Commander and Staff Organization and Operations

May 2015

United States Government
US Army

1. Change 1 to FM 6-0, 5 March 2014, adds the supersession statement to the cover.

2. Modifies figure 7-2.

3. Modifies figure 9-5.

4. Adds joint command relationships to appendix B.

5. Modifies table B-2.

6. Modifies table B-3.

7. Adds definitions of close support, direct liaison authorized, direct support, and mutual support.

8. A number sign (+) marks new material.

9. FM 6-0, 5 May 2014, is changed as follows:

Remove Old Pages	Insert New Pages
front cover	front cover
pages i through vi	pages i through vi
pages 7-1 through 7-2	pages 7-1 through 7-2
pages 9-23 through 9-45	pages 9-23 through 9-46
pages B-1 through B-7	pages B-1 through B-7
pages Glossary-1 through Glossary-9	pages Glossary-1 through Glossary-9
pages Index-1 through Index-9	pages Index-1 through Index-9

7.   File this transmittal sheet in front of the publication for reference purposes.

DISTRUBUTION RESTRICTION: Approved for public release; distribution is unlimited.

By Order of the Secretary of the Army

RAYMOND T. ODIERNO
General, United States Army
Chief of Staff

Official:

GERALD B. O'KEEFE
Administrative Assistant to the
Secretary of the Army
1511202

DISTRIBUTION:
Active Army, Army National Guard, and U.S. Army Reserve: To be distributed in accordance with
the initial distribution number (IDN) 116056, requirements for FM 6-0.

PIN: 104216-001

Field Manual
No. 6-0

Headquarters
Department of the Army
Washington, DC, 5 May 2014

Commander and Staff Organization and Operations

Contents

Distribution Restriction: Approved for public release; distribution is unlimited.

*This publication supersedes ATTP 5-0.1, dated 14 September 2011.

Figures

Tables

This page intentionally left blank.

Preface

FM 6-0, *Commander and Staff Organization and Operations*, provides commanders and their staffs with tactics and procedures for exercising mission command. This publication supersedes ATTP 5-0.1, *Commander and Staff Officer Guide*.

To comprehend the doctrine contained in this publication, readers must first understand the nature of unified land operations as described in ADP 3-0 and ADRP 3-0, *Unified Land Operations*. In addition, readers must also fully understand the principles of mission command as described in ADP 6-0 and ADRP 6-0, *Mission Command*, and the fundamentals of the operations process found in ADP 5-0 and ADRP 5-0, *The Operations Process*.

The principal audience for FM 6-0 includes Army commanders, leaders, and unit staffs (officers, noncommissioned officers, and Soldiers). Commanders and staffs of Army headquarters serving as a joint +headquarters or multinational headquarters should also refer to applicable joint or multinational doctrine concerning the range of military operations as well as the employment of joint or multinational forces. Trainers and educators throughout the Army will also use this publication.

Commanders, staffs, and subordinates ensure their decisions and actions comply with applicable United States, international, and, in some cases, host-nation laws and regulations. Commanders at all levels ensure their Soldiers operate in accordance with the law of war and the rules of engagement. (See FM 27-10.)

FM 6-0 uses joint terms where applicable. Selected joint and Army terms and definitions appear in both the glossary and the text. Terms for which FM 6-0 is the proponent publication (the authority) are marked with an asterisk (*) in the glossary. Terms and definitions for which FM 6-0 is the proponent publication are boldfaced in the text. For other definitions shown in the text, the term is italicized and the number of the proponent publication follows the definition.

FM 6-0 applies to the Active Army, the Army National Guard/Army National Guard of the United States, and the United States Army Reserve unless otherwise stated.

The proponent of FM 6-0 is the United States Army Combined Arms Center. The preparing agency is the Combined Arms Doctrine Directorate, United States Army Combined Arms Center. Send comments and recommendations on a DA Form 2028 (Recommended Changes to Publications and Blank Forms) to Commander, United States Army Combined Arms Center, Fort Leavenworth, ATTN: ATZL-MCD (FM 6-0), 300 McPherson Avenue, Fort Leavenworth, KS 66027-1300; submit an electronic DA Form 2028; or by an e-mail to usarmy.leavenworth.mccoe.mbx.cadd-org-mailbox@mail.mil.

This page intentionally left blank.

Introduction

FM 6-0, *Commander and Staff Organization and Operations*, provides commanders and staffs with many of the tactics and procedures associated with exercising mission command. *Mission command* is the exercise of authority and direction by the commander using mission orders to enable disciplined initiative within the commander's intent to empower agile and adaptive leaders in the conduct of unified land operations (ADP 6-0). Mission command is both a philosophy and a warfighting function.

As the Army's philosophy of command, mission command emphasizes that command is essentially a human endeavor. Successful commanders understand that their leadership guides the development of teams and helps to establish mutual trust and shared understanding throughout the force.

Mission command is also a warfighting function. The *mission command warfighting function* is the related tasks and systems that develop and integrate those activities enabling a commander to balance the art of command and the science of control in order to integrate the other warfighting functions. As a warfighting function, mission command consists of the related tasks and a mission command system that support the exercise of authority and direction by the commander. As a warfighting function, mission command assists commanders in blending the art of command with the science of control, while emphasizing the human aspects of mission command. (See ADRP 6-0 for more details.)

FM 6-0 is intended to serve several purposes. First, it provides commanders and staffs specific information they will need in the exercise of mission command. Second, the manual provides multiple templates and examples of products that commanders and staffs routinely use in the conduct of operations. Finally, FM 6-0 discusses roles and responsibilities that should be understood to facilitate ease of communication among various members of different organizations. It should be noted that although FM 6-0 provides tactics and procedures, commanders may modify products as necessary to meet mission requirements. Local standard operating procedures (SOPs) may also provide examples of products more suitable to specific situations.

FM 6-0 reflects Army leadership decisions to replace the mission command staff task of *conduct inform and influence activities* with *synchronize information-related capabilities*. As a result, FM 6-0 does not use the term *inform and influence activities*. However, commanders remain responsible for the mission command commander task of *inform and influence audiences inside and outside their organizations*. Other changes resulting from this decision include—

- The *assistant chief of staff, G-7 (S-7), inform and influence activities* is replaced by the *information operations officer*. The information operations officer is a special staff officer, coordinated by the G-3 (S-3) operations officer.
- The *inform and influence activities staff section* is replaced by an *information operations element* located in the movement and maneuver cell within each echelon.
- The *inform and influence activities working group* is replaced by the *information operations working group*.
- The military information support operations officer is a special staff officer coordinated by the G-3 (S-3) operations officer.

FM 6-0 contains 16 chapters and 4 appendixes. The chapters are organized by topic and have been updated to reflect changes to doctrine formats (Doctrine 2015) and changes in ADP 3-0 and ADRP 3-0, ADP 6-0 and ADRP 6-0, and ADP 5-0 and ADRP 5-0. The following is a brief introduction and summary of changes by chapter and appendix.

Chapter 1 addresses and provides an update to command post organization and operations as part of the facilities and equipment component of the mission command system described in ADRP 6-0. This chapter describes how commanders organize their headquarters into command posts and cross-functionally organize their staffs. This chapter defines the different types of command posts and describes their purposes. The chapter concludes by providing guidelines for command post operations. Updates to this material include the deletion of a mission command functional cell.

Chapter 2 discusses, updates and describes staff duties, responsibilities, and characteristics as part of the personnel component of the mission command system. It also explains staff relationships and the importance of building staff teams.

Chapter 3 is a new chapter and discusses and expands on the staff task of "conduct knowledge management and information management" found in ADRP 6-0.

Chapter 4 provides an updated discussion of problem solving. Problem solving is a daily activity for leaders and underpins the commander task of drive the operations process and the staff task of conduct the operations process discussed in ADRP 5-0. The major change in the problem solving process is in the first and second steps. Step 1 is now "Gather Information and Knowledge" and Step 2 is now "Identify the Problem." This aligns the problem solving process with the military decisionmaking process found in Chapter 9.

Chapter 5 provides information, instruction, and an annotated example of how to prepare and write a formal report in the form of a staff study.

Chapter 6 provides information, instruction, and an annotated example of how to prepare and write a decision paper.

Chapter 7 provides information, instruction, and annotated examples of how to prepare and conduct the four types of military briefings.

Chapter 8 discusses and provides an annotated example of running estimates. It supports the principle of "commanders drive the operations process" found in ADRP 5-0. This chapter defines running estimates and describes how the commander and staff build and maintain their running estimates throughout the operations process.

Chapter 9 addresses, defines, and provides updated graphics to better organize and explain the military decisionmaking process. Chapter 9 provides two updates to this material:

- In step 2 of the military decisionmaking process, "develop the initial information collection plan" replaces "develop initial intelligence collection synchronization tools."
- In "develop initial intelligence collection tools", step 7 of the military decisionmaking process is now titled "orders production, dissemination, and transition."

Chapter 10 provides information on troop leading procedures, one of the Army's planning methodologies found in ADRP 5-0. While this chapter explains troop leading procedures from a ground-maneuver perspective, it applies to all types of small units.

Chapter 11 is a new chapter and provides information on military deception, one of the additional mission command warfighting functions tasks described in ADRP 6-0. Initially this chapter addresses the principles of military deception. It then discusses how commanders use military deception to shape the operational environment in support of decisive action. The chapter concludes with a discussion of how to plan, prepare, execute, and assess military deception.

Chapter 12 provides information on rehearsals. Rehearsals are a preparation activity as described in ADRP 5-0. This chapter describes types of rehearsals, lists responsibilities of personnel involved, and contains guidelines for conducting rehearsals.

Chapter 13 provides information on liaison, a planning and preparation activity as described in ADRP 5-0. This chapter discusses responsibilities of liaison officers and teams. It includes liaison checklists and an example outline for a liaison officer handbook.

Chapter 14 is a new chapter and describes decisionmaking during execution. It expands on execution activities found in ADRP 5-0. This chapter discusses how commanders, supported by their staffs, assess the operation's progress, make decisions and direct the application of combat power to seize, retain, and exploit the initiative.

Chapter 15 provides information on assessment and its role in the operations process as described in ADRP 5-0. It describes the assessment process and key terms. This chapter concludes with guidelines and details to assist commanders and their staffs in developing formal assessment plans.

Chapter 16 provides information on after action reviews and new information on after action reports as part of assessment described in ADRP 5-0. This chapter provides an annotated after action report format for use by all levels of command.

Appendix A is a new appendix and provides information on operational and mission variables.

Appendix B provides information on and definitions of command and support relationships.

Appendix C provides updated information and annotated examples of the operation order, operation plan, warning order, fragmentary order, and attachments. This appendix contains a complete list of all operation plan or order attachments: annexes, appendixes, tabs, and exhibits.

Paragraph 3 of the base operation order or operation plan now contains a new subparagraph titled "Cyber Electromagnetic Activities."

Appendix D provides updated information and annotated examples of all the annex formats (from A to Z) for the operation plan or order. This appendix does not contain annotated examples of the subordinate attachments to the annex: appendixes, tabs, or exhibits. Changes to formats include—

- Appendix 1 (Design Products) to Annex C (Operations) is now titled Appendix 1 (Army Design Methodology Products) to Annex C (Operations).
- The cyber electromagnetic activities appendix formerly found in Annex D (Fires) is now Appendix 12 (Cyber Electromagnetic Activities) to Annex C (Operations).
- Appendix 13 to Annex C (Operations) is now titled "Military Information Support Operations."
- Appendix 14 to Annex C (Operations) is now titled "Military Deception."
- Appendix 15 to Annex C (Operations) is a new appendix titled "Information Operations."
- The air and missile defense appendix has been moved from Annex E (Protection) to Annex D (Fires) and is titled Appendix 7 (Air and Missile Defense) to Annex D (Fires).
- Annex E (Protection) now has fourteen appendixes and is updated in accordance with ADRP 3-37.
- Annex E (Protection) retains an appendix titled Appendix 12 (Coordinate Air and Missile Defense).
- Appendix 13 (Detainee and Resettlement) to Annex C (Operations) is now Appendix 14 (Detainee and Resettlement) to Annex E (Protection).
- Annex J is now "Public Affairs."
- New annexes include Annex Q (Knowledge Management) and Annex W (Operational Contract Support).

This version of FM 6-0 does not discuss the tactics or procedures related to—

- Information operations.
- Cyber electromagnetic activities.
- Civil affairs operations.
- Installing, operating, or maintaining the network.
- Information protection.

Users should see the appropriate doctrinal publication for details.

FM 6-0 adds the following terms. (See introductory table-1.)

Introductory table-1. New Army terms

Term	Acronym	Remarks
Fragmentary order	FRAGORD	Adopts joint definition and acronym
Warning order	WARNORD	Adopts joint definition and acronym

This page intentionally left blank.

Chapter 1

Command Post Organization and Operations

This chapter describes how commanders organize their headquarters into command posts during the conduct of operations. This chapter defines the different types of command posts and describes their purposes. Next, this chapter discusses the effectiveness and survivability factors commanders consider when organizing their command posts. This chapter also describes how commanders cross-functionally organize their staffs within command posts into functional and integrating cells. The chapter concludes by providing guidelines for command post operations, including the importance of establishing standard operating procedures (SOPs) for the headquarters. (See the corresponding proponent publications for specific guidance on command post organization by echelon or type of unit. See JP 3-33 for more information on an Army headquarters serving as a joint headquarters.)

COMMAND POST ORGANIZATION

1-1.   In operations, effective mission command requires continuous close coordination, synchronization, and information sharing across staff sections. To promote this, commanders cross-functionally organize elements of staff sections in command posts (CPs) and CP cells. Additional staff integration occurs in meetings, including working groups and boards. (See paragraphs 1-65 through 1-71.)

1-2.   **A *command post* is a unit headquarters where the commander and staff perform their activities**. The headquarters design, combined with robust communications, gives commanders a flexible mission command structure consisting of a main CP, a tactical CP, and a command group for brigades, divisions, and corps. Combined arms battalions are also resourced with a combat trains CP and a field trains CP. Theater army headquarters are resourced with a main CP and a contingency CP. (See appropriate echelon publications for doctrine on specific types of CPs and headquarters organizations.)

1-3.   Each CP performs specific functions by design as well as tasks the commander assigns. Activities common in all CPs include, but are not limited to—

- Maintaining running estimates.
- Controlling operations.
- Assessing operations.
- Developing and disseminating orders.
- Coordinating with higher, lower, and adjacent units.
- Conducting knowledge management and information management.
- Conducting network operations.
- Providing a facility for the commander to control operations, issue orders, and conduct rehearsals.
- Maintaining the common operational picture.
- Performing CP administration (examples include sleep plans, security, and feeding schedules).
- Supporting the commander's decisionmaking process.

MAIN COMMAND POST

1-4.   **The *main command post* is a facility containing the majority of the staff designed to control current operations, conduct detailed analysis, and plan future operations.** The main CP is the unit's principal CP. It includes representatives of all staff sections and a full suite of information systems to plan, prepare, execute, and assess operations. It is larger, has more staff members, and is less mobile than the tactical CP. The chief of staff (COS) or executive officer (XO) leads and provides staff supervision of the main CP. Functions of the main CP include, but are not limited to—

- Controlling and synchronizing current operations.
- Monitoring and assessing current operations (including higher and adjacent units) for their impact on future operations.
- Planning operations, including branches and sequels.
- Assessing the overall progress of operations.
- Preparing reports required by higher headquarters and receiving reports for subordinate units.

TACTICAL COMMAND POST

1-5. **The *tactical command post* is a facility containing a tailored portion of a unit headquarters designed to control portions of an operation for a limited time.** Commanders employ the tactical CP as an extension of the main CP to help control the execution of an operation or a specific task, such as a gap crossing, a passage of lines, or an air assault operation. Commanders may employ the tactical CP to direct the operations of units close to each other, such as during a relief in place. The tactical CP may also control a special task force or a complex task, such as reception, staging, onward movement, and integration.

1-6. The tactical CP is fully mobile and includes only essential Soldiers and equipment. The tactical CP relies on the main CP for planning, detailed analysis, and coordination. A deputy commander or operations officer generally leads the tactical CP.

1-7. When employed, tactical CP functions include, but are not limited to—

- Monitoring and controlling current operations.
- Monitoring and assessing the progress of higher and adjacent units.
- Performing short-range planning.
- Providing input to targeting and future operations planning.

1-8. When the commander does not employ the tactical CP, the staff assigned to it reinforces the main CP. Unit SOPs should address the specifics for this, including procedures to quickly detach the tactical CP from the main CP.

COMMAND GROUP

1-9. **A *command group* consists of the commander and selected staff members who assist the commander in controlling operations away from a command post.** The command group is organized and equipped to suit the commander's decisionmaking and leadership requirements. It does this while enabling the commander to accomplish critical mission command warfighting function tasks anywhere in the area of operations.

1-10. Command group personnel include staff representation that can immediately affect current operations, such as maneuver, fires (including the air liaison officer), and intelligence. The mission and available staff, however, dictate the command group's makeup. For example, during a deliberate breach, the command group may include an engineer and an air defense officer. When visiting a dislocated civilians' collection point, the commander may take a translator, a civil affairs operations officer, a medical officer, and a chaplain.

1-11. Divisions and corps headquarters are equipped with a mobile command group. The mobile command group serves as the commander's mobile CP. It consists of ground and air components equipped with information systems. The mobile command group's mobility allows commanders to move to critical locations to personally assess a situation, make decisions, and influence operations. The mobile command group's information systems and small staff allow commanders to do this while retaining communications with the entire force.

EARLY-ENTRY COMMAND POST

1-12. While not part of the unit's table of organization and equipment, commanders can establish an early-entry command post to assist them in controlling operations during the deployment phase of an operation. **An *early-entry command post* is a lead element of a headquarters designed to control operations until the remaining portions of the headquarters are deployed and operational.** The early-entry command post normally consists of personnel and equipment from the tactical CP with additional intelligence analysts, planners, and other staff officers from the main CP based on the situation.

1-13. The early-entry command post performs the functions of the main and tactical CPs until those CPs are deployed and fully operational. A deputy commander, COS (XO), or operations officer normally leads the early-entry command post.

COMMAND POST ORGANIZATION CONSIDERATIONS

1-14. When organizing the CP, commanders must consider effectiveness and survivability. However, effectiveness considerations may compete with survivability considerations, making it difficult to optimize either. Commanders balance survivability and effectiveness considerations when organizing CPs.

EFFECTIVENESS CONSIDERATIONS AND FACTORS

1-15. CP staff and equipment are arranged to facilitate coordination, information exchange, and rapid decisionmaking. CPs must effectively communicate with all subordinate units and the higher headquarters. An effective CP organization enables quick deployment, employment, and displacement throughout the unit's area of operations. Five factors contribute to CP effectiveness: design and layout, standardization, continuity, deployability, and capacity and range.

Design and Layout

1-16. Many design considerations affect CP effectiveness. At a minimum, commanders position CP cells and staff elements to facilitate communication and coordination. Other design considerations include, but are not limited to—

- Efficient facilitation of information flow.
- Connectivity to information systems and the network.
- Positioning information displays for ease of use.
- Integrating information on maps and displays.
- Adequate workspace for the commander and staff.
- Ease of deployment, employment, and displacement (setup, teardown, and mobility).
- Effective and efficient power generation and distribution.

1-17. Well-designed CPs integrate command and staff efforts. Meeting this requirement requires matching the CP's manning, equipment, information systems, and procedures against its internal layout and utilities. Organizing the CP into functional and integrating cells promotes efficiency and coordination. (See paragraphs 1-28 through 1-46.)

Standardization

1-18. Standardization increases efficiency and eases CP personnel training. Commanders develop detailed SOPs for all aspects of CP operations. Standard CP layouts, battle drills, and reporting procedures increase efficiency. Units follow and revise SOPs throughout training. Units constantly reinforce standardization using SOPs to make many processes routine. Staffs then effectively execute them in demanding, stressful operations.

Continuity

1-19. Commanders staff, equip, and organize CPs to control and support 24-hour operations. However, duplicating every staff member within a CP is unnecessary. Commanders carefully consider the primary role and functions assigned to each CP and resource it accordingly. Internal CP SOPs address shifts, rest plans, and other CP activities important to operating continuously. Leaders enforce these provisions.

1-20. Maintaining continuity during displacement or catastrophic loss requires designating alternate CPs and procedures for passing control between them. SOPs address providing continuity when units lose communications with the commander, subordinates, or a particular CP. Commanders designate seconds in command and inform them of all critical decisions. Primary staff officers also designate alternates.

Deployability

1-21. CPs deploy efficiently and move within the area of operations as required. Determining the capabilities, size, and sequence of CPs in the deployment flow requires careful consideration. Commanders can configure CP elements as an early-entry command post if needed. CP size directly affects deployment and employment.

Capacity, Connectivity, and Range

1-22. Efficient and effective CP organization allows the commander to maintain the capacity to plan, prepare, execute, and continuously assess operations. CPs require uninterrupted connectivity to effectively communicate with higher and subordinate headquarters. Commanders and staffs must consider various factors that can adversely affect the efficiency of communications systems, such as built-up areas, mountains, and atmospheric conditions.

SURVIVABILITY FACTORS

1-23. CP survivability is vital to mission success. CPs often gain survivability at the price of effectiveness. When concentrated, the enemy can easily acquire and target most CPs. However, when elements of a CP disperse, they often have difficulty maintaining a coordinated staff effort. When developing command post SOPs and organizing headquarters into CPs for operations, commanders use dispersion, size, redundancy, and mobility to increase survivability.

Dispersion

1-24. Dispersing CPs often enhances survivability. Commanders place minimum resources in the deep and close areas and keep more elaborate facilities in security areas. This makes it harder for the enemy to find and attack them. It also decreases support and security requirements in the deep and close areas. Most of the staff is co-located in the main CP; the tactical CP contains only the staff and equipment essential to controlling portions of an operation for a limited time.

Size

1-25. A CP's size affects its survivability. Larger CPs ease face-to-face coordination; however, they are vulnerable to multiple acquisitions and means of attack. Units can hide and protect smaller CPs more easily, but they may not control all force elements. Striking the right balance provides a responsive yet agile organization. For example, commanders require information for decisions; they do not need every subject matter expert located with them.

Redundancy

1-26. Some personnel and equipment redundancy is required for continuous operations. Redundancy allows CPs to continue operating when mission command systems are lost, damaged, or fail under stress.

Mobility

1-27. CP mobility improves CP survivability, especially at lower echelons. Successful lower-echelon CPs move quickly and often. A smaller size and careful movement planning allow CPs to displace rapidly to avoid the enemy.

COMMAND POST CELLS, STAFF SECTIONS, AND ELEMENTS

1-28. Within CPs, commanders cross-functionally organize their staffs into CP cells and staff sections to assist them in the exercise of mission command. A *command post cell* **is a grouping of personnel and equipment organized by warfighting function or by planning horizon to facilitate the exercise of mission command.** Staff sections are groupings of staff members by areas of expertise under a coordinating, special, or personal staff officer. Elements are groupings of staff members subordinate to specific staff sections. Staff sections and elements of staff sections are the building blocks for CP cells. (See chapter 2 for a detailed discussion on the duties and responsibilities of staffs.)

1-29. Commanders organize their CPs by functional and integrating cells. Functional cells group personnel and equipment by warfighting function (minus mission command). Integrating cells group personnel and equipment by planning horizon. Not all staff sections permanently reside in one of the functional or integrating cells. The G-6 (S-6) signal and G-9 (S-9) civil affairs sections are examples. These staff sections do, however, provide representation to different CP cells as required, and they coordinate their activities in the various meetings established in the unit's battle rhythm. (See figure 1-1.)

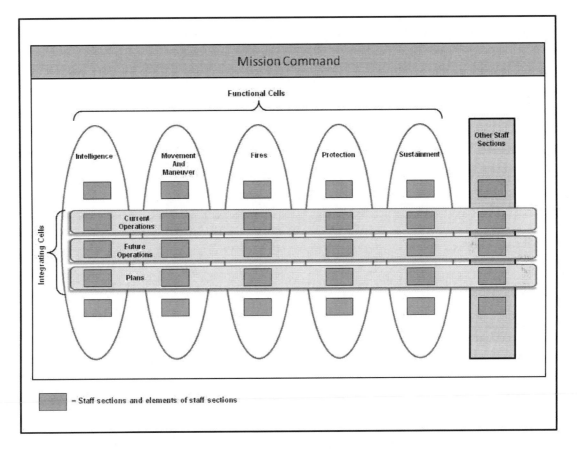

Figure 1-1. Command post organization

Note: Figure 1-1 represents the standard command post organizational design. However, the standard design is tailorable. Commanders organize and reorganize their command post to meet changing situations and the requirements of their specific operations.

MISSION COMMAND

1-30. The entire command post (depicted as the mission command box in figure 1-1) assists the commander in the exercise of mission command. Therefore, commanders do not form a specific mission command functional cell. All of the various command post cells and staff sections assist the commander with specific tasks of the mission command warfighting function. For example, all functional and integrating cells assist the commander with conducting the operations process. As such, the command post as a whole, including the commander, deputy commanders, and command sergeants major, represents the mission command warfighting function.

FUNCTIONAL CELLS

1-31. The functional cells within a CP are intelligence, movement and maneuver, fires, protection, and sustainment. Echelons above brigade are resourced to establish all five functional cells described in paragraphs 1-32 through 1-36. (See appropriate brigade and battalion publications for specifics on the functional cells at those levels.)

Intelligence Cell

1-32. The intelligence cell coordinates activities and systems that facilitate understanding of the threats, terrain and weather, and other relevant aspects of the operational environment. The intelligence cell requests, receives, and analyzes information from multiple sources to produce and distribute intelligence products. The intelligence cell consists of the majority of the intelligence staff and an attached U.S. Air Force weather team. Higher headquarters may augment this organization with additional capabilities to meet mission requirements. The unit's G-2 (S-2) intelligence officer leads this cell.

Movement and Maneuver Cell

1-33. The movement and maneuver cell coordinates activities and systems that move forces to achieve a position of advantage. This includes tasks related to gaining a positional advantage by combining forces with direct fire or fire potential (maneuver) and force projection (movement). Elements of the operations, airspace control, aviation, engineer, geospatial information and service, and space support element form this cell. Staff elements in the movement and maneuver cell also form the core of the current operations integration cell. The unit's operations officer leads this cell. (See paragraphs 1-37 through 1-50 for a discussion of the integrating cells.)

Fires Cell

1-34. The fires cell coordinates, plans, integrates, and synchronizes the employment and assessment of fires in support of current and future operations. The fires cell develops high payoff targets and selects targets for attack. The fires cell recommends targeting guidance to the commander. The fires cell plans, synchronizes, coordinates, and integrates adaptable fires matched to a wide range of targets and target systems. The fires cell coordinates target acquisition, target dissemination, and target engagement functions for the commander. At the division level, the air and missile defense section is integrated within the fires cell to ensure coordination of sense and warning systems, synchronization of fires, and airspace integration. The fires cell coordinates activities and systems that provide collective and coordinated use of Army indirect fires, joint fires, and air and missile defense through the targeting process. The fires cell includes elements of fire support, the Air Force (or air component), the air and missile defense section, and liaison officers from joint or multinational fire support agencies. Additional augmentation to the fires cell includes the naval surface fire support liaison officer and Army space support team(s). The unit's chief of fires (or fire support officer at brigade and below) leads this cell.

Protection Cell

1-35. The protection cell coordinates the activities and systems that preserve the force through risk management. This includes tasks associated with protecting personnel and physical assets. Elements of the following staff sections form this cell: chemical, biological, radiological, and nuclear; engineer; personnel recovery; and provost marshal. Additionally, a safety officer is assigned at theater army and, with augmentation, as required down to the brigade level. The protection cell coordinates with the signal staff section to further facilitate the information protection task. The chief of protection leads this cell.

Sustainment Cell

1-36. The sustainment cell coordinates activities and systems that provide support and services to ensure freedom of action, extend operational reach, and prolong endurance. It includes those tasks associated with logistics, personnel services, and health service support. The following staff sections form this cell: personnel, sustainment, financial management, and surgeon. The chief of sustainment (or logistics officer at brigade and below) leads this cell.

INTEGRATING CELLS

1-37. Whereas functional cells are organized by warfighting functions, integrating cells are organized by planning horizons. They coordinate and synchronize forces and warfighting functions within a specified planning horizon and include the plans cell, future operations cell, and current operations integration cell. A *planning horizon* is a point in time commanders use to focus the organization's planning efforts to shape future events (ADRP 5-0). The three planning horizons are long, mid, and short (generally associated with the plans cell, future operations cell, and current operations integration cell, respectively).

1-38. Planning horizons are situation-dependent and are influenced by events and decisions. For example, the plans cell normally focuses its planning effort on the development of sequels—the subsequent next operation or phase of the operation based on possible outcomes (success, stalemate, or defeat) of the current operation or phase. The future operations cell normally focuses its efforts on branch plans—options built into the base plan that changes the concept of operations based on anticipated events, opportunities, or threats. Planning guidance and decisions by the commander or that of the higher headquarters influence the planning horizons.

1-39. Not all echelons and types of units are resourced for all three integrating cells. Battalions, for example, combine their planning and operations responsibilities in one integrating cell. The brigade combat team has a small, dedicated plans cell but it is not resourced for a future operations cell. Divisions and higher echelons are resourced for all three integrating cells as shown in figure 1-2 on page 1-8.

Plans Cell

1-40. The plans cell is responsible for planning operations for the long-range planning horizons. It prepares for operations beyond the scope of the current order by developing plans and orders, including branch plans and sequels. The plans cell also oversees military deception planning.

1-41. The plans cell consists of a core group of planners and analysts led by the G-5 (S-5) plans officer (or the operations officer at battalion level). All staff sections assist as required. Since a brigade has a small, dedicated plans cell, the majority of its staff sections balance their efforts between the current operations integration and plans cells. Battalions are not resourced for a plans cell. Planning in combined arms battalions occurs in the current operations integration cell.

Future Operations Cell

1-42. The future operations cell is responsible for planning operations in the mid-range planning horizon. It focuses on adjustments to the current operation—including the positioning or maneuvering of forces in depth—that facilitate continuation of the current operation. The cell consists of a core group of planners led by an assistant operations officer (the chief of future operations). All staff sections assist as required. Divisions and higher echelon headquarters have a future operations cell. Battalion and brigade headquarters do not.

1-43. In many respects, the future operations cell serves as a bridge between the plans and current operations integration cells. The future operations cell monitors current operations and determines implications for operations within the mid-range planning horizon. In coordination with the current operations integration cell, the future operations cell assesses whether the ongoing operation must be modified to achieve the current phase's objectives. Normally, the commander directs adjustments to the operation, but the cell may also recommend options to the commander. Once the commander decides to adjust the operation, the cell develops the fragmentary order (FRAGORD) necessary to implement the change. The future operations cell also participates in the targeting working group since the same planning horizons normally concern them both. The future operations cell updates and adds details to the branch plans foreseen in the current operation and prepares any orders necessary to implement a sequel to the operation. (See figure 1-2.)

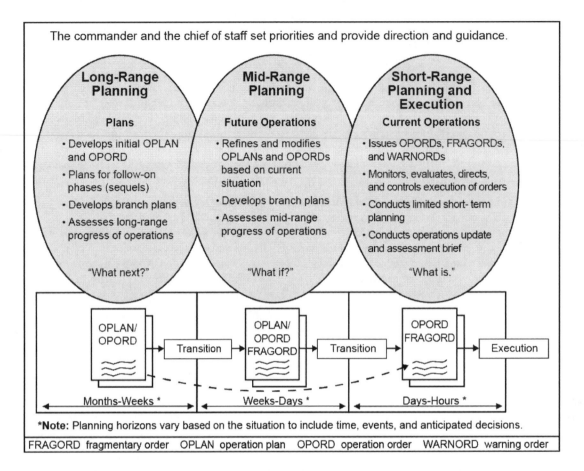

Figure 1-2. Integration of plans, future operations, and current operations

Current Operations Integration Cell

1-44. The current operations integration cell is the focal point for the execution of operations. This involves assessing the current situation while regulating forces and warfighting functions in accordance with the mission, commander's intent, and concept of operations.

1-45. The current operations integration cell displays the common operational picture and conducts shift changes, assessments, and other briefings, as required. It provides information on the status of operations to all staff members and to higher, subordinate, and adjacent units. The operations synchronization meeting is the most important event in the battle rhythm in support of the current operation.

1-46. The operations officer leads the current operations integration cell and is aided by an assistant operations officer (the chief of operations). Elements or watch officers from each staff section and liaison officers from subordinate and adjacent units form this cell. All staff sections are represented in the current operations integration cell, either permanently or on call.

COMMAND POST OPERATIONS

1-47. Units must man, equip, and organize command posts to control operations for extended periods. Effective CP personnel use information systems and equipment to support 24-hour operations while they continuously communicate with all subordinate, higher, and adjacent units. Commanders arrange CP personnel and equipment to facilitate internal coordination, information sharing, and rapid decisionmaking. They also ensure that they have procedures to execute the operations process within the headquarters to enhance how they exercise mission command. Commanders use the battle rhythm, SOPs, and meetings to assist them with effective CP operations.

STANDARD OPERATING PROCEDURES

1-48. SOPs that assist with effective mission command serve two purposes. Internal SOPs standardize each CP's internal operations and administration. External SOPs developed for the entire force standardize interactions among CPs and between subordinate units and CPs. Effective SOPs require that all Soldiers know their duties and train to standards. (See FM 7-15 for more information on the tasks of command post operations.)

1-49. Each CP should have SOPs that address the following:
- Organization and setup.
- Staffing and shift plans, including eating and sleeping plans.
- Physical security and defense.
- Priorities of work.
- Equipment and vehicle maintenance, including journals and a maintenance log.
- Load plans and equipment checklists.
- Orders production and dissemination procedures.
- Plans for handling, storing, and cleaning up hazardous materials.
- Battle rhythm.
- Use of Army Battle Command Systems, such as Command Post of the Future, Advanced Field Artillery Tactical Data System, and Blue Force Tracker.

1-50. In addition to these SOPs, each CP requires—
- CP battle drills.
- Shift-change briefings.
- Reports and returns.
- Operations update and assessment briefings.
- Operations synchronization meetings.
- Procedures for transferring control between CPs.

Command Post Battle Drills

1-51. Each CP requires procedures to react to a variety of situations. Specific actions taken by a CP should be defined in its SOPs and rehearsed during training and operations. Typical CP battle drills include, but are not limited to—
- React to an air, ground, or chemical, biological, radiological, or nuclear attack.
- React to indirect fire.
- React to jamming or suspected communications compromise.
- Execute dynamic targeting.
- Execute a close air support mission.
- React to a cyber intrusion or attack.
- React to a mass casualty incident.
- React to a civil riot or incident.
- React to significant collateral damage.
- React to incorrect information affecting an operational environment.
- React to a degraded network.
- React to a duty status and whereabouts unknown incident.

Shift-Change Briefings

1-52. During continuous operations, CPs operate in shifts. To ensure uninterrupted operations, staffs execute a briefing when shifts change. Depending on the situation, this briefing may be formal or informal and include the entire staff or selected staff members. Normally key CP leaders meet face-to-face. The COS (XO) oversees the briefing, with participants briefing their areas of expertise. The briefing's purpose is to inform the incoming shift of—

- Current unit status.
- Significant activities that occurred during the previous shift.
- Significant decisions and events anticipated during the next shift.

The commander may attend and possibly change the focus of the briefing. If the commander issues guidance or makes a decision, issuing a FRAGORD may be necessary.

1-53. The shift-change briefing format and emphasis change based on the situation. For example, the format for a force supporting civil authorities in a disaster area differs from a force conducting offensive tasks abroad. To facilitate a quick but effective shift-change briefing, unit SOPs should contain tailored formats.

1-54. The shift-change briefing provides a mechanism to formally exchange information periodically among CP staff members. CP staff members coordinate activities and inform each other continuously. They immediately give information to the commander that answers a commander's critical information requirement or is exceptional information. They disseminate information that potentially affects the entire force to the commander, higher headquarters, and subordinate units as the situation dictates. Situational understanding for CP staff members includes knowing who needs what relevant information and why they need it. CP staff members exercise initiative when they ensure relevant information gets to the people who need it. (See table 1-1 for a sample shift change briefing.)

Table 1-1. Sample shift-change briefing

Current mission and commander's intent (COS [XO])

Enemy situation (G-2 [S-2])
- *Significant threat or local populace attitudes and actions during the last shift.*
- *Current enemy situation and changes in the most likely enemy courses of action.*
- *Anticipated significant threat or undesired local populace activity in the next 12/24/48 hours.*
- *Changes in priority intelligence requirements (PIRs).*
- *Weather update and weather effects on operations in the next 12/24/48 hours.*
- *Changes to information collection priorities.*
- *Status of information collection units and capabilities.*

Civil Situation (G-2 [S-2] and G-9 [S-9])
- *Significant actions by the population during the last shift.*
- *Current civil situation.*
- *Disposition and status of civil affairs units and capabilities.*
- *Significant activities involving the population anticipated during the next shift.*

Friendly situation (G-3 [S-3])
- *Significant friendly actions during the last shift.*
- *Subordinate units' disposition and status.*
- *Higher and adjacent units' disposition and status.*
- *Major changes to the task organization and tasks to subordinate units that occurred during the last shift.*
- *Answers to CCIRs and changes in CCIRs.*
- *Changes to information collection.*
- *Disposition and status of selected information collection units and capabilities.*
- *Answers to EEFIs and changes in EEFIs.*
- *Significant activities and decisions scheduled for next shift (review of the decision support matrix).*
- *Anticipated planning requirements.*
- *Liaison officer update.*

Running estimate summaries by warfighting function and staff section. Briefers include—
- *Fires*
- *Air liaison officer*
- *Aviation officer*
- *Air and missile defense officer*
- *Information operations officer*
- *Engineer officer*
- *CBRN officer*
- *Provost marshal*
- *G-1 (S-1)*
- *G-4 (S-4)*
- *G-6 (S-6)*
- *Electronic warfare officer*

Briefings include—
- *Any significant activities that occurred during the last shift.*
- *The disposition and status of units within their area of expertise.*
- *Any changes that have staff wide implications (for example, "higher headquarters changed the controlled supply rate for 120 mm HE, so that means...").*
- *Upcoming activities and anticipated changes during the next shift.*

CP operations and administration (headquarters commandant or senior operations NCO)—
- *CP sustainment issues.*
- *CP security.*
- *CP displacement plan and proposed new locations.*
- *Priority of work.*

COS (XO) guidance to the next shift, including staff priorities and changes to the battle rhythm.

CBRN	chemical, biological, radiological, and nuclear	HE	high explosive
CCIR	commander's critical information requirement	mm	millimeter
COS	chief of staff	NCO	noncommissioned officer
CP	command post	PIR	priority intelligence requirement
EEFI	essential element of friendly information	S-1	personnel staff officer
G-1	assistant chief of staff, personnel	S-2	intelligence staff officer
G-2	assistant chief of staff, intelligence	S-3	operations staff officer
G-3	assistant chief of staff, operations	S-4	logistics staff officer
G-4	assistant chief of staff, logistics	S-6	signal staff officer
G-6	assistant chief of staff, signal	S-9	civil affairs operations staff officer
G-9	assistant chief of staff, civil affairs operations	XO	executive officer

Reports and Returns

1-55. A unit's reporting system facilitates timely and effective information exchange among CPs and higher, lower, and adjacent headquarters. An established SOP for reports and returns drives effective information management. These SOPs state the writer, the frequency and time, and recipient of each report. (List nonstandard reports in Annex R [Reports] of the operation plan and operation order.)

Operation Update and Assessment Briefing

1-56. An operation update and assessment briefing may occur daily or any time the commander calls for one. Its content is similar to the shift-change briefing, but it has a different audience. The staff presents it to the commander and subordinate commanders. It provides all key personnel with common situational understanding. Often commanders require this briefing shortly before an operation begins to summarize changes made during preparation, including changes resulting from information collection efforts.

1-57. During the briefing, staff sections present a summary of their running estimates. Subordinate commanders brief their current situation and planned activities. Rarely do all members conduct this briefing in person. All CPs and subordinate commanders participate using available communications equipment, including radios, conference calls, and video teleconferences. The briefing follows a sequence and format specified by SOPs. This keeps transmissions short, ensures completeness, and eases note taking. This briefing normally has a format similar to a shift-change briefing. However, it omits CP administrative information and includes presentations by subordinate commanders in an established sequence.

Operations Synchronization Meeting

1-58. The operations synchronization meeting is the key event in the battle rhythm in support of the current operation. Its primary purpose is to synchronize all warfighting functions and other activities in the short-term planning horizon. It is designed to ensure that all staff members have a common understanding of current operations, including upcoming and projected actions at decision points.

1-59. The operations synchronization meeting does not replace the shift-change briefing or operation update and assessment briefing. Chaired by the G-3 (assistant chief of staff, operations) or S-3 (operations staff officer), representatives of each CP cell and separate staff section attend the meeting. The operations synchronization meeting includes a FRAGORD addressing any required changes to maintain synchronization of current operations, and any updated planning guidance for upcoming working groups and boards. All warfighting functions are synchronized and appropriate FRAGORDs are issued to subordinates based on the commander's intent for current operations.

Transferring Control of Operations Between Command Posts

1-60. The employment and use of CPs are important decisions reflected in the operation order. Often, a particular CP may control part or all of the operation for a specific time. Effectively transferring control between CPs requires a well-understood SOP and clear instructions in the operation order.

1-61. While all CPs have some ability to exercise control on the move, they lose many capabilities they have when stationary. Therefore, CPs normally control operations from a static location. During moves, they transfer control responsibilities to another CP. Transfer of control requires notifying subordinates, since many network operations change to route information to the new controlling CP. SOPs establish these requirements to minimize interruptions when transferring control.

BATTLE RHYTHM

1-62. A headquarters' battle rhythm consists of a series of meetings (to include working groups and boards), briefings, and other activities synchronized by time and purpose. **The *battle rhythm* is a deliberate daily cycle of command, staff, and unit activities intended to synchronize current and future operations.** The COS (XO) oversees the unit's battle rhythm. The COS (XO) ensures activities are logically sequenced so that the output of one activity informs another activity's inputs. Not only is this important internally within the headquarters, the unit's battle rhythm must nest with the higher headquarters. This ensures that the information pertinent to decisions and the recommendations on decisions made in the headquarters are provided in a timely manner to influence the decisionmaking of the higher headquarters, where appropriate. Understanding the purpose and potential decisions of each meeting and activity is equally important. This understanding allows members of the staff and subordinate commanders to provide appropriate input to influence decisions. The COS (XO) balances other staff duties and responsibilities with the time required to plan, prepare for, and hold meetings and conduct briefings. The COS (XO) also critically examines attendance requirements. Some staff sections and CP cells may lack the personnel to attend all events. The COS (XO) and staff members constantly look for ways to combine meetings and eliminate unproductive ones.

1-63. The battle rhythm enables—

- Establishing a routine for staff interaction and coordination.
- Facilitating interaction between the commander and staff.
- Synchronizing activities of the staff in time and purpose.
- Facilitating planning by the staff and decisionmaking by the commander.

1-64. The battle rhythm changes during execution as operations progress. For example, early in the operation a commander may require a daily plans update briefing. As the situation changes, the commander may only require a plans update every three days. Some factors that help determine a unit's battle rhythm include the staff's proficiency, higher headquarters' battle rhythm, and current mission. In developing the unit's battle rhythm, the chief COS (XO) considers—

- Higher headquarters' battle rhythm and report requirements.
- Subordinate headquarters' battle rhythm requirements.
- The duration and intensity of the operation.
- Integrating cells' planning requirements.

MEETINGS

1-65. Meetings are gatherings to present and exchange information, solve problems, coordinate action, and make decisions. They may involve the staff; the commander and staff; or the commander, subordinate commanders, staff, and others as necessary (including unified action partners). Who attends depends on the issue. Commanders establish meetings to integrate the staff and enhance planning and decisionmaking within the headquarters. Commanders also identify staff members to participate in the higher commander's meeting, including working groups and boards. (JP 3-33 discusses the various working groups and boards used by joint force commanders.) Decisions made during meetings must be shared internally within the headquarters, with the higher headquarters, and with subordinate units.

1-66. The number of meetings and the subjects they address depend on the situation and echelon. While numerous informal meetings occur daily within a headquarters, meetings commonly included in a unit's battle rhythm and the cells responsible for them include—

- A shift-change briefing (current operations integration cell).
- An operation update and assessment briefing (current operations integration cell).
- An operations synchronization meeting (current operations integration cell).
- Planning meetings and briefings (plans or future operations cells).
- Working groups and boards (various functional and integrating cells).

1-67. Often, the commander establishes and maintains only those meetings required by the situation. Commanders—assisted by the COS (XO)—establish, modify, and dissolve meetings as the situation evolves. The COS (XO) manages the timings of these events through the unit's battle rhythm.

1-68. For each meeting, a unit's SOPs address—

- Purpose.
- Frequency.
- Composition (chair and participants).
- Inputs and expected outputs.
- Agenda.

1-69. Boards and working groups are types of meetings and are included in the unit's battle rhythm. A *board* **is a grouping of predetermined staff representatives with delegated decision authority for a particular purpose or function.** Boards are similar to working groups. However, commanders appoint boards to make decisions. When the process or activity being synchronized requires command approval, a board is the appropriate forum.

1-70. **A** *working group* **is a grouping of predetermined staff representatives who meet to provide analysis, coordinate, and provide recommendations for a particular purpose or function.** Their cross-functional design enables working groups to synchronize contributions from multiple CP cells and staff sections. For example, the targeting working group brings together representatives of all staff elements concerned with targeting. It synchronizes the contributions of all staff elements with the work of the fires cell. It also synchronizes fires with future operations and current operations integration cells. (See table 1-2 for a sample SOP for a working group.)

Table 1-2. Sample SOP for a division civil affairs operations working group

Purpose and frequency	**Purpose:** • *Establish policies, procedures, priorities, and overall direction for all civil-military operations projects* • *Provide update on ongoing civil-military operations projects* • *Identify needs within the area of operations* • *Present suggested future projects* **Frequency:** *Weekly*
Composition	**Chair:** G-9 **Attendees:** • *Civil affairs battalion representative*　• *Military information support element representative* • *G-2 representative* • *G-3 operations representative*　• *Provost marshal or force protection representative* • *G-5 planner* • *Information operations representative*　• *Special operations forces liaison officer* • *Staff judge advocate representative*　• *Surgeon* • *Host-nation liaison officers*　• *Chaplain* • *Engineer planner*　• *Project manager and contractor representatives* • *Public affairs brigade combat team and Marine Corps liaison officer*
Inputs and outputs	**Inputs:** • *Project management status* • *Information operations working group (last week's)* • *Targeting board* • *Higher headquarters operation order* **Outputs:** • *Updated project status matrix* • *Proposed project matrix* • *Long-range civil-military operation plan adjustment*
Agenda	• *G-2 update or assessment* • *Operations update* • *Public perception update* • *Civil affairs project update* • *Engineer project update* • *Staff judge advocate concerns* • *Discussion or issues* • *Approval of information operations working group inputs*

G-2	assistant chief of staff, intelligence	G-5	assistant chief of staff, plans
G-3	assistant chief of staff, operations	G-9	assistant chief of staff, civil affairs operations

1-71. Working groups address various subjects depending on the situation and echelon. Battalion and brigade headquarters normally have fewer working groups than higher echelons have. Working groups may convene daily, weekly, monthly, or intermittently depending on the subject, situation, and echelon. Typical working groups and the lead cell or staff section at division and corps headquarters include the following:

- Assessment working group (plans or future operations cell).
- Operations and intelligence working group (intelligence cell).

- Targeting working group (fires cell).
- Protection working group (protection cell)
- Civil affairs operations working group (civil affairs operations staff section).
- Information operations working group (movement and maneuver cell).
- Cyber electromagnetic activities working group (electronic warfare element).

This page intentionally left blank.

Chapter 2
Staff Duties and Responsibilities

This chapter describes staffs, including their responsibilities, characteristics, and relationships, and explains the importance of building staff teams. This chapter also outlines the basic staff structure common to all headquarters and provides a discussion of the common duties and responsibilities of all staff sections. This chapter concludes by describing the duties and responsibilities of specific coordinating, special, and personal staff officers by area of expertise.

PRIMARY STAFF RESPONSIBILITIES

2-1.   The staff is a key component of the mission command system. In addition to executing the mission command staff tasks (see ADRP 6-0), the primary responsibilities of any staff are to—
- Support the commander.
- Assist subordinate commanders, staffs, and units.
- Inform units and organizations outside the headquarters.

SUPPORT THE COMMANDER

2-2.   Staffs support the commander in understanding, visualizing, and describing the operational environment; making and articulating decisions; and directing, leading, and assessing military operations. Staffs make recommendations and prepare plans and orders for the commander. Staff products consist of timely and relevant information and analysis. Staffs use knowledge management to extract that information from the vast amount of available information. (See chapter 3 for more information on knowledge management.) Staffs synthesize this information and provide it to commanders in the form of running estimates to help commanders build and maintain their situational understanding. (See chapter 8 for more information on running estimates.)

2-3.   Staffs support and advise the commander within their area of expertise. While commanders make key decisions, they are not the only decisionmakers. Trained and trusted staff members, given decisionmaking authority based on the commander's intent, free commanders from routine decisions. This enables commanders to focus on key aspects of operations.

2-4.   Staffs support the commander in communicating the commander's decisions and intent through plans and orders. (See appendixes C and D for more information on plans and orders formats and annexes.)

ASSIST SUBORDINATE COMMANDERS, STAFFS, AND UNITS

2-5.   Effective staffs establish and maintain a high degree of coordination and cooperation with staffs of higher, lower, supporting, supported, and adjacent units. Staffs help subordinate headquarters understand the larger context of operations. They do this by first understanding their higher headquarters' operations and commander's intent, and nesting their own operations with higher headquarters. They then actively collaborate with subordinate commanders and staffs to facilitate a shared understanding of the operational environment. Examples of staffs assisting subordinate units include performing staff coordination, staff assistance visits, and staff inspections.

INFORM UNITS AND ORGANIZATIONS OUTSIDE THE HEADQUARTERS

2-6.   The staff keeps its units well informed. The staff also keeps civilian organizations informed with relevant information according to their security classification, as well as their need to know. As soon as a staff receives information and determines its relevancy, that staff passes that information to the appropriate headquarters. The key is relevance, not volume. Masses of data are worse than meaningless data; they inhibit mission command by distracting staffs from relevant information. Effective knowledge management helps staffs identify the information the commander and each staff element need, and its relative importance. (See chapter 3 for more details on knowledge management.)

2-7. Information should reach recipients based on their need for it. Sending incomplete information sooner is better than sending complete information too late. When forwarding information, the sending staff highlights key information for each recipient and clarifies the commander's intent. Such highlighting and clarification assists receivers in analyzing the content of the information received in order to determine that information that may be of particular importance to the higher and subordinate commanders. The sending staff may pass information directly, include its analysis, or add context to it. Common, distributed databases can accelerate this function; however, they cannot replace the personal contact that adds perspective.

COMMON STAFF DUTIES AND RESPONSIBLITIES

2-8. In addition to the mission command staff tasks, each staff element has specific duties and responsibilities by area of expertise. However, all staff sections share a set of common duties and responsibilities:

- Advising and informing the commander.
- Building and maintaining running estimates.
- Providing recommendations.
- Preparing plans, orders, and other staff writing.
- Assessing operations.
- Managing information within area of expertise.
- Identifying and analyzing problems.
- Conducting staff assistance visits.
- Performing risk management.
- Performing intelligence preparation of the battlefield.
- Conducting staff inspections.
- Conducting staff research.
- Performing staff administrative procedures.
- Exercising staff supervision over their area of expertise.
- Consulting and working with the servicing legal representative.

STAFF CHARACTERISTICS

2-9. In addition to the leader attributes and core competencies addressed in Army leadership doctrine, a good staff officer is competent, exercises initiative, applies critical and creative thinking, is adaptable, is flexible, has self-confidence, is cooperative, is reflective, and communicates effectively. (See ADRP 6-22 for more details.) Effective staff officers seek a shared understanding of the operational environment with their commander, as well as the commanders of both higher and subordinate headquarters. This shared understanding includes the commander's visualization of the operational approach, to include his intent. Staffs continually reassess that understanding as changes occur within the operational environment.

2-10. Effective staff officers are competent in all aspects of their area of expertise. They are experts in doctrine and the processes and procedures associated with the operations process, and they understand the duties of other staff members enough to accomplish coordination both vertically and horizontally.

2-11. Staff officers exercise individual initiative. They anticipate requirements rather than waiting for instructions. They anticipate what the commander needs to accomplish the mission and prepare answers to potential questions before they are asked.

2-12. Staffs apply critical and creative thinking throughout the operations process to assist commanders in understanding and decisionmaking. As critical thinkers, staff officers discern truth in situations where direct observation is insufficient, impossible, or impractical. They determine whether adequate justification exists to accept conclusions as true based on a given inference or argument. As creative thinkers, staff officers look at different options to solve problems. They use adaptive approaches (drawing from previous similar circumstances) or innovative approaches (coming up with completely new ideas). In both instances, staff officers use creative thinking to apply imagination and depart from the old way of doing things.

2-13. Effective staff officers are adaptive. They recognize and adjust to changing conditions in the operational environment with appropriate, flexible, and timely actions. They rapidly adjust and continuously assess plans, tactics, techniques, and procedures.

2-14. Staff officers are flexible. They avoid becoming overwhelmed or frustrated by changing requirements and priorities. Commanders may change their minds or redirect the command after receiving additional information or a new mission and may not inform the staff of the reason for a change. Staff officers remain flexible and adjust to any changes. They set priorities when there are more tasks to accomplish than time allows. They learn to manage multiple commitments simultaneously.

2-15. Staff officers possess discipline and self-confidence. They understand that all staff work serves the commander, even if the commander rejects the resulting recommendation. Staff officers do not give a "half effort" even if they think the commander will disagree with their recommendations. Alternative and possibly unpopular ideas or points of view assist commanders in making the best possible decisions.

2-16. Staff officers are team players. They cooperate with other staff members within and outside the headquarters. This practice contributes to effective collaboration and coordination.

2-17. Staff officers are reflective in their actions. While conducting actions, they are able to quickly assess and implement corrective measures that lead to successful outcomes. Upon completion of actions, they analyze and assess events to implement measures that maximize efficiencies in the future.

2-18. Staff officers communicate clearly and present information orally, in writing, and visually (with charts, graphs, and figures). Staff officers routinely brief individuals and groups. They know and understand briefing techniques that convey complex information in easily understood formats. They can write clear and concise orders, plans, staff studies, staff summaries, and reports.

STAFF RELATIONSHIPS

2-19. Staff effectiveness depends in part on relationships of the staff with commanders and other staff. Collaboration and dialogue aids in developing shared understanding and visualization among staffs at different echelons. A staff acts on behalf of, and derives its authority from, the commander. Although commanders are the principal decisionmakers, individual staff officers make decisions within their authority based on broad guidance and unit standard operating procedures (SOPs). Commanders insist on frank dialogue between themselves and their staff officers. A staff gives honest, independent thoughts and recommendations, so commanders can make the best possible decisions. Once the commander makes a decision, staff officers support and implement the commander's decision even if the decision differs from their recommendations.

2-20. Teamwork within a staff and between staffs produces the staff integration essential to synchronized operations. A staff works efficiently with complete cooperation from all staff sections. A force operates effectively in cooperation with all headquarters. Commanders and staffs contribute to foster this positive climate during training and sustain it during operations. However, frequent personnel changes and augmentation to the headquarters adds challenges to building and maintaining the team. While all staff sections have clearly defined functional responsibilities, none can operate effectively in isolation. Therefore, coordination is extremely important. Commanders ensure staff sections are properly equipped and manned. This will allow staffs to efficiently work within the headquarters and with their counterparts in other headquarters. Commanders ensure staff integration through developing the unit's battle rhythm, including synchronizing various meetings, working groups, and boards.

STAFF ORGANIZATION

2-21. The basis for staff organization depends on the mission, each staff's broad areas of expertise, and regulations and laws. While staffs at every echelon and type of unit are structured differently, all staffs share some similarities. (See paragraphs 2-22 to 2-38.)

CONSIDERATIONS

2-22. The mission determines which activities to accomplish. These activities determine how commanders organize, tailor, or adapt their individual staffs to accomplish the mission. The mission also determines the size and composition of a staff, including staff augmentation.

2-23. Regardless of mission, every Army staff has common broad areas of expertise that determine how the commander divides duties and responsibilities. The duties and responsibilities inherent in an area of expertise are called functional responsibilities. Grouping related activities allows an effective span of control and unity of effort. Areas of expertise may vary slightly, depending on the echelon of command and mission. For example, at battalion level there is no financial manager, while certain sustainment units combine the intelligence and operations functions.

2-24. Army regulations and laws establish special relationships between certain staff officers and the commander. For example, regulations require the inspector general (AR 20-1), staff judge advocate (AR 27-10), and chaplain (AR 165-1) to be members of the commander's personal staff.

2-25. Every organization requires an authorization document that states a headquarters' approved structure and resources. It is the basis and authority for personnel assignments and equipment requisitions. This document is a table of organization and equipment (TOE), a modified TOE, or a table of distribution and allowances (known as TDA). Commanders establish authorizations by developing a modified TOE from the TOE for their individual units. Commanders prescribe in more detail the organization, personnel, and equipment to be authorized to accomplish missions in specific operational environments. Commanders can change their individual modified TOEs with Department of the Army approval.

STRUCTURE

2-26. The basic staff structure includes a COS (XO) and various staff sections. **A *staff section* is a grouping of staff members by area of expertise under a coordinating, special, or personal staff officer**. A principal staff officer—who may be a coordinating, special, or personal staff officer for the commander—leads each staff section. The number of coordinating, special, and personal principal staff officers and their corresponding staff sections varies with different command levels.

Commander

2-27. Commanders are responsible for all their staffs do or fail to do. A commander cannot delegate this responsibility. The final decision, as well as the final responsibility, remains with the commander. When commanders assign a staff member a task, they delegate the authority necessary to accomplish it. Commanders provide guidance, resources, and support. They foster a climate of mutual trust, cooperation, and teamwork.

Deputy Commander and Assistant Division Commanders

2-28. The commander determines the duties and responsibilities of the deputy and assistant commanders. These duties and responsibilities are formally declared and outlined in a terms of reference memorandum signed by the commander. In a corps or division, the deputy or assistant commander extends the commander's span of control in areas and functions as the commander designates. The deputy or assistant commander's specific duties vary from corps to corps and division to division.

2-29. The corps deputy commander serves as the commander's primary assistant and second-in-command of the corps. The corps deputy commander has specific duties directed by the commander and described in corps SOPs. The corps deputy commander and assistant division commanders do not have their own staffs, but they can request staff assistance at any time. They may supervise or control certain staff elements based on responsibilities assigned by the commander.

2-30. A division has two assistant division commanders who support the commander. The division commander specifies and assigns responsibility for tasks to the assistant division commanders to achieve the commander's intent. Normally one assistant division commander is the senior officer in the main command post and is responsible for supervising the execution of current operations. Both assistant division commanders prepare to execute operations from the tactical command post or mobile command group, as directed by the commander, to help control the execution of all division operations.

2-31. The corps deputy commander and assistant division commanders interact with the chief of staff and staff principal advisors based on duties the commanding general assigns. The deputy commander and assistant division commanders maintain situational understanding to enable them to assume command at any time. Because of this requirement, the corps deputy commander normally remains at the main command post (CP) to co-locate physically or virtually with the commanding general. The deputy commander and assistant division commanders have two general responsibilities:

- Temporarily assume the commanding general's duties.
- Assume certain delegated authorities.

Chief of Staff (Executive Officer)

2-32. The Chief of Staff (Executive Officer) (COS [XO]) is the commander's principal assistant. Commanders normally delegate executive management authority to the COS (XO). As the key staff integrator, the COS (XO) frees the commander from routine details of staff operations and the management of the headquarters. Division and higher units are assigned a COS. Brigade and battalions are assigned an XO. The COS (XO) ensures efficient and prompt staff actions. The COS (XO) duties include, but are not limited to—

- Coordinating and directing the work of the staff.
- Establishing and monitoring the headquarters battle rhythm and nesting with higher and subordinate headquarters battle rhythms for effective planning support, decisionmaking, and other critical functions.
- Representing the commander when authorized.
- Formulating and disseminating staff policies.
- Ensuring effective liaison exchanges with higher, lower, and adjacent units and other organizations as required.
- Supervising the sustainment of the headquarters and activities of the headquarters and headquarters battalion or company.
- Supervising staff training.
- Supervising the special staff sections in division through Army Service component command headquarters.

Principal Staff Officers

2-33. The principal staff officers consist of officers from the coordinating and special staff sections, as well as personal staff officers. Paragraphs 2-36 through 2-72 discuss coordinating staff officers. Paragraphs 2-73 through 2-104 discuss special staff officers. Paragraphs 2-105 through 2-114 discuss personal staff officers.

Noncommissioned Officers

2-34. Noncommissioned officers (NCOs) serve alongside their staff officer counterparts in all staff sections. They execute similar duties as those of their staff officer. NCOs often provide the experience and continuity in their particular staff section. They are to be counted upon to provide expert advice to the staff officer and other members of the staff section. NCOs display the same characteristics as good staff officers as described in paragraphs 2-9 to 2-18.

Augmentation

2-35. Often, Army headquarters receive augmentation teams to assist with mission command. Commanders integrate these teams and detachments into their command posts. For example, divisions commonly receive a civil affairs battalion when deployed. A civil affairs planning team within that battalion augments the civil affairs staff section and plans cell in the division headquarters. In other instances, commanders may request staff augmentation. Augmentation teams include, but are not limited to—

- Army space support team.
- Army cyberspace operations support team.
- Civil affairs planning team.
- Combat camera team.
- Legal support teams.
- Mobile public affairs detachment.
- Military history detachment.
- Military information support operations units.
- Army information operations field support team.
- Individual augmentation by specialty (for example, assessment, or economic development).

COORDINATING STAFF OFFICERS

2-36. Coordinating staff officers are the commander's principal assistants who advise, plan, and coordinate actions within their area of expertise or a warfighting function. Commanders may designate coordinating staff officers as assistant chiefs of staff, chiefs of a warfighting function, or staff officers. Coordinating staff officers may also exercise planning and supervisory authority over designated special staff officers.

Note: The commander's rank determines whether the staff is a G staff or an S staff. Organizations commanded by a general officer have G staffs. Other organizations have S staffs. Most battalions and brigades do not have plans or financial management staff sections.

2-37. The coordinating staff consists of the following positions:

- Assistant chief of staff (ACOS), G-1 (S-1)—personnel.
- ACOS, G-2 (S-2)—intelligence.
- ACOS, G-3 (S-3)—operations.
- ACOS, G-4 (S-4)—logistics.
- ACOS, G-5—plans.
- ACOS, G-6 (S-6)—signal.
- ACOS, G-8—financial management.
- ACOS, G-9 (S-9)—civil affairs operations.
- Chief of fires.
- Chief of protection.
- Chief of sustainment (see paragraph 2-55).

2-38. A chief of fires, a chief of protection, and a chief of sustainment are authorized at division and corps levels. They coordinate their respective warfighting functions for the commander through functional cells within the main command post. (See chapter 1.)

ASSISTANT CHIEF OF STAFF, G-1 (S-1), PERSONNEL

2-39. The ACOS, G-1 (S-1) is the principal staff officer for all matters concerning human resources support (military and civilian). The G-1 (S-1) also serves as the senior adjutant general officer in the command. Specific responsibilities of the G-1 (S-1) include manning, personnel services, personnel support, and headquarters management. The G-1 (S-1) has coordinating staff responsibility for the civilian personnel officer and the equal opportunity advisor. The G-1 (S-1) prepares a portion of Annex F (Sustainment) to the operation order or operation plan. (See FM 1-0 for more details.)

Man the Force

2-40. Manning the force impacts the effectiveness of all Army organizations, regardless of size, and affects the ability to successfully accomplish all other human resource core competencies and key functions. Manning includes five functional tasks: personnel readiness management, personnel accountability, personnel strength reporting, retention operations, and personnel information management. Corps and division G-1s maintain overall responsibility for personnel readiness management of subordinate elements. Corps and division G-1s maintain the responsibility to assist brigade S-1s and the national provider in shaping the force to meet mission requirements. Personnel accountability is the by-name management of the location and duty status of every person assigned or attached to a unit. Personnel strength reporting is a numerical product of the accountability process. The Army Retention Program is the long-term answer for maintaining end strength. Personnel information management is a process to collect, process, store, display, and disseminate information about Soldiers, Army civilians, units, and other personnel as required.

Provide Human Resources Services (Essential Personnel Services)

2-41. Essential personnel services are initiated by the Soldier, unit commanders, unit leaders, G-1s (S-1s), or from the top of the human resource command. Typical actions initiated by the Soldier are personnel action requests, requests for leaves or passes, changes to record of emergency data or life insurance elections, changes to dependent information, allotments, saving bonds, and direct deposit information. Typical actions initiated by commanders include requests for awards or decorations, promotions, reductions, and bars to reenlistment. Normally, the supervisor at all levels initiates evaluation reports (such as change of rater and complete the record reports). The military postal system operates as an extension of the United States Postal Service. Casualty operations record, report, verify, and process casualty information from the unit level to the casualty and mortuary affairs operations center, notify appropriate individuals, and provide casualty assistance to the next-of-kin.

Coordinate Personnel Support

2-42. Personnel support activities encompass those functions and activities that contribute to unit readiness by promoting fitness, building morale and cohesion, enhancing quality of life, and providing recreational, social, and other support services for Soldiers, Army civilians, and other personnel who deploy with the force. Personnel support encompasses the following functions: morale, welfare, and recreation, command interest programs, and Army band operations. Commanders at all levels are responsible for the morale, welfare, and recreation support provided to their Soldiers and civilians. Command interest programs include family readiness, Army substance abuse program, suicide prevention program, and other programs as directed. Army bands provide music for ceremonial and morale support in all operations to sustain Soldiers and to inspire leaders.

Headquarters Management

2-43. Headquarters management includes, but is not limited to—
- Managing the organization and administration of the headquarters.
- Providing administrative support for military and civilian personnel, including leaves, passes, counseling, transfers, awards, and personal affairs.
- Providing information services, including publications, printing, distribution, and material for the Freedom of Information Act.
- Providing administrative support for non-U.S. forces, foreign nationals, and civilian internees.
- Administering discipline, law, and order (with the provost marshal), including desertion, court-martial offenses, punishments, and straggler dispositions.

ASSISTANT CHIEF OF STAFF, G-2 (S-2), INTELLIGENCE

2-44. The ACOS, G-2 (S-2) is the chief of the intelligence warfighting function and the principal staff officer responsible for providing intelligence to support current and future operations and plans. This officer gathers and analyzes information on enemy, terrain, weather, and civil considerations for the commander. The G-2 (S-2) is responsible for the preparation of Annex B (Intelligence) and assists the assistant chief of staff, operations (G-3 [S-3]) in the preparation of Annex L (Information Collection). (See FM 2-0 for additional information on the G-2 [S-2], and see ADP 2-0 and ADRP 2-0 for more details.)

2-45. The G-2 (S-2), together with the G-3 (S-3), helps the commander coordinate, integrate, and supervise the execution of information collection plans and operations. The G-2 (S-2) helps the commander focus and integrate these assets and resources to satisfy the battalion through corps intelligence requirements. Some of the specific responsibilities of the G-2 (S-2) include, but are not limited to–

- Overseeing the intelligence functional cell, specifically situation development, target development, support to lethal and nonlethal targeting, support to indications and warnings, support to assessment, and support to protection.
- Providing the commander and staff with assessments of threat capabilities, intentions, and courses of action (COAs) as they relate to the division or corps and its mission.
- Identifying gaps in intelligence and developing collection strategies.
- Disseminating intelligence products throughout the unit (battalion through corps) as well as to higher and subordinate headquarters.
- Answering requests for information from subordinate commanders, staffs, and higher and adjacent units.
- Coordinating the units' intelligence requirements with supporting higher, lateral, and subordinate echelons.
- Overseeing the intelligence cell's contributions to planning requirements and assessing collection.
- Participating with staff in performing intelligence preparation of the battlefield (IPB).
- Monitoring intelligence operations.
- Ensuring ongoing intelligence operations are collecting information needed for anticipated decisions or other priority intelligence requirements (PIRs).
- Ensuring information concerning the PIRs is processed and analyzed first.
- Recommending changes to the information collection plan based on changes in the situation and weather.
- Counterintelligence responsibilities including, but not limited to—
  - Coordinating counterintelligence activities.
  - Identifying enemy intelligence collection capabilities, such as efforts targeted against the unit.
  - Evaluating enemy intelligence capabilities as they affect operations security, signals security, countersurveillance, security operations, military deception planning, military information support operations, and protection.
  - Vetting all contractors and their employees to deter the subversive nature of insurgent activities.
- Support to security programs includes—
  - Supervising the command and personnel security programs.
  - Evaluating physical security vulnerabilities to support the G-3 (S-3) and assistant chief of staff, signal (G-6 [S-6]).
  - Performing staff planning and supervising the special security office.

ASSISTANT CHIEF OF STAFF, G-3 (S-3), OPERATIONS

2-46. The ACOS, G-3 (S-3) operations officer's responsibilities are unique within the coordinating staff. The G-3 (S-3) is the chief of the movement and maneuver warfighting function and the principal staff officer responsible for all matters concerning training, operations and plans, and force development and modernization. In addition to coordinating the activities of the movement and maneuver warfighting function, the operations officer is the primary staff officer for integrating and synchronizing the operation as a whole for the commander. While the COS (XO) directs the efforts of the entire staff, the operations officer ensures warfighting function integration and synchronization across the planning horizons in current operations integration, future operations, and plans integrating cells. (See chapter 1.) Additionally, the operations officer authenticates all plans and orders for the commander to ensure the warfighting functions are synchronized in time, space, and purpose in accordance with the commander's intent and planning guidance.

2-47. The G-3 (S-3) has coordinating staff responsibility for the G-5, aviation officer, engineer officer, military information support officer, information operations officer, force management officer, and space operations officer, as well as other staff officers residing in the movement and maneuver cell.

2-48. The G-3 (S-3) is responsible for and prepares Annex L (Information Collection) and Annex V (Interagency Coordination). In conjunction with the G-5 (S-5), the G-3 (S-3) prepares Annex A (Task Organization), Annex C (Operations), and Annex M (Assessment) to the operation order or operation plan. In conjunction with the knowledge management officer (KMO), the G-3 (S-3) prepares Annex R (Reports) and Annex Z (Distribution).

Training

2-49. G-3 (S-3) training responsibilities include, but are not limited to—
- Conducting training within the command.
- Preparing training guidance for the commander's approval.
- Identifying training requirements, based on the unit mission essential task list and training status.
- Determining requirements for and allocation of training resources.
- Organizing and conducting internal schools, and obtaining and allocating quotas for external schools.
- Conducting training inspections, tests, and evaluations.
- Maintaining the unit readiness status of each unit in the command.
- Compiling training records and reports.

Plans and Operations

2-50. The G-3 (S-3) has responsibilities for plans and operations. Overall, this officer prepares, coordinates, authenticates, reviews, publishes, and distributes written operation orders and plans. This includes the command SOP, plans, orders (including fragmentary orders [FRAGORDs] and warning orders [WARNORDs]), exercises, terrain requirements, and products involving contributions from other staff sections. The G-3 (S-3) provides coordination, integrates information collection, and allocates resources.

2-51. The G-3 (S-3) coordinates with other staff officers during plans and operations. This list is not all-inclusive. This officer coordinates with the G-1 (S-1) for civilian personnel involvement in tactical operations and with the assistant chief of staff, civil affairs operations (G-9 [S-9]) on using Army forces to establish or reestablish civil government. By coordinating with the commander, the COS (XO), G-6 (S-6), and the G-3 (S-3) can establish, oversee, and supervise staff activities of the command post. Coordinating with the engineer officer, G-2 (S-2), chief of protection, G-9 (S-9), and surgeon, the G-3 (S-3) establishes environmental vulnerability protection levels. Coordinating with the chief of protection and operations security officer, the G-3 (S-3) establishes operations security priorities, plans, and guidance.

2-52. The G-3 (S-3) integrates information collection during plans and operations. This officer integrates information collection into the concept of operations and manages the information collection effort through integrated staff processes and procedures. The G-3 (S-3) also synchronizes information collection with the overall operation throughout the operations process (with the rest of the staff). By developing the information collection plan (with rest of the staff) to support the commander's visualization, the information collection plan produces an initial information collection order.

2-53. The G-3 (S-3) allocates resources during plans and operations and ensures units provide necessary support requirements when and where required. The G-3 (S-3) retasks and refocuses collection assets during execution (considering recommendations from the rest of the staff). This officer recommends use of resources, including resources required for military deception, and sustainment requirements (with the G-1 [S-1]) and the assistant chief of staff, logistics G-4 [S-4]).

2-54. During plans and operations, the G-3 (S-3) also—

- Develops the information collection annex to plans and orders (with the rest of the staff).
- Allocates information collection tasks (considering recommendations from the rest of the staff).
- Integrates fires into operations.
- Plans tactical troop movements, including route selection, priority of movement, timing, security, bivouacking, quartering, staging, and preparing movement orders.
- Develops the ammunition required supply rate (with the G-2 [S-2], chief of fires [fire support officer], and G-4 [S-4]).
- Requisitions replacement units (through operations channels).
- Participates in course of action and decision support template development (with the G-2 [S-2] and the chief of fires [fire support officer]).
- Recommends general command post locations.
- Recommends task organizations and assigns missions to subordinate elements.
- Supports linguist requirements, to include consolidating linguist requirements and establishing priorities for using linguists.

ASSISTANT CHIEF OF STAFF, G-4 (S-4), LOGISTICS

2-55. The ACOS, G-4 (S-4) is the principal staff officer for sustainment plans and operations, supply, maintenance, transportation, services, and operational contract support. At division and corps level the G-4 is titled the chief of sustainment. At brigade level and below the S-4 serves as the principal staff officer coordinating sustainment. The G-4 (S-4) helps the support unit commander maintain logistics visibility with the commander and the rest of the staff. As the chief of sustainment, the G-4 has coordinating staff responsibility for the G-1, G-8, transportation officer, and the surgeon. The G-4 (S-4) prepares Annex F (Sustainment), Annex P (Host-Nation Support) and Annex W (Operational Contract Support) to the operation order or operation plan. (See ADRP 4-0 for more details.)

Sustainment Plans and Operations (General)

2-56. The G-4 (S-4) responsibilities for sustainment plans and operations include, but are not limited to—

- Developing the logistic plan to support operations (with the G-3 [S-3]).
- Coordinating with the G-3 (S-3), G-2 (S-2), and engineer officer to requisition cataloged topographic foundation data and existing mission-specific data sets from the Defense Logistics Agency.
- Coordinating with the G-3 (S-3) and G-1 (S-1) on equipping replacement personnel and units.
- Coordinating with the support unit commander on the current and future support capability of that unit.
- Coordinating the selection of main supply routes and logistic support areas (with the engineer officer) and recommending them to the G-3 (S-3).
- Performing logistic preparation of the battlefield (with the support command).
- Recommending command policy for collecting and disposing of excess property and salvage.

Supply

2-57. The G-4 (S-4) responsibilities for supply include, but are not limited to—

- Determining supply requirements, except medical (with the support unit commander and the G-3 [S-3]).
- Coordinating all classes of supply except Class VIII (which is coordinated through medical supply channels).
- Coordinating the requisition, acquisition, and storage of supplies and equipment and the maintenance of materiel records.
- Recommending sustainment priorities and controlled supply rates.
- Ensuring that accountability and security of supplies and equipment are adequate (with the provost marshal).
- Calculating and recommending to the G-3 (S-3) basic and prescribed loads, and helping the G-3 (S-3) determine required supply rates.

Maintenance

2-58. The G-4 (S-4) responsibilities for maintenance include, but are not limited to—

- Monitoring and analyzing the equipment readiness status.
- Determining maintenance workload requirements, except medical (with the support command).
- Coordinating equipment recovery and evacuation operations (with the support command).
- Determining maintenance timelines.

Transportation

2-59. The G-4 (S-4) responsibilities for transportation include, but are not limited to—

- Conducting operational and tactical planning to support mode and terminal operations, and movement control.
- Planning administrative troop movements (with the G-3 [S-3]).
- Coordinating transportation assets for other Services.
- Coordinating with the G-9 (S-9) for host-nation support.
- Coordinating special transport requirements to move the command post.
- Coordinating with the G-1 (S-1) and the provost marshal to transport replacement personnel and enemy prisoners of war.
- Coordinating with the G-3 (S-3) for sustainment of tactical troop movements.

Services

2-60. The G-4 (S-4) responsibilities for services include, but are not limited to—

- Coordinating the construction of facilities and installations, except for fortifications and signal systems.
- Coordinating field sanitation.
- Coordinating organizational clothing and individual equipment exchange and replacement.
- Coordinating unit spill-prevention plans.
- Coordinating or providing food preparation, water purification, mortuary affairs, aerial delivery, laundry, shower, and clothing and light textile repair.
- Coordinating the transportation, storage, handling, and disposal of hazardous material or hazardous waste.

Staff Planning and Supervision

2-61. The G-4 (S-4) has the following staff planning and supervisory responsibilities:

- Identifying requirements the unit can meet through contracting.
- Identifying requirements and restrictions, in conjunction with the staff judge advocate, for using local civilians, enemy prisoners of war, civilian internees, and detainees in sustainment operations.
- Coordinating with the staff judge advocate on legal aspects of contracting.
- Coordinating with financial managers on the financial resources availability.
- Coordinating real property control and fire protection for facilities.

2-62. A support operations officer or materiel officer is authorized in support commands and battalions. As the principal staff officer for coordinating logistics, the support operations officer or materiel officer provides technical supervision for the sustainment mission of the support command and is the key interface between the supported unit and the support command. The responsibilities of the support operations officer or materiel officer include, but are not limited to—

- Advising the commander on support requirements versus support assets available.
- Coordinating external support requirements for supported units.
- Synchronizing support requirements to ensure they remain consistent with current and future operations.
- Planning and monitoring support operations and making adjustments to meet support requirements.
- Coordinating with other staff.
- Preparing and distributing the external service support SOP that provides guidance and procedures to supported units.

ASSISTANT CHIEF OF STAFF, G-5, PLANS

2-63. The ACOS, G-5 (S-5) is the principal staff officer for planning operations for the mid- to long-range planning horizons at division echelon and higher. In conjunction with the G-3 (S-3), the G-5 prepares Annex A (Task Organization), Annex C (Operations), and Annex M (Assessment) to the operation order or operation plan. (See ADRP 5-0 for more details.)

2-64. Plans and orders consist of—

- Preparing, coordinating, authenticating, publishing, and distributing operation plans, concept plans, and operation orders.
- Conducting mission analysis of higher headquarters plans and orders.
- Reviewing subordinate supporting plans and orders.
- Coordinating and synchronizing warfighting functions in all plans and orders.

2-65. The G-5 has staff planning and supervisory responsibility for—

- Overseeing operations beyond the scope of the current order (such as the next operation or the next phase of the current operation).
- Developing plans, orders, branches, and sequels.
- Conducting military deception planning.
- Developing policies and other coordinating or directive products, such as memorandums of agreement.

ASSISTANT CHIEF OF STAFF, G-6 (S-6), SIGNAL

2-66. The ACOS G-6 (S-6) is the principal staff officer for all matters concerning network operations (jointly consisting of Department of Defense Information Network Operations and applicable portions of the Defensive Cyberspace Operations), network transport, information services, and spectrum management operations within the unit's area of operations. The G-6 (S-6) prepares Annex H (Signal) and participates in preparation of Appendix 12 (Cyberspace Electromagnetic Activities) to Annex C (Operations) with input from the G-2 (S-2) and in coordination with the G-3 (S-3), to the operation order or operation plan. (See FM 6-02.70 and FM 6-02.71 for more details.) G-6 (S-6) responsibilities include, but are not limited to—

- Preparing and maintaining network operations estimates, plans, and orders.
- Overseeing Department of Defense Information Network Operations related functions that engineer and install the network to support operational requirements.
- Directing and managing the operation of the network to ensure network and information system availability and information delivery.
- Managing the execution of Defensive Cyberspace Operations for the network in coordination with other staff sections.
- Overseeing or participating in the development of plans and orders for cyber electromagnetic activities in conjunction with other staff sections.
- Overseeing or participating in the development and maintenance of the cyberspace common operational picture with assistance from the G-2 (S-2) and other staff sections.
- Coordinating and managing spectrum management operations and communications security within the area of operations.
- Recommending CP locations, based on operational requirements and the information environment.
- Recommending network-related essential elements of friendly information.
- Coordinating contractor and maintenance support for all network operations, information services, and electromagnetic spectrum management.

Network Operations

2-67. The network operations officer oversees the operation and defense of the warfighting information network and ensures the confidentiality, integrity, and availability of information critical to mission command and the establishment of a cyber-related common operation picture. At the operational and tactical level, network operations capabilities include network and systems management, information assurance and computer network defense (to include response actions), information dissemination management, and content staging. G-6 (S-6) responsibilities related to network operations include, but are not limited to—

- Coordinating, planning, and directing the integration of the mission-related networks and information systems with those of unified action partners.
- Ensuring the effective and efficient operations of information systems, elements of systems, and services (including operating systems, databases, and hosts of the end-users).
- Provisioning networked system services with the desired level of quality and guaranteed availability.
- Coordinating unit commercial and military satellite communications requirements with the space operations officer.
- Managing radio frequency allocations and assignments and providing electromagnetic spectrum management within the area of operations.
- Planning, coordinating, and directing all measures that protect information and information systems by ensuring their availability, integrity, authentication, confidentiality, and nonrepudiation.
- Monitoring, detecting, analyzing, and responding to unauthorized activity (malicious or non-malicious) occurring within information networks and systems.
- Ensuring that information dissemination management and content staging capabilities used to deliver, discover, and store information meet the command's critical information requirements.
- Coordinating, planning, and directing all command information assurance activities.
- Providing operational and technical support to all assigned or attached units.

ASSISTANT CHIEF OF STAFF, G-8, FINANCIAL MANAGEMENT

2-68. The ACOS, (G-8) is the principal staff officer singularly responsible for all financial management (resource management and finance operations). As the principal financial management advisor to the commander, this officer directs, prioritizes, and supervises the operations and functions of the G-8 staff sections assigned to the G-8 and the contingency command post. In coordination with the financial management center and through the theater sustainment command, the G-8 establishes and implements command finance operations policy. The G-8 is responsible for those operational financial management tasks supporting the theater. This officer works with the servicing legal representative for advice regarding laws and financial management regulations governing obligations, expenditures, and limitations on the use of public funds. The G-8 coordinates financial management policies and practices with the expeditionary contracting command to ensure guidance is executed in accordance with Department of the Army (DA) mandates. The G-8 prepares a portion of Annex F (Sustainment). (See FM 1-06 for more details.)

2-69. The financial management center is a modular and tailorable operational financial management unit. Its mission is inextricably linked to the theater army G-8, but it is assigned to a theater sustainment command. The financial management center supports the Army Service component command, theater sustainment command, and the expeditionary sustainment command by providing cash management, internal control measures, accounting, automation, and technical guidance for financial management companies and financial management detachments. To provide adequate theater and national-provider responsiveness and support, the financial management center maintains oversight of all financial management operations and placement of all operational and tactical financial management units in theater. The financial management center provides technical coordination of all theater finance operations and collected advice to the theater army G-8 and the theater sustainment command commander on all aspects of theater finance operations.

ASSISTANT CHIEF OF STAFF, G-9 (S-9), CIVIL AFFAIRS OPERATIONS

2-70. The ACOS, G-9 (S-9) is the principal staff officer responsible for all matters concerning civil affairs. The G-9 (S-9) establishes the civil-military operations center, evaluates civil considerations during mission analysis, and prepares the groundwork for transitioning the area of operations from military to civilian control. The G-9 (S-9) advises the commander on the military's effect on civilians in the area of operations, relative to the complex relationship of these people with the terrain and institutions over time. The G-9 (S-9) is responsible for enhancing the relationship between Army forces and the civil authorities and people in the area of operations. The G-9 (S-9) is required at all echelons from battalion through corps, but it is normally authorized only at division and corps. Once deployed, units below division level may be authorized an S-9. The G-9 (S-9) prepares Annex K (Civil Affairs Operations) to the operation order or operation plan. (See FM 3-57 for more information.) G-9 (S-9) responsibilities include, but are not limited to—

- Operating a civil-military operations center to maintain liaison with other U.S. government agencies, host-nation civil and military authorities, and nongovernmental and international organizations in the area of operations.
- Coordinating with the chief of fires or fire support officer on the restricted target list and the no-strike list.
- Planning community relations programs to gain and maintain public understanding and goodwill and to support military operations.
- Providing the G-2 (S-2) information gained from civilians in the area of operations.
- Coordinating with the surgeon on the military use of civilian medical treatment facilities, materials, and supplies.
- Coordinating with the information operations officer to ensure disseminated information is not contradictory.
- Coordinating with the public affairs officer on supervising public information media under civil control.

- Providing instruction to units, officials (friendly, host-nation civil, or host-nation military), and the population on identifying, planning, and implementing programs to support civilian populations and strengthen host-nation internal defense and development.
- Identifying and assisting the G-6 (S-6) with coordinating military use of local information systems.
- Coordinating with the provost marshal to control civilian traffic in the area of operations.
- Helping the G-4 (S-4) coordinate facilities, supplies, and other materiel resources available from the civil sector to support operations.

CHIEF OF FIRES OR FIRE SUPPORT OFFICER

2-71. The chief of fires is the principal staff officer responsible for the fires warfighting function at division through theater army. At brigade and below, the fire support officer serves as a special staff officer for fires. This officer synchronizes and coordinates fire support for the S-3 who integrates fire support into plans and operations. The chief of fires has coordinating responsibility for the air and missile defense officer and the air liaison officer. The chief of fires or fire support officer prepares Annex D (Fires) to the operation order or operation plan. (See ADRP 3-09 for more details.) The chief of fires or fire support officers' responsibilities include, but are not limited to—

- Planning, preparing, executing, and assessing all fires tasks in support of offensive, defensive and stability tasks and providing inputs for preparation of the operation plan and operation order.
- Developing, with the commander and G-3 (S-3), a scheme of fires to support the operation.
- Planning and coordinating fire support tasks.
- Developing a proposed high-payoff target list, target selection standards, and an attack guidance matrix.
- Identifying named and target areas of interest, high-payoff targets, and additional events that may influence the positioning of fire support assets.
- Coordinating the positioning of fires assets.
- Providing information on the status of Army, joint, and multinational fires and their systems, including target acquisition assets and munitions.
- Recommending fire support coordination measures to support current and future operations and managing changes to them.
- Recommending and implementing the commander's counterfire and target engagement priorities.
- Recommending to the commander the establishment, responsibilities, authorities, and duties of a force field artillery headquarters as necessary.
- Conducting the tasks associated with integrating and synchronizing joint fires and multinational fires with the other warfighting functions.
- Training fires cell personnel to perform all of their functions.
- Advising the commander and staff of available fires capabilities and limitations.
- Leading the targeting working group.
- Working with the COS (XO), and G-3 (S-3) to integrate all types of fires into the commander's concept of operations.
- Accompanying the commander in the command group during execution of tactical operations (when directed).

CHIEF OF PROTECTION

2-72. The chief of protection is the principal staff officer responsible for the protection warfighting function at division through theater army. The chief of protection is the principal advisor to the commander on all matters relating to the protection warfighting function. Brigade and lower echelon headquarters are not assigned a chief of protection. The chief of protection has coordinating staff responsibilities for the chemical, biological, radiological, and nuclear officer; the explosive ordnance disposal officer; the operations security officer; the personnel recovery officer; the provost marshal; and the safety officer. At brigade and lower echelon headquarters, the S-3 has coordinating staff responsibility for these staff officers. The chief of protection prepares Annex E (Protection) to the operation order or operation plan. (See ADRP 3-37 for more details.) The chief of protection's responsibilities include, but are not limited to—

- Directing analysis, planning, and coordinating protection functions and missions.
- Advising the commander on the allocation and employment of all assigned or attached protection assets.
- Chairing the protection working group.
- Coordinating input and making recommendations to the commander on the assets to be included in the critical and defended asset lists.
- Monitoring and assessing the protection effort.
- Conducting staff coordination with other headquarters cells, nodes, and functional groupings.
- Managing protection support for major operations.
- Synchronizing protection operations between CPs.
- Managing training and materiel enhancements.
- Providing guidance on protection systems and the execution of protection tasks.

SPECIAL STAFF OFFICERS

2-73. Every staff has special staff officers. This section addresses the specific duties of each special staff officer. The number of special staff officers and their responsibilities vary with authorizations, the desires of the commander, and the size of the command. If a special staff officer is not assigned, the officer with coordinating staff responsibility for the area of expertise assumes those functional responsibilities. During operations, special staff officers work in parts of the CP designated by the commander, COS, or their supervising coordinating staff officer. In general, the COS (XO) exercises coordinating staff responsibility over those special staff officers without a coordinating staff officer.

AIR AND MISSILE DEFENSE OFFICER

2-74. The air and missile defense officer is responsible for coordinating air and missile defense activities and plans with the area air and missile defense commander, joint force air component commander, and airspace control authority. The air and missile defense officer coordinates the planning and use of all joint air and missile defense systems, assets, and operations. A key role of the air and missile defense officer is to advise the commander on developing a critical asset list and a defended asset list of assets that can be defended by air and missile defense forces. Army forces air and missile defense plans are synchronized with the area air defense commander's area air defense plan, the joint force air component commander's joint air operations plan and daily air tasking order, and the airspace control authority's airspace control plan, and daily airspace control order. The air and missile defense officer prepares a portion of Annex D (Fires) to the operation order or operation plan. (See FM 3-01, FM 3-27, and FM 3-52 for more details.)

2-75. The air and missile defense officer is the senior air defense artillery officer in the command and the commander of an air defense artillery unit supporting it. An air and missile defense officer is authorized at the division, corps, and theater army levels. Examples of air and missile defense officer responsibilities include, but are not limited to—

- Disseminating air tasking order and airspace control order information to air defense artillery units.
- Integrating airspace coordinating measures to support air and missile defense operations.
- Recommending active and passive air defense measures.
- Determining requirements and recommending assets to support air and missile defense.
- Planning and coordinating airspace use with other staffs.
- Providing information on the status of air and missile defense systems, air and missile attack early warning radars, and air defense artillery ammunition.

- Estimating the adequacy of the air defense artillery ammunition controlled supply rate.
- Coordinating and synchronizing Army forces air and missile defense with joint force air and missile defense.

AIR LIAISON OFFICER

2-76. The air liaison officer is responsible for coordinating aerospace assets and operations, such as close air support, air interdiction, air reconnaissance, airlift, and joint suppression of enemy air defenses. The air liaison officer is the senior Air Force officer with each tactical air control party. Air liaison officer responsibilities include, but are not limited to—

- Advising the commander and staff on employing aerospace assets.
- Operating and maintaining the Air Force tactical air direction radio network and Air Force air request network.
- Transmitting requests for immediate close air support and reconnaissance support.
- Transmitting advance notification of impending immediate airlift requirements.
- Acting as liaison between air and missile defense units and air control units.
- Planning the simultaneous employment of air and surface fires.
- Coordinating air support missions with the chief of fires or fire support officer, and the appropriate airspace control element.
- Supervising joint terminal attack controllers and the tactical air control party.
- Integrating air support sorties with the Army concept of operations.
- Participating in targeting team meetings.
- Directing close air support missions.
- Providing Air Force input into airspace control.

AVIATION OFFICER

2-77. The aviation officer is responsible for coordinating Army aviation assets and operations at division, corps, and theater army levels. (See FM 3-52 for more details.) The aviation officer's responsibilities include, but are not limited to—

- Exercising staff supervision and training over Army aviation operations.
- Monitoring aviation flying-hour, standardization, and safety programs.
- Planning and supervising Army aviation operations.
- Providing technical advice and assistance on using Army aviation for evacuation (medical or other).
- Participating in targeting meetings.

CHEMICAL, BIOLOGICAL, RADIOLOGICAL, AND NUCLEAR OFFICER

2-78. The chemical, biological, radiological, and nuclear (CBRN) officer is responsible for CBRN operations, obscuration operations, and CBRN asset use. The CBRN officer prepares a portion of Annex C (Operations) and a portion of Annex E (Protection) to the operation order or operation plan. (See FM 3-11.21 for more details.) The CBRN officer's responsibilities include, but are not limited to—

- Recommending courses of action to minimize friendly and civilian vulnerability.
- Coordinating across the entire staff while assessing the effect of enemy CBRN-related attacks and hazards on current and future operations.
- Coordinating Army health system support requirements for CBRN operations with the surgeon.
- Coordinating with other staff for CBRN-related operations.
- Planning, supervising, and coordinating CBRN decontamination (except patient decontamination) operations.
- Assessing weather and terrain data to determine environmental effects on potential CBRN hazards and threats.

- Overseeing construction of CBRN shelters.
- Planning and recommending integration of obscuration into tactical operations.
- Advising the commander on CBRN threats and hazards and passive defense measures.

CIVILIAN PERSONNEL OFFICER

2-79. The civilian personnel officer manages and administers the civilian employee personnel management program. This civilian employee has a permanent position on the staff at divisions and corps. (See Army Regulation [AR] 690-11 and other regulations in this series for more details.) Specific duties include, but are not limited to—

- Advising the commander and staff concerning the civilian employee personnel management program and supervising the management and administration of that program within the command.
- Administering civilian personnel management laws and regulations.
- Participating, when appropriate, in negotiations with host nations on labor agreements.
- Developing plans and standby directives for procuring, using, and administering the civilian labor force and using local labor in foreign areas during emergencies (with other staff members).

DENTAL SURGEON

2-80. The dental surgeon coordinates dental activities within the command. All dental activities are planned at the medical brigade (support), medical command (deployment support), or Army Service component command. (See FM 4-02.19 for more details.) Dental surgeon responsibilities include, but are not limited to—

- Coordinating dental activities with the surgeon.
- Exercising staff supervision over and providing technical assistance to dental activities.
- Planning and supervising dental functions.
- Developing a program for dental support of foreign humanitarian assistance.
- Providing advice and technical assistance in constructing, rehabilitating, and using dental facilities.

ELECTRONIC WARFARE OFFICER

2-81. The electronic warfare officer is a specially trained officer who performs electronic warfare duties and integrates cyber electromagnetic activities. The electronic warfare officer prepares a portion of Annex C (Operations) to the operation order or operation plan and contributes to any section that has a cyber electromagnetic activities subparagraph such as Annex N (Space Operations). Electronic warfare officer responsibilities include, but are not limited to—

- Leading the electronic warfare element.
- Integrating and synchronizing cyber electromagnetic activities.
- Coordinating, preparing, and maintaining the electronic warfare target list, electronic attack taskings, and electronic attack requests.
- Coordinating with other staff when conducting electronic warfare.
- Assessing opponent strength and vulnerabilities, friendly capabilities, and friendly missions in electronic warfare terms.
- Developing a prioritized adversary target list based on high-value targets and high-payoff targets (with the chief of fires and fire support officer).
- Coordinating the electronic attack target list with organic military intelligence units and with adjacent and higher commands, including joint and multinational commands when appropriate.
- Coordinating with the information operations officer to deconflict cyber electromagnetic activities and information operations.
- Leading the cyber electromagnetic activities working group.
- Participating in working groups and targeting meetings.

ENGINEER OFFICER

2-82. The engineer officer resides in the protection cell and is responsible for planning and assessing survivability operations. The engineer officer is involved in planning and operations with more than just the protection warfighting function. For example, mobility and countermobility are part of movement and maneuver, general engineering is part of sustainment, and geospatial engineering supports the intelligence warfighting function. The engineer officer prepares Annex G (Engineer) to the operation order or operation plan. (See FM 3-34 for more details.) Specific duties include, but are not limited to—

- Advising the chief of protection on survivability operations.
- Coordinating and synchronizing survivability operations.
- Synchronizing and integrating engineer operations (combat and construction) between multiple command posts and organizations.
- Writing engineer FRAGORDs, WARNORDs, and related products.
- Providing real-time reachback linkage to United States Army Corps of Engineers knowledge centers and supporting national assets.
- Updating the running estimate.

EXPLOSIVE ORDNANCE DISPOSAL OFFICER

2-83. The explosive ordnance disposal officer is responsible for coordinating the detection, identification, recovery, evaluation, rendering safe, and final disposal of explosive ordnance. An explosive ordnance disposal officer is authorized at corps and divisions. The explosive ordnance disposal officer prepares a portion of Annex E (Protection) to the operation order or operation plan. (See AR 75-15 for more details.) Explosive ordnance disposal officer responsibilities include, but are not limited to—

- Establishing and operating an explosive ordnance disposal incident reporting system.
- Establishing, operating, and supervising technical intelligence reporting procedures.
- Coordinating requirements for explosive ordnance disposal support with requesting units, other Army commands, other Services, federal agencies, and multinational partners. Coordination may include arranging for administrative and logistic support for subordinate explosive ordnance disposal units.
- Monitoring the supply status of and expediting requests for special explosive ordnance disposal tools, equipment, and demolition materials.

EQUAL OPPORTUNITY ADVISOR

2-84. The equal opportunity advisor coordinates matters concerning equal opportunity for Soldiers and their families. Commanders at every echelon are authorized to appoint an equal opportunity advisor. (See AR 600-20 for more details.) The duties and responsibilities of the equal opportunity advisor include, but are not limited to—

- Advising and assisting the commander and staff on all equal opportunity matters, including sexual harassment, discrimination, and affirmative action.
- Consulting with the servicing legal representative during all informal and formal investigations.
- Recognizing and assessing indicators of institutional and individual discrimination and sexual harassment.
- Recommending, developing, and monitoring affirmative action and equal opportunity plans and policies to reduce or prevent discrimination and sexual harassment.
- Collecting and processing demographic data concerning all aspects of equal opportunity climate assessment.
- Managing or conducting all equal opportunity education and training programs within the command, to include conducting ethnic observances.
- Receiving and helping process complaints.

FORCE MANAGEMENT OFFICER

2-85. The force management officer is responsible for accounting for the force and its resources. This officer evaluates the organizational structure, functions, and workload of military and civilian personnel to ensure their proper use and requirements. By conducting formal, on-site manpower and equipment surveys, this officer ensures documents for the modified TOE and the tables of distribution and allowances reflect the minimum essential and most economical equipment needed for the mission.

FOREIGN DISCLOSURE OFFICER

2-86. The foreign disclosure officer is responsible for the oversight and coordination of specific disclosure of or access to classified military information or controlled unclassified information to representatives of foreign governments and international organizations. (See AR 380-10 for more details.)

HISTORIAN

2-87. The historian is the special staff officer responsible for implementing the commander's history program. This includes collecting, preserving, and expressing the accurate historical record of the command. The historian can be either an officer or a civilian professional historian. Historian responsibilities include, but are not limited to—

- Preparing the Command Report and Annual History.
- Collecting and preservating unit historical documents such as plans, orders, and after action reviews and reports.
- Providing historical perspective and institutional memory to the planning and decisionmaking process.
- Preparing special studies and reports, based on assembled historical material.
- Maintaining a historical research collection to support the institutional memory of the command.
- Supervising the command's historical program.

HUMAN TERRAIN TEAMS

2-88. The human terrain team's mission is to conduct operationally relevant social science research and provide commanders and staffs with an embedded knowledge capability to establish a coherent, analytical, socio-cultural framework for operational planning, preparation, execution, and assessment.

2-89. When deployed with a unit, human terrain teams support the commander and staff in the planning, preparation, execution, and assessment of operations. Human terrain team members may also assist the commander in building relationships with local leaders and power brokers. This will assist the supported commander in gaining local, regional, socio-cultural, economic, and political insight. Human terrain team responsibilities include, but are not limited to—

- Conducting social science research in support of the commander and staff to assist in the development of commander's critical information requirements (CCIRs) as well as perceived gaps in the unit's cultural knowledge.
- Gathering data and conducting analysis to make that data operationally relevant to the commander and staff for given situations.
- Ensuring the analytic cultural framework for operational planning, decisionmaking, and assessment is incorporated into the continuous planning processes conducted by the commander and staff.

INFORMATION OPERATIONS OFFICER

2-90. The information operations officer is the special staff officer responsible for synchronizing and deconflicting information-related capabilities employed in support of unit operations. The information operations officer is authorized at theater army through brigade. Coordinated by the G-3, the information operations officer leads the information operations element located in the movement and maneuver cell at echelons above brigade. The responsibilities of the information operations officer include, but are not limited to—

- Analyzing the information environment to discern impacts it will have on unit operations and to exploit opportunities to gain an advantage over threat forces.
- Identifying opportunities to employ information-related capabilities to enable achieving friendly operational objectives.
- Assessing the risk associated with the use of information-related capabilities as part of supported unit operations.
- Providing input to the synchronization matrix for the use of available information-related capabilities in support of unit operations.
- Identifying information-related capabilities support gaps not resolvable at the unit level.
- Coordinating with other Army, Service, or joint forces to use information-related capabilities to augment existing unit capability shortfalls.
- Providing information as required in support of operations security at the unit level.
- Providing information as required in support of military deception at the unit level.
- Leading the information operations working group.
- Assessing the effectiveness of employed information-related capabilities.

KNOWLEDGE MANAGEMENT OFFICER

2-91. The knowledge management officer directs the knowledge management section. Knowledge management officers ensure that units understand knowledge management processes and procedures. They demonstrate how these processes and procedures can improve efficiency and shared understanding during training and enhance operational effectiveness during operations, especially in time-constrained environments. Knowledge management officers need not remain in the CP. Commanders may require their knowledge management officer to move with them. The knowledge management officer's duties and responsibilities include, but are not limited to—

- Developing knowledge management techniques, policies, and procedures and ensuring command-wide dissemination.
- Performing staff planning and coordination of knowledge management functions and activities to improve shared understanding, learning, and decisionmaking.
- Creating an organizational knowledge network and metrics for evaluating its effectiveness.
- Leading the staff in assessing unit performance.
- Integrating and synchronizing knowledge management functions and activities with higher echelon and subordinate commands.
- Monitoring emerging knowledge management trends for incorporation into unit operations.
- Chairing the knowledge management working group.

LIAISON OFFICER

2-92. Liaison officers are the commander's representative at the headquarters or agency to which they are sent. They promote coordination, synchronization, and cooperation between their parent unit and higher headquarters, interagency, coalition, host-nation, adjacent, and subordinate organizations as required. As subject matter experts from their assigned headquarters, liaison officers are usually embedded in another organization to provide face-to-face coordination. (See chapter 13 and unit SOPs.)

MILITARY DECEPTION OFFICER

2-93. The military deception officer is responsible for coordinating military deception assets and operations. The military deception officer works within the G-5 plans movement and maneuver cell. A military deception officer is authorized at corps and theater army levels. At division and lower echelons, the commander will designate a military deception officer when necessary. Usually the individual designated by the commander is well versed in the use of Army information-related capabilities that are employed to influence an enemy decisionmaker. The military deception officer provides input to various annexes to the operation order or operation plan as warranted, including preparing a portion of Annex C. Military deception officer responsibilities include, but are not limited to—

- Exercising staff supervision over military deception activities.
- Providing expertise in military deception operations planning.
- Managing information required to conduct military deception operations and civil considerations analysis to better determine the effects of ambiguity.
- Determining requirements or opportunities for military deception operations (with the G-2) through red teaming the adversaries' most probable courses of action.
- Coordinating with the military information support operations planner for support to the deception targets, deception objectives, and deception story.
- Coordinating with the information operations officer to ensure themes, messages, and actions support and enable the military deception plan.
- Coordinating with the cyber electromagnetic activities staff lead to ensure that cyber electromagnetic activities support and enable the military deception plan.
- Producing, distributing, briefing, and coordinating the military deception plan on a need-to-know basis.
- Coordinating operations security measures to shield the military deception plan with the operations security officer.
- Integrating military deception assets (both conventional and unconventional).
- Assessing the execution and effects of military deception operations.

MILITARY INFORMATION SUPPORT OFFICER

2-94. The military information support officer is responsible for synchronizing military information support operations (MISO) in support of unit operations. A military information support officer is authorized at division, corps, and theater army. A military information support NCO is authorized at the brigade level. If no military information support NCO is assigned, the commander of an attached military information support element may assume the military information support staff officer's responsibilities. Coordinated by the G-3, the military information support officer prepares a portion of Annex C (Operations) to the operation order or operation plan. (See MISO doctrine for more details.) Responsibilities include, but are not limited to—

- Coordinating with the information operations officer to ensure synchronization of MISO.
- Planning, coordinating, and synchronizing MISO to support the overall operation.
- Recommending prioritization of the efforts of attached military information support forces.
- Evaluating enemy information efforts and the effectiveness of friendly MISO on target groups (with the G-2, information operations officer, and G-9).
- Assessing MISO effectiveness.
- Assessing the potential effects of adversary information, misinformation, and propaganda on command objectives and determining the best response, if any, in conjunction with the information operations officer.
- Providing military information products and support to an approved military deception plan.

OPERATIONS RESEARCH AND SYSTEMS ANALYSIS OFFICER

2-95. The operations research and system analysis officer conducts analysis in support of operations, across staff sections and employed forces. The operations research and system analysis officer is the principal advisor to the COS and the commander on analytical techniques. The operations research and system analysis officer provides a flexible capability to the commander to use mathematical rigor to confirm or deny theories, to analyze problems and determine their true cause, and to project current trends. The operations research and system analysis officer is often tasked to be the action officer for the assessment working group. This officer's duties and responsibilities include, but are not limited to—

- Managing, analyzing, and visualizing data using statistical information, geospatial information, spreadsheets, and graphics software.
- Developing customized tools for staff sections.
- Providing quality control.
- Supporting COA analyses and operations planning.
- Conducting assessments to determine effectiveness of an operation.
- Conducting analyses to support the staff's military decisionmaking process.

OPERATIONS SECURITY OFFICER

2-96. The operations security officer is responsible for the command's operations security program. The operations security officer prepares a portion of Annex E (Protection) to the operation order or operation plan. (See AR 530-1 for information on operations security policy and procedures.) The operations security officer's responsibilities include, but are not limited to—

- Identifying and recommending the essential elements of friendly information (EEFIs).
- Conducting analysis of adversaries as part of the IPB process.
- Conducting analysis of vulnerabilities as part of the IPB process.
- Assessing operations security risk.
- Developing, coordinating, and applying operations security measures across the staff.
- Writing the running estimate for operations security.
- Writing the operations security appendix to the protection annex.
- Monitoring, assessing, and adjusting operations security as required.
- Reviewing internal staff documents, information system logs, and news releases for sensitive information and potential compromise of EEFIs.
- Searching news sources, web logs (blogs), and other web sites for sensitive information and compromise of EEFIs.
- Attending the information operations working group as required.

PERSONNEL RECOVERY OFFICER

2-97. The personnel recovery officer is responsible for the coordination of all personnel recovery related matters. The personnel recovery officer prepares a portion of Annex E (Protection) to the operation order or operation plan. (See FM 3-50.1 for more details.) Personnel recovery officer duties include, but are not limited to—

- Developing and maintaining the organization's personnel recovery program.
- Recommending recovery COAs to the commander.
- Coordinating personnel recovery issues, both vertically and horizontally.
- Developing personnel recovery SOPs, plans, and annexes.
- Supporting joint personnel recovery or establishing a joint personnel recovery center as required.
- Assisting personnel recovery officers in developing subordinate recovery programs.

PROVOST MARSHAL

2-98. The provost marshal is responsible for planning, coordinating, and employing all organic, assigned, or attached military police assets. Usually the senior military police officer in the command, the provost marshal augments the staff with a small planning cell that typically works within the G-5. A provost marshal is authorized at corps and division headquarters. The provost marshal prepares a portion of Annex C (Operations) and a portion of Annex E (Protection) to the operation order or operation plan. (See ADRP 3-37 and FM 3-39 for more details.) Provost marshal responsibilities include, but are not limited to—

- Conducting maneuver and mobility support operations, including route reconnaissance, surveillance, circulation control, dislocated civilian and straggler control, and information dissemination.
- Directing components of area security operations, including activities associated with antiterrorism operations, zone and area reconnaissance, checkpoint access control, and physical security of critical assets, nodes, and sensitive materials.
- Managing detainee operations and resettlement operations.
- Coordinating and directing law and order operations, including liaison with local civilian law enforcement authorities.
- Conducting police intelligence operations, including activities related to the collection, assessment, development, and dissemination of police intelligence products.
- Coordinating customs and counterdrug activities.
- Providing physical security guidance for commanders.
- Assisting with area damage control and CBRN detection and reporting.
- Helping the commander administer discipline, law, and order.

RED TEAM OFFICER

2-99. Red teaming enables commanders to fully explore alternative plans and operations in the context of the operational environment and from the perspective of partners, adversaries, and others. Red teams assist commanders and staffs with critical and creative thinking and help them avoid groupthink, mirror imaging, cultural missteps, and tunnel vision throughout the conduct of operations. Red teams are part of the commander's staff at division headquarters through the theater army headquarters. Brigades may be augmented with a red team as required. Commanders use red teams to provide alternatives during planning, execution, and assessment. The red team officer's duties and responsibilities include, but are not limited to—

- Broadening the understanding of the operational environment.
- Assisting the commander and staff in framing problems and defining end state conditions.
- Challenging assumptions.
- Ensuring the perspectives of the enemies and adversaries and others are appropriately considered.
- Aiding in identifying friendly and enemy vulnerabilities and opportunities.
- Assisting in identifying areas for assessment.
- Anticipating cultural perceptions of partners, adversaries, and others.
- Conducting independent critical reviews and analyses of plans and concepts to identify potential weaknesses and vulnerabilities.

SECRETARY OF THE GENERAL STAFF

2-100. The secretary of the general staff is the special staff officer who acts as XO for the COS. Corps, divisions, major support commands, and general officers with a staff are authorized a secretary of the general staff. Secretary of the general staff responsibilities include, but are not limited to—

- Planning and supervising conferences chaired by the commander, deputy or assistant commanders, or the COS.
- Directing preparation of itineraries for distinguished visitors to the headquarters and monitoring their execution.
- Monitoring preparation and execution of all official social events and ceremonies involving the commander, deputy or assistant commanders, or the COS.
- Monitoring and disseminating command correspondence.
- Acting as the informal point of contact for liaison officers.

STAFF WEATHER OFFICER

2-101. The G-2 (S-2) staff, with the support of the staff weather officer, is responsible for providing the commander with a thorough understanding of terrestrial and solar weather effects and their impact on friendly and threat systems and operations, as well as civil considerations. The G-2 (S-2) staff provides this information during the planning process and incorporates significant weather effects into all of the primary intelligence products (intelligence estimates, intelligence summaries, and the intelligence portion of the common operational picture [COP]). Weather effects are analyzed based on the military aspects of weather (visibility, wind, precipitation, cloud cover, temperature, and humidity). The staff weather officer is responsible for coordinating operational weather support and weather services through the G-2 (S-2). The staff weather officer, an Air Force officer or NCO, leads a combat weather team of two or more personnel. (See AR 115-10 for more information.) Staff weather officer responsibilities include—

- Coordinating weather support procedures for garrison and before and during deployments with its supported Army command.
- Advising the Army commander on Air Force weather capabilities, limitations, and the ways in which weather support can enhance operations.
- Helping the G-2 (S-2) arrange weather support for subordinate units.
- Helping the G-2 (S-2) and staff produce weather displays, graphic COP overlays, and weather effects tactical decision aids displaying weather effects on friendly and threat forces, weapons systems and sensor payloads, and information collection units and assets.
- Evaluating and disseminating weather products and data.
- Advising the Air Force on Army operational weather support requirements.
- Helping the G-2 (S-2) monitor the weather support mission, identify responsibilities, and resolve weather support deficiencies.

SPACE OPERATIONS OFFICER

2-102. The space operations officer is in charge of the space support element and is responsible for providing space-related tactical support and coordination of space-based capabilities available to the command. An Army space support team is often placed under operational control to a command, if the workload or product requirements exceed the capacity of the organic space support element. The team's officer in charge fulfills the space operations officer's responsibilities. The space operations officer prepares Annex N (Space Operations) to the operation order or operation plan. (See FM 3-14 for more details.) Space operations officer responsibilities include, but are not limited to—

- Advising the commander on the space architectures, capabilities, limitations, and use of theater, strategic, national, and commercial space assets.
- Calculating, analyzing, and disseminating global positioning system satellite coverage and accuracy data.
- Facilitating the dynamic retasking of space-based assets to support current and future operations.
- Assisting in acquiring Department of Defense and commercial satellite terrain and weather imagery (classified and unclassified) to enhance mapping, mission analysis, and other actions requiring near real-time imagery from denied areas.
- Advising the G-2 and the information operations officer on capabilities and vulnerabilities of threat and commercial space systems.
- Providing estimates on the effects of space weather activities on current and future operations and the effects of terrestrial weather on space-based capabilities.
- Nominating threat or foreign ground stations for targeting (with the G-3 chief of fires).
- Coordinating the activities of the Army space support team supporting the command.
- Integrating into special technical operations to maximize all the unique and specialized space-related technical capabilities into operations.

TRANSPORTATION OFFICER

2-103. The transportation officer coordinates transportation assets and operations and also prepares a portion of Annex F (Sustainment). (See ATP 4-16 for more details.) Transportation officer responsibilities include, but are not limited to—

- Planning and directing administrative movements, including onward movement from ports of debarkation, sustainment movements, and other movements as directed.
- Planning movement scheduling and regulations of main supply routes.
- Planning the mode of operations (truck, rail, air, and water).
- Planning the movement of materiel and personnel.
- Monitoring movements on routes two echelons lower than the unit.

VETERINARY OFFICER

2-104. The veterinary officer is responsible for coordinating assets and activities concerning veterinary services within the command. All veterinarian activities are planned at the medical brigade (support), medical command (deployment support), or Army Service component command. (See ADRP 4-0 and FM 4-02.18 for more details.) Veterinary officer responsibilities include, but are not limited to—

- Coordinating veterinary activities with the surgeon and other staff members.
- Determining requirements for veterinary supplies and equipment.
- Ensuring safety of food and food sources.
- Advising on health and operational risks of animal disease, including possible biological warfare events.
- Monitoring the sanitation of food storage facilities and equipment.
- Managing veterinary equipment and facilities.
- Coordinating animal housing.
- Participating in civil-military operations.
- Coordinating the use of medical laboratory services by veterinary personnel.
- Preparing reports on command veterinary activities.

PERSONAL STAFF OFFICERS

2-105. Personal staff officers work under the immediate control of, and have direct access to, the commander. By law and regulation, personal staff officers have a unique relationship with the commander. The commander establishes guidelines or gives guidance on when a personal staff officer informs or coordinates with the COS (XO) or other staff members. Some personal staff officers have responsibilities as special staff officers and work with a coordinating staff officer. They do this on a case-by-case basis, depending on the commander's guidance or the nature of the task. Personal staff officers also may work under the supervision of the COS (XO). Although there are other members in the commander's personal staff, this section discusses only staff officers and the command sergeant major.

AIDE-DE-CAMP

2-106. The aide-de-camp serves as a personal assistant to a general officer. An aide-de-camp is authorized for general officers in designated positions. The rank of the aide-de-camp depends on the rank of the general officer. No officer exercises coordinating staff responsibility over the aide-de-camp. Aide-de-camp responsibilities include, but are not limited to—

- Providing for the general officer's personal well-being and security, and relieving the general officer of routine and time-consuming duties.
- Preparing and organizing schedules, activities, and calendars.
- Preparing and executing trip itineraries.
- Coordinating protocol activities.
- Acting as an executive assistant.

- Meeting and hosting the general officer's visitors.
- Supervising other personal staff members (secretaries, assistant aides, enlisted aides, and drivers).
- Performing varied duties, according to the general officer's desires.

CHAPLAIN

2-107. The mission of the chaplain is to perform or provide religious support to the unit by assisting the commander in providing for the free exercise of religion and religious, moral, and ethical leadership. Chaplains execute their distinct religious support mission for Soldiers, members of other military services, family members, and authorized civilians in a variety of geographical locations, operational situations, and circumstances. Chaplains provide religious support and advise commanders on the impact of religion, reflecting the dual roles of religious leaders and religious staff advisors. Chaplains and chaplain assistants are assigned at all battalion and higher echelons. The chaplain prepares a portion of Annex F (Sustainment). (See FM 1-05 for more details.)

COMMAND SERGEANT MAJOR

2-108. The command sergeant major is the senior noncommissioned officer of the command at battalion and higher echelons. Command sergeants major carry out policies and enforce standards for the performance, training, and conduct of enlisted Soldiers. They give advice and initiate recommendations to the commander and staff in matters pertaining to enlisted Soldiers. In operations, a commander employs the command sergeant major throughout the area of operations to extend command influence, assess the morale of the force, and assist during critical events.

INSPECTOR GENERAL

2-109. The inspector general is responsible for advising the commander on the command's overall welfare and state of discipline. The inspector general is a confidential advisor to the commander. An inspector general is authorized for general officers in command positions. The inspector general prepares Annex U (Inspector General) to the operation order or operation plan. (See AR 20-1 for more details.) Inspector general responsibilities include, but are not limited to—

- Advising commanders and staffs on inspection policy.
- Advising commanders on the effectiveness of the organizational inspection program.
- Conducting inspections as the commander requires and monitoring corrective actions.
- Receiving allegations and conducting investigations and investigative inquiries.
- Monitoring and informing the commander of trends, both positive and negative, in all activities.
- Consulting with staff sections, as appropriate, to obtain items for the special attention of inspectors and to arrange for technical assistance.
- Providing the commander continuous, objective, and impartial assessments of the command's operational and administrative effectiveness.
- Assisting Soldiers, Army civilians, family members, retirees, and other members of the force who seek help with Army-related problems.
- Identifying and helping to resolve systemic problems.

INTERNAL REVIEW OFFICER

2-110. The internal review officer provides professional internal audit capability and delivers pertinent, timely, and reliable information and advice to the commander. This information and advice evaluates risk, assesses management control measures, fosters stewardship, and improves the quality, economy, and efficiency of business practices. (See AR 11-7 for more details.) Internal review officer responsibilities include, but are not limited to—

- Completing internal audits of functions or organizational entities in the command with known or suspected problems, determining the nature and cause of problems, and suggesting resolutions.
- Providing troubleshooting—quick reaction efforts to prevent serious problems from developing.

- Providing an audit compliance function by serving as the point of contact with external audit groups. In addition, the internal review office facilitates the external audit reply and response process and performs follow-up audits.
- Coordinating with higher headquarters and other agencies to ensure units properly follow all standards and policies.

PUBLIC AFFAIRS OFFICER

2-111. The public affairs officer understands and coordinates the flow of information to Soldiers, the Army community, and the public. The public affairs officer prepares Annex J (Public Affairs) to the operation order or operation plan. (See FM 3-61 for more details.) Public affairs officer responsibilities include, but are not limited to—

- Planning and supervising the command public affairs program.
- Advising and informing the commander of the public affairs impact and implications of planned or current operations.
- Preparing themes and messages for the commander for public communications.
- Serving as the command representative for all communications with external media.
- Assessing the information requirements and expectations of the Army and the public, monitoring media and public opinion, and evaluating the effectiveness of public affairs plans and operations.
- Coordinating with other information-related capabilities to ensure synchronization through the information operations working group.
- Coordinating logistics and administrative support of civilian journalists under unit administrative control.
- Conducting liaison with media representatives to provide accreditation, mess, billet, transport, and escort as authorized and appropriate.
- Developing and educating the command on policies and procedures for protecting against the release of information detrimental to the mission, national security, and personal privacy.
- Informing Soldiers, family members, and Army civilians of their rights under the Privacy Act, operations security responsibilities, and roles as implied representatives of the command when interacting with news media.
- Recommending news, entertainment, and information for Soldiers and home station audiences.

SAFETY OFFICER

2-112. The safety officer coordinates safety activities throughout the command and advises the commander on matters relating to the Army safety program, including its implementation and effectiveness. Commanders at every echelon from battalion through corps appoint a safety officer. An aviation safety officer is authorized for corps staffs and all aviation units. The safety officer prepares a portion of Annex E (Protection) to the operation order or operation plan. (See AR 385-10 for more details.) Safety officer responsibilities include, but are not limited to—

- Implementing the command safety and occupational health program.
- Implementing the accident prevention program.
- Coordinating with the inspector general and provost marshal on correcting unsafe trends identified during inspections.
- Providing input to the G-1 (S-1) on projected accident losses.
- Providing safety training to the local civilian labor force.
- Reviewing risk assessments and recommending risk-reduction control measures for all operations.

STAFF JUDGE ADVOCATE

2-113. The staff judge advocate is the senior legal advisor in the command and the primary legal advisor to the commander. The commander and the staff judge advocate shall, at all times, communicate directly on matters relating to the administration of military justice, including, but not limited to, all legal matters affecting the morale, good order, and discipline of the command. The staff judge advocate also provides legal advice and support to the staff and coordinates actions with other staff sections to ensure the timely and accurate delivery of legal services throughout the command. The staff judge advocate is a member of the commander's personal and special staff. A legal support element—typically composed of three judge advocates—deploys in direct support of each brigade-level task force. The staff judge advocate provides complete legal support encompassing the six core legal disciplines—military justice, international and operational law, administrative and civil law, contract and fiscal law, claims, legal assistance. The staff judge advocate prepares a portion of Annex C (Operations) and Annex F (Sustainment) to the operation order or operation plan. (See AR 27-1 and FM 1-04 for more details.)

SURGEON

2-114.   The surgeon is responsible for coordinating health assets and operations within the command. This officer provides and oversees medical care to Soldiers, civilians, and enemy prisoners of war. Organizations from battalion through Army Service component command level are authorized a surgeon. The surgeon prepares a portion of Annex E (Protection) and Annex F (Sustainment) to the operation order or operation plan. (See FM 4-02.21 for more details.) Surgeon responsibilities include, but are not limited to—

- Advising the commander on the health of the command.
- Providing health education and training.
- Coordinating medical evacuation, including Army dedicated medical evacuation platforms (air and ground).
- Ensuring the establishment of a viable veterinary services program (including inspection of subsistence and outside the continental U. S. food production and bottled water facilities, veterinary preventive medicine, and animal medical care).
- Ensuring an area medical laboratory capability or procedures for obtaining this support from out of theater resources are established for the identification and confirmation of the use of suspect biological and chemical warfare agents by opposition forces. This includes the capability for specimens and samples packaging, and establishing handling requirements and escort and chain of custody requirements.
- Establishing clinical laboratory capabilities, including blood banking.
- Planning for and implementing preventive medicine operations (including preventive medicine programs, initiating preventive medicine measures to counter the health threat, and establishing medical and occupational and environmental health surveillance).
- Supervising and preparing health-related reports and statistics.
- Advising on the effects of the medical threat on personnel, rations, and water.
- Advising how operations affect the public health of personnel and the indigenous populations.

This page intentionally left blank.

Chapter 3

Managing Knowledge and Information

This chapter expands on the mission command warfighting function tasks "conduct knowledge management and information management." First, it presents a model used to explain the progressive transformation of data into understanding. Then the chapter discusses the specifics of knowledge management followed by a discussion on information management. Finally, it describes the interrelationship between knowledge management and information management.

KNOWLEDGE AND UNDERSTANDING

3-1. Success in operations demands timely and effective decisions based on applying judgment to available information and knowledge. Throughout the conduct of operations, commanders (supported by their staffs, subordinate commanders, and unified action partners) seek to build and maintain situational understanding. *Situational understanding* is the product of applying analysis and judgment to relevant information to determine the relationships among the operational and mission variables to facilitate decisionmaking (ADP 5-0).

3-2. Commanders also strive to create shared understanding within the force and with unified action partners. Essential to mission command is the shared understanding and appreciation of the operational context and the commander's intent by multiple participants. The shared understanding of an operational environment, the operation's purpose, the problem, and approaches to solving the problem form the basis for unity of effort and trust. This shared understanding is the context within which decentralized actions can be performed by units as if they were centrally coordinated. Knowledge management helps create shared understanding through the alignment of people, processes, and tools within the organizational structure and culture in order to increase collaboration and interaction between leaders and subordinates. Knowledge management facilitates situational understanding and acts as a catalyst for enhanced shared understanding. (See Army doctrine for more information on shared understanding.)

3-3. Knowledge management and information management assist commanders with progressively adding meaning at each level of processing and analyzing to help build and maintain their situational understanding. They are interrelated activities that support the commander's decisionmaking. There are four levels of meaning. From the lowest level to the highest level, they include data, information, knowledge, and understanding. At the lowest level, processing transforms data into information. Analysis then refines information into knowledge. Commanders and staffs then apply judgment to transform knowledge into understanding. Commanders and staffs continue a progressive development of learning, as organizations and individuals assign meaning and value at each level. (See figure 3-1.)

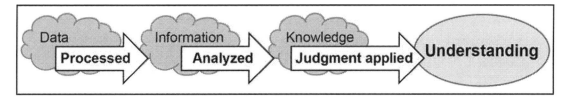

Figure 3-1. Achieving understanding

3-4. *Data* consist of unprocessed signals communicated between any nodes in an information system, or sensing from the environment detected by a collector of any kind (human, mechanical, or electronic) (ADRP 6-0). In typical organizations, data often flows to command posts from subordinate units.

Subordinate units push data to inform higher headquarters of events that facilitate situational understanding. Data can be quantified, stored, and organized in files and databases; however, data only becomes useful when processed into information.

3-5. Information is the meaning that a human assigns to data by means of the known conventions used in their representation. Information alone rarely provides an adequate basis for deciding and acting. Effective mission command requires further developing information into knowledge so commanders can achieve understanding.

3-6. *Knowledge* is information that has been analyzed to provide meaning or value or evaluated as to implications for the operation. (FM 6-01.1) It is also comprehension gained through study, experience, practice, and human interaction that provides the basis for expertise and skilled judgment. Staffs work to improve and share two types of knowledge, tacit and explicit.

3-7. Tacit knowledge resides in an individual's mind. It is the domain of individuals, not technology. All individuals have a unique, personal store of knowledge gained from life experiences, training, and formal and informal networks of friends and professional acquaintances. This knowledge includes learned nuances, subtleties, and work-arounds. Intuition, mental agility, effective responses to crises, and the ability to adapt are also forms of tacit knowledge. Leaders use tacit knowledge to solve complex problems and make decisions. They also routinely engage subordinates' tacit knowledge to improve organizational learning and enhance unit innovation and performance.

3-8. Explicit knowledge consists of written or otherwise documented information that can be organized, applied, and transferred using digital (such as computer files) or non-digital (such as paper) means. Explicit knowledge lends itself to rules, limits, and precise meanings. Examples of explicit knowledge include dictionaries, official department publications (field manuals, technical manuals, tactics, techniques, and procedures publications, and Department of the Army pamphlets) and memorandums. Explicit knowledge is primarily used to support situational awareness and shared understanding as it applies to decisionmaking.

3-9. Understanding is knowledge that has been synthesized and had judgment applied to it to comprehend the situation's inner relationships. Judgment is based on experience, expertise, and intuition. Ideally, true understanding should be the basis for decisions. However, commanders and staffs realize that uncertainty and time preclude achieving perfect understanding before deciding and acting.

KNOWLEDGE MANAGEMENT

3-10. *Knowledge management* is the process of enabling knowledge flow to enhance shared understanding, learning, and decisionmaking (ADRP 6-0). Knowledge flow refers to the ease of movement of knowledge within and among organizations. Knowledge must flow to be useful. The purpose of knowledge management is to create shared understanding through the alignment of people, processes, and tools within the organizational structure and culture in order to increase collaboration and interaction between leaders and subordinates. This results in better decisions and enables improved flexibility, adaptability, integration, and synchronization to achieve a position of relative advantage. Effective and efficient use of knowledge in conducting operations and supporting organizational learning are essential functions of knowledge management. (See figure 3-2 for a depiction of knowledge management flow.) Sound knowledge management practices include—

- Collaboration among personnel at different places.
- Rapid knowledge transfer between units and individuals.
- Reachback capability to Army schools, centers of excellence, and other resources.
- Leader and Soldier agility and adaptability during operations.

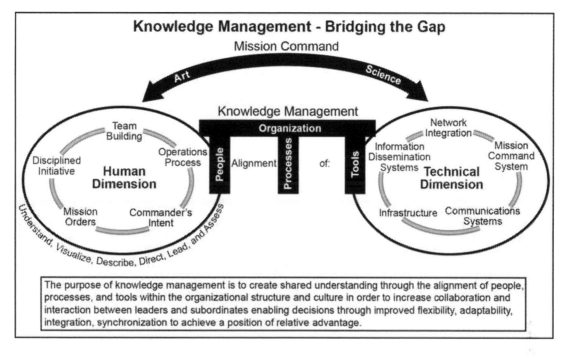

Figure 3-2. Knowledge management flow

3-11. Knowledge management provides the methods and means to efficiently share knowledge among individuals and distribute relevant information where and when it is needed. Knowledge management creates, organizes, applies, transfers, collects, codifies, and exchanges knowledge and information between people. It seeks to align people and processes with appropriate tools to continuously capture, maintain, and re-use key information, decisions, and lessons learned to help units learn and adapt and improve mission performance.

3-12. Knowledge management leverages knowledge that resides in individuals and organizations. It facilitates the flow of that knowledge across the organization and between organizations so units can apply that knowledge to mission or operational requirements. Every Soldier must understand and practice knowledge management. It enables the Army and its subordinate commands at every level to be learning organizations.

KNOWLEDGE MANAGEMENT COMPONENTS

3-13. Knowledge management is organized into four components that enable the understanding and visualization of leaders and develops shared understanding of subordinates. The four components are—
- People.
- Processes.
- Tools.
- Organization.

People

3-14. Of the four components, people are the most vital for successful knowledge management. People include the commander and staff; higher echelon, lower echelon, and adjacent commanders and staffs; other Army leaders; and other agencies that might contribute to answering information requirements. They include those inside and outside the organization that create, organize, apply, and transfer knowledge; and the leaders who act on that knowledge. Knowledge only has meaning in a human context. It moves between and benefits people, not machines.

3-15. The knowledge management officer (KMO) plays a special role in knowledge management. Working through the chief of staff (COS) or executive officer (XO), the KMO is responsible for developing the knowledge management plan that integrates and synchronizes knowledge and information management. The KMO synchronizes knowledge and information management to facilitate the commander's situational understanding for any problem set and to provide the staff shared understanding. The KMO accomplishes this by using the tools, processes, and people available to facilitate an environment of shared understanding. For instance, the KMO assists the COS (XO) in the command post design, the reporting procedures standard operating procedure (SOP), the battle rhythm, and the working groups' structure. Moreover, the KMO is concerned with sharing and disseminating knowledge and information inside and outside the organization. The KMO is the principal advisor in developing the plan, or blueprint, to facilitate situational and shared understanding of the operational environment, the problem, and approaches to solving the problem.

3-16. Staffs execute the knowledge management plan developed by the KMO. Staffs develop and provide knowledge from which commanders and other decisionmakers achieve situational awareness and shared understanding, make decisions, and execute those decisions. Staffs are involved in both directing actions and assessing progress. The structure of personnel, units, and activities creates explicit communication channels for knowledge transfer within and between organizations.

Processes

3-17. The five-steps in the knowledge management process are assess, design, develop, pilot, and implement. Its activities are integrated into the operations process used in the planning, preparing, executing, and assessing of operations. This integration enables the transfer of knowledge between and among individuals and organizations. Soldiers, groups, teams, and units employ them. Knowledge exchange occurs both formally—through established processes and procedures—and informally—through collaboration and dialogue. The knowledge management process also seeks to ensure that knowledge products and services are relevant, accurate, timely, and usable to commanders and decisionmakers. (See FM 6-01.1 for more information on the knowledge management process.)

3-18. Assessment precedes all other steps. This step begins with determining what information leaders need to make decisions, and how the unit provides information for those leaders. Knowledge management officers identify gaps in the process, and as they establish objectives, they begin to consider possible solutions to address those gaps. As solutions are implemented, they are assessed to evaluate their effectiveness and to recommend new solutions or improvements.

3-19. Design is the second step in the knowledge management process. Design is identifying tailored frameworks for knowledge management products or services that effectively and efficiently answer information requirements and meet the objectives established in the assessment step. Services created in the design step are often shared through virtual communities that are established to share information on a certain topic. The knowledge management products or services could be refinements of an existing process or a new solution identified after the assessment. For example, when conducting a relief in place the incoming unit members may reorganize the command post that they assume from the outgoing unit based on their knowledge management assessment. In this example, the KMO assists in reorganizing the command post to develop situational and shared understanding.

Note: The design step of the knowledge management process differs from and should not be confused with Army design methodology. See ADRP 5-0 for more information on Army design methodology.

3-20. Develop is the step that actually builds the solution derived from the assessment and design steps. First, the KMO and the staff collaborate to establish the social framework for the virtual communities and other knowledge-sharing venues designed in the previous step. A social framework is the means by which individuals and organizations with a common interest are able to communicate with each other (such as through video and audio teleconferencing and messaging). Knowledge management representatives provide insight and advice on the social frameworks best suited to the organization. Once the social frameworks are established for the virtual communities, signal staff section personnel, usually portal administrators or designers, assist in connecting them to the technical network. The knowledge management section works with the unit on both aspects of this step. (See knowledge management doctrine for information on knowledge management representatives.)

3-21. Pilot is the phase that deploys the knowledge management solution and tests and validates it with the unit. This aspect is an incremental test of a modification to an existing process or procedure. Important considerations of the pilot step include communicating the proposed knowledge management plan to the commander and staff and ensuring acceptance or discussing alternatives as needed. The knowledge management section must be prepared to train and coach unit personnel as needed in order to successfully deploy and test the solution.

3-22. Implement is the phase that executes the validated knowledge management plan and integrates it into the unit information systems. Training and coaching personnel on their specific roles and tasks continues as needed. Knowledge managers monitor the initial implementation of the knowledge management plan and make any necessary adjustments. Once the knowledge management plan is fully implemented and integrated into the operations process, knowledge managers continue to monitor and assess results.

3-23. The steps of the knowledge management process and their associated activities are not ends in themselves. The knowledge management section uses them to improve knowledge management within the organization before operations, throughout the operations process, and after operations. Furthermore, the KMO synchronizes them with the unit's battle rhythm.

Tools

3-24. Throughout the process, the KMO uses tools that include information systems and various software tools used to put knowledge products and services into organized frameworks. Knowledge management tools are anything that is used to share and preserve information. The mission determines the tool. Commonly used tools include—

- Information systems: The equipment and facilities that collect, process, store, display, and disseminate information. This includes computers—hardware and software—and communications, as well as policies and procedures for their use.
- Collaboration tools: These tools are information systems that include online capabilities that make team development and collaboration possible. Examples include chat, white-boarding, professional forums, communities of interest, communities of practice, and virtual teaming.
- Expertise-location tools: These tools support finding subject matter experts.
- Data-analysis tools: These tools support data synthesis that identifies patterns and establishes relationships among data elements.
- Search-and-discover tools: These tools include search engines that look for topics, recommend similar topics or authors, and show relationships to other topics.
- Expertise-development tools: These tools use simulations and experiential learning to support developing experience, expertise, and judgment.

Organization

3-25. An organization is the matrix in which people, processes, and tools function to integrate individual learning and organizational learning strategies. The commander is responsible for establishing a culture of shared understanding and knowledge, which is critical for learning organizations. The commander creates a mindset of shared understanding and has the greatest influence on the organization. Knowledge management is effective in climates that are conducive to openly sharing ideas. Organizations bring their attitudes, feelings, values, and behaviors together, creating a system of processes facilitated by tools that will characterize that group. The KMO must consider this dynamic when advising and assisting organizations regarding the knowledge management plan. The KMO cannot plan in a vacuum and must have the commander's commitment in executing the knowledge management plan.

KNOWLEDGE MANAGEMENT TASKS

3-26. Knowledge management components are supported by four knowledge management tasks that bring an organization closer to situational and shared understanding. The four knowledge management tasks are creating knowledge, organizing knowledge, applying knowledge, and transferring knowledge.

Creating Knowledge

3-27. *Knowledge creation* is the process of developing new knowledge or combining, restructuring, or repurposing existing knowledge in response to identified knowledge gaps. (FM 6-01.1) Knowledge comes from a variety of sources, including new technology, answering the commander's critical information requirements, or the sharing of information that others need to know. Knowledge is also created when organizations learn, which in turn enables organizations to adapt. The output of collaborative planning from working groups is an example of creating knowledge.

Organizing Knowledge

3-28. Organizing knowledge includes archiving, labeling, and identifying. These are specific tasks of knowledge managers under the implement step of the knowledge management process. Organizing knowledge ensures that users can discover and retrieve knowledge that is relevant, and knowledge managers can track knowledge products throughout their life cycle. Archiving consists of moving outdated and irrelevant knowledge from active status to an inactive status, based on rules and policies. Labeling takes content that is no longer relevant, archives it, and keeps it separate from current knowledge products. Identifying involves determining whether to archive or dispose of content. Subject matter experts do this by reviewing content that exceeds a specified date or does not meet usage benchmarks. Based on this review, they determine whether regulations require retaining content or destroying it.

Applying Knowledge

3-29. Applying knowledge refers to making knowledge accessible to those who need to use it. KMOs seek to create conditions so users can retrieve and apply the knowledge they need. This is the primary purpose of content management, and it occurs during the implement step of the knowledge management process. (See FM 6-01.1 for information on content management.) A key aspect of knowledge management is ensuring that multiple users can easily retrieve knowledge products, which enables collaboration in applying knowledge. One example is posting the decision template on the unit portal so subordinate units can access it is applying knowledge.

Transferring Knowledge

3-30. *Knowledge transfer* is the movement of knowledge—including knowledge based on expertise or skilled judgment—from one person to another (FM 6-01.1). It describes how knowledge is passed between individuals and groups. It includes knowledge developed within the unit and received from other sources. Effective knowledge transfer allows all involved to build on each other's knowledge in ways that strengthen not only individual Soldiers but also the entire organization. It is more than simply moving or transferring files and data. Since knowledge transfer occurs between people, knowledge management includes creating techniques and procedures to develop knowledge skills in leaders, build experience, and transfer expertise.

INFORMATION MANAGEMENT

3-31. *Information management* is the science of using procedures and information systems to collect, process, store, display, disseminate, and protect data, information, and knowledge products (ADRP 6-0). Information management supports, underpins, and enables knowledge management. The two are linked to facilitate understanding and decisionmaking. Information management is a technical discipline that involves the planning, storage, manipulating, and controlling of information throughout its life cycle in support of the commander and staff. Information management employs both staff management and processes to make information available to the right person at the right time. Information management provides a structure so commanders and staffs can process and communicate relevant information and make decisions. Effective information management contributes to the knowledge management tasks of knowledge creation and supports shared understanding for all unit members.

3-32. Generally, information management relates to the tasks of collection, processing, display, storage, distribution, and protection of data and information. In contrast, knowledge management uses information to create, organize, apply, and transfer knowledge to support achieving understanding, making decisions, and ultimately taking effective action. The assistant chief of staff, signal (G-6 [S-6]) enables knowledge management by providing network architecture and the technological tools necessary to support content management and knowledge sharing.

INFORMATION MANAGEMENT COMPONENTS

3-33. Information management is organized into two components, procedures and information systems. These two components facilitate the collection, processing, storing, displaying, disseminating, and protecting of knowledge and information. Information management provides the timely and protected distribution of relevant information to commanders and staff elements. It supports and is a component of knowledge management.

Procedures

3-34. The two primary procedures for information management are establishing information requirements and information categories. An *information requirement* is any information element the commander and staff require to successfully conduct operations (ADRP 6-0). Information management begins by identifying information gaps and developing information requirements. Commanders and staffs may use tools such as mission variables and operational variables to categorize information. (See appendix A for a discussion of mission and operational variables.) All information given to commanders should be relevant information for answering information requirements. Effective information management identifies and organizes relevant information and processes data into information for development into and use as knowledge. Information management then quickly routes information to those who need it. That is, commanders should only receive information that they need for exercising mission command. Forces determine the relevance of information based on the following characteristics:

- Accurate—conveys the true situation.
- Timely—is available in time to make decisions.
- Useable—is portrayed in common, easily understood formats and displays.
- Complete—provides all necessary data.
- Precise—has the required level of detail.
- Secure—affords required protection.

3-35. As operations progress, commanders and staffs require additional information to gain further understanding and support decisionmaking. Effective commanders and staffs prioritize the collection of required information. Doctrine organizes information into two categories: the commander's critical information requirements (CCIRs) and the essential elements of friendly information (EEFIs).

3-36. A *commander's critical information requirement* is an information requirement identified by the commander as being critical to facilitating timely decision making (JP 3-0). The two key elements are friendly forces information (FFIR) and priority intelligence requirements (PIRs). A CCIR is—

- Specified by the commander for a specific operation.
- Applicable only to the commander who specifies it.
- Situation dependent-directly linked to a current or future mission.
- Time-sensitive.

3-37. *A friendly force information requirement* is information the commander and staff need to understand the status of friendly force and supporting capabilities (JP 3-0). FFIRs identify the information about the mission, troops and support available, and time available for friendly forces that the commander considers most important. In coordination with the staff, the operations officer manages FFIRs for the commander.

3-38. *A priority intelligence requirement* is an intelligence requirement, stated as a priority for intelligence support, that the commander and staff need to understand the adversary or other aspects of the operational environment (JP 2-01). PIRs identify the information about the enemy, terrain and weather, and civil considerations that the commander considers most important. The intelligence officer manages PIRs for the commander.

3-39. An effective mission command system anticipates and answers a commander's information requirements. Commanders carefully allocate collection resources to obtain data and information for critical tasks. They set priorities for collection by establishing the CCIRs. Commanders widely distribute their CCIRs and revise them as the situation changes. Information is continuously collected. It may be delivered on a routine schedule or as requested. An information-push system pushes information from the source to the user as it becomes available or according to a schedule (such as through routine reports). An information-pull system supplies information as requested. Commanders and staffs determine how to use and integrate both types of systems.

3-40. An *essential element of friendly information* is a critical aspect of a friendly operation that, if known by the enemy, would subsequently compromise, lead to failure, or limit success of the operation and therefore should be protected from enemy detection (ADRP 5-0). Although EEFIs are not CCIRs, they have the same priority. EEFIs establish an element of information to protect rather than one to collect. EEFIs identify those elements of friendly force information that, if compromised, would jeopardize mission success. EEFIs help commanders protect vital friendly information. Their identification is the first step in the operations security process and central to the protection of information.

Information Systems

3-41. Information systems are the physical dimension of information management. Staffs use automated systems for efficient processing, storing, and disseminating of information. Information systems—especially when merged into a single, integrated network—enable extensive information sharing. Effective information systems and processes make relevant information easy to share and easy for commanders to use. The goal is not to process vast amounts of information but to enable commanders to develop an accurate situational understanding as quickly as possible.

3-42. An effective way to communicate relevant information is the *common operational picture*—a single display of relevant information within a commander's area of interest tailored to the user's requirements and based on common data and information shared by more than one command (ADRP 6-0). The common operational picture (COP) integrates many digital information systems to display relevant information. Initially, commanders and staffs analyze their mission using operational and mission variables. They begin to develop the COP. Commanders determine their information requirements, and additional information is collected based on those requirements. Commanders and staffs continue to refer to and refine the COP as the situation evolves. They use the COP as a tool for developing knowledge and understanding. Commanders and staffs are obligated to share their understanding of the COP to subordinate and higher commands to facilitate synchronized operations and parallel understanding.

INFORMATION MANAGEMENT TASKS

3-43. The information management components are supported by six primary tasks to enable knowledge management and facilitate situational understanding and decisionmaking. The six tasks are collect, process, store, display, disseminate and protect. The network operations (signal) essential task of content management enables information producers and consumers to more efficiently perform information management. Content management consists of information dissemination management and content staging capabilities and is defined as the technologies, techniques, processes, policies, and procedures necessary to provide Soldiers an awareness of relevant, accurate information through automated access to newly discovered or recurring information in a timely, efficient, and usable format. Information dissemination management officers seek to facilitate the retrieval or distribution of the right information, to the right place, at the right time, and in the right format. Content staging is a capability by which information can be compiled, cataloged, and cached. (See JP 6-0 and FM 6-02.71 for more information on content management, information dissemination management, and content staging.)

Collect

3-44. Data collection operations work to answer the commander's critical information requirements. Collection describes compiling data based on those requirements. Through information dissemination management and content staging capabilities, producers of information can quickly accumulate data sensed or generated at the beginning of the information management process.

Process

3-45. Processing information describes the act of cataloging data via established and usually routine sets of procedures to link or fuse it with other related data in order to create information. Information dissemination management and content staging capabilities enable the automated registering of data in order for it to be developed into information and stored until needed.

Store

3-46. Storage of information describes the caching of information using any medium necessary. Storage provides the physical and virtual staging of information. Information regarding storage locations may be listed in unit SOPs or operation orders (OPORDs). Information dissemination management and content staging capabilities enable the digital caching of processed information in a secure system.

Display

3-47. Display of information describes the visual presentation of collected information, data, or knowledge. Examples of displays include graphic control images and map boards, as well as the various electronic displays used in command posts. These displays serve to enhance understanding of the status of operations.

Disseminate

3-48. Disseminating information is a primary task for information management. Information dissemination management capabilities allow for the efficient distribution or retrieval of relevant information of any kind from one person or place to another, in a usable form, by any means to improve understanding or to initiate or govern action. Information dissemination takes the following two basic forms: broadcast or point-to-point dissemination. Information dissemination management activities should exhibit a judicious combination of broadcast and point-to-point forms of dissemination.

3-49. Broadcast dissemination allows senders to distribute information simultaneously to a large number of users. Anyone with access to the network can receive the information. The greatest advantage of this method is that information managers can disseminate information to the widest audience in the shortest amount of time. Since the information is sent to a variety of users with varying relevant information requirements, the information cannot be tailored to a specific commander's needs. Another major drawback of broadcast dissemination is that undisciplined use of this method can quickly lead to information overload.

3-50. Point-to-point dissemination directs information to a specific user or users. Using this method, information can be easily passed from one commander to the next. Dissemination methods should be tailored to meet specific relevant information needs of each recipient with built-in mechanisms that are not present in broadcast dissemination.

Protect

3-51. For information producers and users, protection involves actions taken to counter threats and vulnerabilities during all steps of the information management process. Protection activities include encryption of data at rest, granting access to information on a need-to-share basis, and using secure, authorized processes to disseminate and display information.

3-52. Protection ensures the confidentiality, integrity, and availability of information traversing networks and residing on information systems from the time it is collected, processed, and stored until it is discovered, distributed, and used by the users, systems, and decisionmakers.

KNOWLEDGE AND INFORMATION MANAGEMENT IN PRACTICE

3-53. The staff task of "conduct knowledge management and information management" is essential to the mission command warfighting function and entails the continuous application of the knowledge management process of assess, design, develop, pilot, and implement activities designed to capture and distribute knowledge throughout the organization. The knowledge management process is used throughout the operations process to put the knowledge management plan into practice. Paragraphs 3-54 through 3-62 describe the activities involved in the conduct of knowledge and information management.

3-54. Assessments are critical to conducting knowledge and information management. They provide the feedback to the organization on what is effective. The assessment leads to an organization that learns and adapts. Assessment, the continuous monitoring and evaluation of the current situation, precedes and guides every operations process activity and concludes each operation or phase of an operation. Staffs use several tools to assess progress. Running estimates, the COP, and after action reviews are the three most prevalent.

3-55. Preparation activities help commanders, staffs, and subordinates understand the situation and their roles in upcoming operations. Knowledge management and information management support preparation by working to enable learning across the organization through various venues such as forums and working groups. Preparation consists of activities that units perform to improve their ability to execute an operation. Knowledge management and information management supports staff preparatory activities such as plan refinement, training battle drills, rehearsals, liaison, inspections, confirmation briefs, and coordination. Knowledge management staff sections have the essential function of facilitating this learning. Based on their improved situational understanding, commanders refine the plan, as required, prior to execution.

3-56. Knowledge management and information management support mission execution through reporting, refinement of communications, and collaborative processes. Knowledge management and information management support learning during execution by capturing (through reports and after action reviews) the available data, information, and knowledge from individuals and organizations. Knowledge that is needed for daily work is collated, standardized, and transferred for use to create value through learning. This knowledge is a result of cross-functional staff analysis coordination and subsequent recommendations and decisions. These are outputs from boards and workgroups, as well as from individuals.

3-57. Staffs support the commander in understanding situations, making and implementing decisions, controlling operations, and assessing progress by providing timely and relevant information and analysis. Staffs use knowledge management and information management to extract knowledge from the vast amount of available information. This enables staffs to provide knowledge to commanders in the form of recommendations and running estimates to help commanders build and maintain their situational understanding.

3-58. Effective staffs establish and maintain a high degree of coordination and cooperation with staffs of higher echelon, lower echelon, supporting, supported, and adjacent units. They do this by actively collaborating with commanders and staffs of other units to solve problems. During coordination, the KMO helps align the processes to the people, leading to an identification of the proper tools, while the information management officer installs, operates, and maintains many of the identified tools.

3-59. Staffs apply the science of control to support the commander's tasks by conducting knowledge management and information management. Knowledge management supports improving organizational learning, innovation, and performance. Conducting information management develops tools for the collection and distribution of timely and relevant information to commanders and staffs. These technical tools help enable knowledge transfer. Tools within the technical dimension can be large and complex like computer networks or simple items like maps and overlay graphics. Conducting knowledge management and information management helps commanders develop a shared understanding with other commanders and staffs.

3-60. Examples of knowledge and information activities include, but are not limited to, the following—

- Maintaining running estimates.
- Completing requests for information.
- Collecting and displaying important information (such as CCIR and execution matrices).
- Synchronizing the unit's battle rhythm with the higher headquarters' battle rhythm.
- Updating and sharing the COP.
- Producing orders, plans, and reports.
- Convening or participating in meetings to include boards, and working groups.
- Collaborating across communities of interest, communities of practice, and professional and functional forums.
- Standardizing knowledge management and information management procedures.
- Conducting after action reviews and applying lessons learned.
- Identifying knowledge gaps in organizations and processes and applying knowledge management processes to develop solutions and streamline processes.
- Supporting learning before, during, and after operations and training events.
- Developing, piloting, establishing, and sustaining collaborative knowledge networks.
- Developing and providing knowledge and information management policies and processes to facilitate knowledge capture, search, and dissemination.
- Developing digital continuity books and Web sites.
- Designing and administering knowledge networks and forums.

3-61. In practice, the KMO and signal officer work in close coordination to fulfill the commander's knowledge management needs. The COS or XO is responsible for the knowledge management program, and the KMO is the principal advisor. The KMO focuses on the knowledge management plan to facilitate situational understanding for the commander and shared understanding for the unit. The signal officer focuses on the technical aspects to support the knowledge management plan.

3-62. The technical element of information management helps enable knowledge creation, transfer, and flow. An illustration of the differences between knowledge and information management functionality is a unit's tactical web portal. The signal officer is normally responsible for the technical management of the tactical portal, while the KMO establishes policies and governance related to knowledge and information management and provides the signal officer with an understanding of requirements related to connectivity, performance, security, and configuration control.

This page intentionally left blank.

Chapter 4

Problem Solving

Problem solving is a daily activity for leaders. This chapter describes types of problems followed by a description of a systematic approach to assist in solving well- and medium-structured problems.

PROBLEMS

4-1.   The ability to recognize and effectively solve problems is an essential skill for leaders. A problem is an issue or obstacle that makes it difficult to achieve a desired goal or end state. The degree of interactive complexity of a given situation is the primary factor that determines that problem's structure. Problems range from well-structured to ill-structured. (See ADRP 5-0 more discussion on types of problems.)

4-2.   Well-structured problems are easy to identify, required information is available, and methods to solve them are somewhat obvious. While often difficult to solve, well-structured problems have verifiable solutions. Problems of mathematics and time and space relationships, as in the case with detailed logistics planning and engineering projects, illustrate well-structured problems. For well-structured problems, leaders may use the problem solving process, troop leading procedures, or the military decisionmaking process (MDMP).

4-3.   Medium-structured problems are more interactively complex than well-structured problems. For example, a field manual describes how a combined arms battalion conducts a defense, but it offers no single solution that applies to all circumstances. Leaders may agree on the problem and the end state for the operation. However, they may disagree about how to apply the doctrinal principles to a specific piece of terrain against a specific enemy. Medium-structured problems may require iterations of the problem solving process, troop leading procedures, or the MDMP.

4-4.   Ill-structured problems are complex, nonlinear, and dynamic; therefore, they are the most challenging to understand and solve. Unlike well- or medium-structured problems, leaders disagree about how to solve ill-structured problems, what the end state should be, and whether the desired end state is even achievable. Army design methodology assists leaders in understanding ill-structured problems and developing operational approaches to manage or solve those problems. (See ADRP 5-0 for more information on the Army design methodology.)

4-5.   Not all problems require lengthy analysis. For well-structured problems, leaders may make quick decisions based on their experiences. For well-structured or medium-structured problems involving a variety of factors, leaders need a systematic problem-solving process. The objective of problem solving is not just to solve near-term problems, but to also do so in a way that forms the basis for long-term success.

THE PROBLEM SOLVING PROCESS

4-6.   Troop leading procedures and the MDMP are specifically designed for planning and problem solving for conducting operations. For situations when operational planning is not appropriate, the Army's approach to problem solving involves the following steps:
- Gather information and knowledge.
- Identify the problem.
- Develop criteria.
- Generate possible solutions.
- Analyze possible solutions.

- Compare possible solutions.
- Make and implement the decision.

GATHER INFORMATION AND KNOWLEDGE

4-7. Gathering information and knowledge and is an important first step in problem solving. Leaders cannot understand or identify the problem without first gathering information and knowledge. While described as a step, gathering information and knowledge continues throughout the problem solving process. It helps leaders understand the situation and determine what the problem is by defining its limitations and scope. Leaders never stop acquiring and assessing the impact of new or additional information relevant to the problem.

4-8. Leaders require facts and assumptions to solve problems. Understanding facts and assumptions is critical to understanding problem solving. In addition, leaders need to know how to handle opinions and organize information.

Facts

4-9. Facts are verifiable pieces of information that have objective reality. They form the foundation on which leaders base solutions to problems. Regulations, policies, doctrinal publications, commander's guidance, plans and orders, and personal experiences are just a few sources of facts.

Assumptions

4-10. An *assumption* is a supposition on the current situation or a presupposition on the future course of events, either or both assumed to be true in the absence of positive proof, necessary to enable the commander in the process of planning to complete an estimate of the situation and make a decision on the course of action (JP 5-0). In other words, an assumption is information that is accepted as true in the absence of facts, but cannot be verified. Appropriate assumptions used in decisionmaking have two characteristics:

- They are valid; that is, they are likely to be true.
- They are necessary; that is, they are essential to continuing the problem solving process.

4-11. If the process can continue without making a particular assumption, leaders discard that assumption. So long as an assumption is both valid and necessary, leaders treat it as a fact. Problem solvers continually seek to confirm or deny the validity of their assumptions.

Opinions

4-12. When gathering information, leaders evaluate opinions carefully. An opinion is a personal judgment that the leader or another individual makes. Opinions cannot be totally discounted. They are often the result of years of experience. Leaders objectively evaluate opinions to determine whether to accept them as facts, include them as opinions, or reject them.

Organizing Information

4-13. Leaders check each piece of information to verify its accuracy. If possible, two individuals should check and confirm the accuracy of facts and the validity of assumptions. Being able to establish whether a piece of information is a fact or an assumption is of little value if those working on the problem do not know the information exists. Leaders share information with the decisionmaker, subordinates, and peers, as appropriate. A proposed solution to a problem is only as good as the information that forms the basis of the solution. Sharing information among members of a problem-solving team increases the likelihood that a team member will uncover the information that leads to the best solution.

4-14. Organizing information includes coordination with units and agencies that may be affected by the problem or its solution. Leaders determine these as they gather information. They coordinate with other leaders as they solve problems, both to obtain assistance and to keep others informed of situations that may affect them. Such coordination may be informal and routine. For an informal example, a squad leader checks with the squad to the right to make sure their fields of fire overlap. For a formal example, a division action officer staffs a decision paper with the major subordinate commands. As a minimum, leaders always coordinate with units or agencies that might be affected by a solution they propose before they present it to the decisionmaker.

IDENTIFY THE PROBLEM

4-15. A problem exists when the current state or condition differs from or impedes achieving the desired end state or condition. Leaders identify problems from a variety of sources. These include—

- Higher headquarters' directives or guidance.
- Decisionmaker's guidance.
- Subordinates.
- Personal observations.

4-16. When identifying a problem, leaders actively seek to identify its root cause, not merely the symptoms on the surface. Symptoms may be the reason that the problem became visible. They are often the first things noticed and frequently require attention. However, focusing on the symptoms of a problem may lead to false conclusions or inappropriate solutions. Using a systematic approach to identifying the real problem helps avoid the "solving symptoms" pitfall.

4-17. Leaders do the following to identify the root cause of a problem:

- Compare the current situation to the desired end state.
- Define the problem's scope or boundaries.
- Answer the following questions:
  - Who does the problem affect?
  - What does the problem affect?
  - When did the problem occur?
  - Where is the problem?
  - Why did the problem occur?
- Determine the cause of obstacles between current and desired end state.
- Write a draft problem statement.
- Focus information collection efforts specific to the problem.
- Redefine the problem as necessary as the staff acquires and assesses new knowledge and information.
- Update facts and assumptions.

4-18. After identifying the root causes, leaders develop a problem statement—a statement that clearly describes the problem to be solved. When the staff bases the problem upon a directive from a higher authority, it is best to submit the problem statement to the decisionmaker for approval. This ensures the problem solver has understood the decisionmaker's guidance before continuing.

4-19. Once leaders develop a problem statement, they make a plan to solve the problem. Leaders make the best possible use of available time and allocate time for each problem-solving step. This allocation provides a series of deadlines to meet in solving the problem. Leaders use reverse planning to prepare their problem-solving timeline. They use this timeline to periodically assess progress. They do not let real or perceived pressure cause them to abandon solving the problem systematically. They change time allocations as necessary, but they do not omit steps.

DEVELOP CRITERIA

4-20. The third step in the problem-solving process is developing criteria. A criterion is a standard, rule, or test by which something can be judged—a measure of value. Problem solvers develop criteria to assist them in formulating and evaluating possible solutions to a problem. Criteria are based on facts or assumptions. Problem solvers develop two types of criteria: screening and evaluation.

SCREENING CRITERIA

4-21. Leaders use screening criteria to ensure solutions they consider can solve the problem. Screening criteria defines the limits of an acceptable solution. They are tools to establish the baseline products for analysis. Leaders may reject a solution based solely on the application of screening criteria. Leaders commonly ask five questions of screening criteria to test a possible solution:

- **Is it suitable?**—Does it solve the problem and is it legal and ethical?
- **Is it feasible?**—Does it fit within available resources?
- **Is it acceptable?**—Is it worth the cost or risk?
- **Is it distinguishable?**—Does it differ significantly from other solutions?
- **Is it complete?**—Does it contain the critical aspects of solving the problem from start to finish?

EVALUATION CRITERIA

4-22. After developing screening criteria, the problem solver develops the evaluation criteria in order to differentiate among possible solutions. (See figure 4-1.) Well-defined evaluation criteria have five elements:

- **Short Title**—the criterion name.
- **Definition**—a clear description of the feature being evaluated.
- **Unit of Measure**—a standard element used to quantify the criterion. Examples of units of measure are U.S. dollars, miles per gallon, and feet.
- **Benchmark**—a value that defines the desired state or "good" for a solution in terms of a particular criterion.
- **Formula**—an expression of how changes in the value of the criterion affect the desirability of the possible solution. The problem solver states the formula in comparative terms (for example, less is better) or absolute terms (for example, a night movement is better than a day movement).

Short Title: Cost

Definition: The maximum total cost of each truck.

Unit of Measure: Dollars

Benchmark: $38,600

Formula: ≤$38,600 is an advantage; >$38,600 is a disadvantage; less is better.

Figure 4-1. Sample evaluation criterion

4-23. A well thought-out benchmark is critical for meaningful analysis. Decisionmakers employ analysis to judge a solution against a standard, determining whether that solution is good in an objective sense. It differs from comparison, in which decisionmakers judge possible solutions against each other, determining whether a solution is better or worse in a relative sense. Benchmarks are the standards used in such analysis. They may be prescribed by regulations or guidance from the decisionmaker. Sometimes, a decisionmaker can infer the benchmark by the tangible return expected from the problem's solution. Often, however, leaders establish benchmarks themselves. Four common methods for doing this are—

- **Reasoning**—based on personal experience and judgment as to what is good.
- **Historical precedent**—based on relevant examples of prior success.

- **Current example**—based on an existing condition, which is considered desirable.
- **Averaging**—based on the mathematical average of the solutions being considered. Averaging is the least preferred of all methods because it essentially duplicates the process of comparison.

4-24. In practice, the criteria by which choices are made are almost never of equal importance. Because of this, it is often convenient to assign weights to each evaluation criterion. Weighting criteria establishes the relative importance of each one with respect to the others. Weighting should reflect the judgment of the decisionmaker or acknowledged experts as closely as possible. For example, a decisionmaker or expert might judge that two criteria are *equal* in importance, or that one criterion is *slightly favored* in importance, or *moderately* or *strongly favored.* If decisionmakers assign these verbal assessments numerical values, from 1 to 4 respectively, they can use mathematical techniques to produce meaningful numerical criteria weights.

GENERATE POSSIBLE SOLUTIONS

4-25. After gathering information relevant to the problem and developing criteria, leaders formulate possible solutions. They carefully consider the guidance provided by the commander or their superiors, and develop several alternatives to solve the problem. Too many possible solutions may result in time wasted on similar options. Experience and time available determine how many solutions leaders consider. Leaders should consider at least two solutions. Limiting solutions enables the problem solver to use both analysis and comparison as problem-solving tools. Developing only one solution to "save time" may produce a faster solution but risks creating more problems from factors not considered.

4-26. When developing solutions, leaders generate options. They then summarize solutions in writing, sketches, or both.

GENERATE OPTIONS

4-27. Leaders must use creativity to develop effective solutions. Often, groups can be far more creative than individuals. However, those working on solutions should have some knowledge of or background in the problem area.

4-28. The basic technique for developing new ideas in a group setting is brainstorming. Brainstorming is characterized by unrestrained participation in discussion. While brainstorming, leaders—

- State the problem and make sure all participants understand it.
- Appoint someone to record all ideas.
- Withhold judgment of ideas.
- Encourage independent thoughts.
- Aim for quantity, not quality.
- Hitchhike ideas—combine one person's thoughts with those of others.

At the conclusion of brainstorming, leaders may discard solutions that clearly miss the standards described by the screening criteria. If this informal screen leaves only one or no solution, then leaders need to generate more options.

SUMMARIZE THE SOLUTION IN WRITING AND SKETCHES

4-29. After generating options, leaders accurately record each possible solution. The solution statement clearly portrays how the action or actions solve the problem. In some circumstances, the solution statement may be a single sentence. For example, it might be "Provide tribal leader with the means to dig a well." In other circumstances, the solution statement may require more detail, including sketches or concept diagrams. For example, if the problem is to develop a multipurpose small-arms range, leaders may choose to portray each solution with a narrative and a separate sketch or blueprint of each proposed range.

ANALYZE POSSIBLE SOLUTIONS

4-30. Having identified possible solutions, leaders analyze each one to determine its merits and drawbacks. If criteria are well defined, including a careful selection of benchmarks, analysis is greatly simplified.

4-31. Leaders use screening criteria and benchmarks to analyze possible solutions. They apply screening criteria to judge whether a solution meets minimum requirements. For quantitative criteria, they measure, compute, or estimate the raw data values for each solution and each criterion. In analyzing solutions that involve predicting future events, they use war-gaming, models, and simulations to visualize events and estimate raw data values for use in analysis. Once raw data values have been determined, the leader judges them against applicable screening criteria to determine if a possible solution merits further consideration. Leaders screen out any solution that fails to meet or exceeds the set threshold of one or more screening criteria.

4-32. After applying the screening criteria to all possible solutions, leaders use benchmarks to judge them with respect to the desired state. Data values that meet or exceed the benchmark indicate that the possible solution achieves the desired end state. Data values that fail to meet the benchmark indicate a poor solution that fails to achieve the desired end state. For each solution, leaders list the areas in which analysis reveals it to be good or not good. Sometimes the considered solutions fail to reach the benchmark. When this occurs, the leader points out the failure to the decisionmaker.

4-33. Leaders carefully avoid comparing solutions during analysis. Comparing solutions during analysis undermines the integrity of the process and tempts problem solvers to jump to conclusions. They examine each possible solution independently to identify its strengths and weaknesses. They are also careful not to introduce new criteria.

COMPARE POSSIBLE SOLUTIONS

4-34. During this step, leaders compare each solution against the others to determine the optimum one. Comparing solutions identifies which solution best solves the problem based on the evaluation criteria. Leaders use any comparison technique that helps reach the best recommendation. The most common technique is a decision matrix. (See paragraphs 9-176 through 9-182 for information on using a decision matrix.)

MAKE AND IMPLEMENT THE DECISION

4-35. After completing their analysis and comparison, leaders identify the preferred solution. If a superior assigned the problem, leaders prepare the necessary products (verbal, written, or both) needed to present the recommendation to the decisionmaker. Before presenting the findings and a recommendation, leaders coordinate their recommendation with those affected by the problem or the solutions. In formal situations, leaders present their findings and recommendations to the decisionmaker as staff studies, decision papers, or decision briefings.

4-36. A good solution can be lost if the leader cannot persuade the audience that it is correct. Every problem requires both a solution and the ability to communicate the solution clearly. The writing and briefing skills a leader possesses may ultimately be as important as good problem-solving skills.

4-37. Based on the decisionmaker's decision and final guidance, leaders refine the solution and prepare necessary implementing instructions. Formal implementing instructions can be issued as a memorandum of instruction, policy letter, or command directive. Once leaders have given instructions, they monitor their implementation and compare results to the measure of success and the desired end state established in the approved solution. When necessary, they issue additional instructions.

4-38. A feedback system that provides timely and accurate information, periodic review, and the flexibility to adjust must also be built into the implementation plan. Leaders stay involved and carefully avoid creating new problems because of uncoordinated implementation of the solution. Army problem solving does not end with identifying the best solution or obtaining approval of a recommendation.

Chapter 5

Staff Studies

This chapter describes staff studies. It provides instructions and a format for a staff study. It then provides an example for preparing a staff study. It concludes with instructions on coordinating staff studies and a list of common problems.

DEVELOPING STAFF STUDIES

5-1. A staff study is a detailed formal report to a decisionmaker requesting action on a recommendation. It provides the information and methodology used to solve a problem. (See chapter 4 for more information on problems.) The staff study includes an official memorandum for the commander's signature that implements the action. The leader coordinates staff studies with all affected organizations. Staff studies include statements of nonconcurrence, if applicable, so that the decisionmaker clearly understands all staff members' support for the recommendation. A staff study is comprehensive; it includes all relevant information needed to solve the problem and a complete description of the methodology used to arrive at the recommended solution.

5-2. The staff study follows the seven-step Army problem-solving process described in chapter 4. This ensures that the staff clearly identifies the problem, follows a logical sequence, and produces a justifiable solution.

5-3. The body of a completed staff study is a stand-alone document. While enclosures are a part of most staff studies, a decisionmaker should not have to refer to them to understand the recommendation and the basis for it. Enclosures contain details and supporting information and help keep the body of the study concise.

THE STAFF STUDY FORMAT

5-4. Staff officers prepare staff studies as informal memorandums in the format found in figure 5-1 on pages 5-2 through 5-4. Units may establish their own format to meet local requirements.

Office Symbol   Date

MEMORANDUM FOR *Address the staff study to the decisionmaker. Include through addressees if required.*

SUBJECT: *Succinctly describe the subject to distinguish it from other documents as a courtesy to the decisionmaker. Do not simply state "staff study" as this does not provide sufficient detail, nor does it convey any information about the subject.*

1. **PROBLEM.** *Concisely state the problem.*

2. **RECOMMENDATION.** *Recommend a solution or solutions based on the conclusion in paragraph 10. If there are several recommendations, state each one in a separate subparagraph.*

3. **BACKGROUND.** *Briefly state why the problem exists. Provide enough information to place the problem in context. This discussion may include the origin of the action and a summary of related events. If a tasking document is the source of the problem, place it in enclosure 2 and refer to it here.*

4. **FACTS.** *State all facts that influence the problem or its solution. List each fact as a separate subparagraph. Make sure to state the facts precisely and attribute them correctly. Facts must stand-alone: either something is a generally accepted fact or it is attributed to a source that asserts it to be true. There is no limit to the number of facts as long as every fact is relevant. Include all facts relevant to the problem, not just facts used to support the study. The decisionmaker must have an opportunity to consider facts that do not support the recommendation. State any guidance given by the decisionmaker. Refer to enclosures as necessary for amplification, references, mathematical formulas, or tabular data.*

5. **ASSUMPTIONS.** *Identify assumptions necessary for a logical discussion of the problem. List each assumption as a separate subparagraph.*

6. **POSSIBLE SOLUTION.** *List all solutions considered. Place each solution in a separate subparagraph. List each solution by number and name or as a short sentence in the imperative (for example, "Increase physical security measures at key assets"). If a solution is not self-explanatory, include a brief description of it. Use enclosures to describe complex solutions.*

7. **CRITERIA.** *List and define, in separate subparagraphs, the screening and evaluation criteria. A fact or an assumption in paragraph 4 or 5 should support each criterion. At a minimum, the number of facts and assumptions should exceed the number of criteria. In a third subparagraph, explain the rationale for how the evaluation criteria are weighted.*

a. *Screening Criteria. List the screening criteria, each in its own sub-subparagraph. Screening criteria define the minimum and maximum characteristics of the solution to the problem. Answer each screening criterion: Is it suitable, feasible, acceptable, distinguishable, and complete? (See chapter 4, paragraph 4-18.) Screening criteria are not weighted. They are required, absolute standards. Reject courses of action that do not meet the screening criteria.*

b. *Evaluation Criteria. List the evaluation criteria, each in its own sub-subparagraph. List them in order of their weight, from most to least important. Define each evaluation criterion in terms of five required elements: short title, definition, unit of measure, benchmark, and formula. (See chapter 4, paragraphs 4-19 to 4-22 and figure 4-1.)*

c. *Weighting of Criteria. State the relative importance of each evaluation criterion with respect to the others. Explain how each criterion compares to each of the other criteria (equal, slightly favored, favored, or strongly favored) or provide the values from the decision matrix and explain why the criterion is measured in that way. (See chapter 4.) This subparagraph explains the order in which the evaluation criteria are listed in subparagraph 7b.*

8. **ANALYSIS.** *List the courses of action that do not meet the screening criteria and the results of applying the evaluation criteria to the remaining ones.*

Figure 5-1. Staff study paper format example

a. Screened Out Courses of Action. *List the courses of action that did not meet the screening criteria, each in its own subparagraph, and the screening criteria each did not meet. This subparagraph is particularly important if a solution the decisionmaker wanted to be considered does not meet the screening*

b. Course of Action 1. *In subsequent subparagraphs, list the courses of action evaluated, each in a separate subparagraph. Discuss the advantages and disadvantages of each solution. For quantitative criteria, include the payoff value. Discuss or list advantages and disadvantages in narratives. Use the form that best fits the information. Avoid using bullets unless the advantage or disadvantage is self-evident.*

(1) Advantages. List the advantages for course of action 1.

(2) Disadvantages. List the disadvantages for course of action 1.

c. *Course of Action 2. (Use the same format as above and continue the analysis.)*

(1) Advantages. *List the advantages for course of action 2.*

(2) Disadvantages. *List the disadvantages for course of action 2.*

9. **COMPARISON.** *Compare the courses of action to each other, based on the analysis outlined in paragraph 8. Develop in a logical, orderly manner the rationale used to reach the conclusion stated in paragraph 10. If leaders use quantitative techniques in the comparison, summarize the results clearly enough that the reader does not have to refer to an enclosure. Include any explanations of quantitative techniques in enclosures. State only the results in this paragraph.*

10. **CONCLUSION.** *State the conclusion drawn based on the analysis (paragraph 8) and comparison (paragraph 9). The conclusion must answer the question or provide a possible solution to the problem. It must match the recommendation in paragraph 2.*

11. **COORDINATION.** *List all organizations with which the study was coordinated ("staffed"). If the list is long and space is a consideration, place it at enclosure 3. If the staffing list is placed in enclosure 3, indicate the number of nonconcurrences with the cross-reference (for example, "See enclosure 3; 2 nonconcurrences"; or "See enclosure 3; no nonconcurrences").*

A representative of each organization with which the study was staffed indicates whether the organization concurs with the study, nonconcurs, or concurs with comment. Representatives place their initials in the blank, followed by their rank, name, position, telephone number, and e-mail address. If separate copies were sent to each organization (rather than sending one copy to each organization in turn), this information may be typed into the final copy of the study and the actual replies placed in enclosure 4. Recommend this technique when using e-mail for staffing.

Place all statements of nonconcurrence and considerations of nonconcurrence in enclosure 3, or in separate enclosures for each nonconcurrence. Concurrences with comments may be placed in enclosure 3 or in a separate enclosure or enclosures.

ACOS, G-1	**CONCUR/NONCONCUR**_____	**CMT**_____	**DATE:**_____
ACOS, G-2	**CONCUR/NONCONCUR**_____	**CMT**_____	**DATE:**_____
ACOS, G-3	**CONCUR/NONCONCUR**_____	**CMT**_____	**DATE:**_____
ACOS, G-4	**CONCUR/NONCONCUR**_____	**CMT**_____	**DATE:**_____

12. **APPROVAL/DISAPPROVAL.** *Restate the recommendation from paragraph 2 and provide a format for the approval authority to approve or disapprove the recommendation.*

a. That the *(state the approving authority and recommended solution).*

APPROVED_____**DISAPPROVED**_____**SEE ME**_____

b. That the *(approving authority)* sign the implementing directive(s) (TAB A).

APPROVED_____**DISAPPROVED**_____**SEE ME**_____

Figure 5-1. Staff study paper format example (continued)

> **13. POINT OF CONTACT.** *Record the point of contact (or action officer) and contact information. Additional contact information may include the action officer's organization, a civilian telephone number, a unit address, and an e-mail address.*
>
> **[Signature Block]** *Prepare the signature block as specified in Chapter 2 of AR 25-50.*
>
> **[#] Encl** *(Tab the enclosures)*
> 1. Implementing document - *Enclosure 1 contains implementing memorandums, directives, or letters submitted for signature or approval. Since a staff study requests a decision, enclosure 1 contains the documents required to implement the decision (Tab A).*
> 2. Tasking document - *Enclosure 2 contains the document that directed the staff study or decision paper. If the requirement was given verbally, include the memorandum for record that documents the conversation. If no record exists, enter "Not used" in the annex list in the body (Tab B).*
> 3. Coordination list - *Enclosure 3 contains the staffing list if the list is too long for paragraph 11. If paragraph 11 contains the entire staffing list, enter "Not used" in the enclosure list in the body (Tab C).*
> 4. Nonconcurrences - *Enclosure 4 contains statements of nonconcurrence and considerations of nonconcurrence. These documents may be placed in separate enclosures. Place concurrences with comment in either enclosure 4 or a separate enclosure. If there are no statements of nonconcurrence, enter "Not used" in the enclosure list in the body (Tab D).*
> 5–[#].   Other supporting documents, listed as separate enclosures - *Other enclosures contain detailed data, lengthy discussions, and bibliographies. Number the pages of each enclosure separately, except when an enclosure contains several distinct documents (such as, concurrences) (Tabs E through Z, if necessary).*

Figure 5-1. Staff study paper format example (continued)

COORDINATING STAFF STUDIES

5-5.   Preparing a staff study normally involves coordinating with other staff officers and organizations. At a minimum, action officers obtain concurrences or nonconcurrences from agencies affected by the study's recommendations. Other aspects of the study may require coordination as well. Coordination should be as broad as time permits but should be limited to agencies that might be affected by possible recommendations or that have expertise in the subject of the study.

5-6.   Action officers anticipate nonconcurrences and try to resolve as many as possible before staffing the final product. An action officer who cannot resolve a nonconcurrence has two options:

- Modify the staff study to satisfy the nonconcurrence, but only if the analysis and comparison supports the change. If this is done after the final draft has been staffed, the officer must re-staff the study.
- Prepare a consideration of nonconcurrence and include it and the statement of nonconcurrence in enclosure 4 to the staff study.

Statements of Nonconcurrence

5-7.   A statement of nonconcurrence is a recommendation for the decisionmaker to reject all or part of the staff study. Statements of nonconcurrence are prepared in the memorandum format; e-mails may be accepted at the commander's discretion. They address specific points in the recommendations or the study, stating why they are wrong or unacceptable. They offer an alternative or a constructive recommendation when possible.

Considerations of Nonconcurrence

5-8.   Action officers prepare considerations of nonconcurrence as a memorandum for record. They present the reasons for the nonconcurrence accurately and assess them objectively. Then they state why the study is correct and why the decisionmaker should reject the nonconcurrence.

5-4   **FM 6-0**   **5 May 2014**

COMMON PROBLEMS WITH STAFF STUDIES

5-9.   These questions identify the most common problems found in staff studies. Leaders should review them before beginning a staff study and periodically thereafter:

- Is the subject too broad?
- Is the problem properly defined?
- Are facts or assumptions clear and valid?
- Are there any unnecessary facts or assumptions?
- Are there any facts that appear for the first time in the discussion?
- Are there a limited number of options or courses of action?
- Are evaluation criteria invalid or too restrictive?
- Is the discussion too long?
- Is the discussion complete?
- Must readers consult the enclosures to understand the staff study?
- Does the conclusion include a discussion?
- Is the logic flawed or incomplete?
- Does the conclusion follow from the analysis?
- Can the solution be implemented within resource and time constraints?
- Do the conclusions and recommendations solve the problem?
- Is there an implementing directive?
- Have new criteria been introduced in the analysis or comparison?

This page intentionally left blank.

Chapter 6

Decision Papers

This chapter explains decision papers. It provides instructions, a format, and an example for preparing decision papers.

PREPARING DECISION PAPERS

6-1. A decision paper is a piece of correspondence that requests the decisionmaker to act on its recommendation and provides the required implementing documents for signature. Action officers use a decision paper when they do not need a formal report or the decisionmaker does not require the details a staff study provides.

6-2. Decision papers are brief. Unlike staff studies, decision papers are not self-contained. For a decision paper, much of the material that would be included in a staff study is kept in the action officer's file. Decision papers contain the minimum information the decisionmaker needs to understand the action and make a decision. The action officer synthesizes the facts, summarizes the issues, presents feasible alternatives, and recommends one of them. Action officers attach essential explanations and other information as enclosures, which are always tabbed.

FORMATTING DECISION PAPERS

6-3. Action officers prepare decision papers as informal memorandums (see AR 25-50) in the format shown in figure 6-1 on page 6-2. This format also parallels the steps of the Army problem-solving process. Commands may establish format standards to meet local requirements. Decision papers should not exceed two pages, excluding the staffing list and supporting documentation. The coordination requirements for a decision paper are the same as those for a staff study.

Office Symbol (Marks Number)   Date

MEMORANDUM FOR *Address the decision paper to the decisionmaker. Include through addressees or on the routing slip, as specified by command policy.*

SUBJECT: *Briefly state the decision's subject. Be specific as the reader should not have to begin reading the body of the decision paper to figure out the subject. "Decision paper" is not an acceptable subject.*

1. For DECISION. *Indicate if the decision is time-sensitive, tied to an event, or has a suspense date to a higher headquarters. Show internal suspense dates on the routing slip, if necessary. However, do not show them in this paragraph. (Paragraph headings may be either underlined or bolded, according to command policy.)*

2. PURPOSE. *State clearly the decision required, as an infinitive phrase. An infinitive phrase uses a verb, but has no subject, for example, "To determine the...," or, "To obtain...." Include in the purpose statement who, what, when, and where, if pertinent.*

3. RECOMMENDATION. *Recommend a solution or solutions to the problem. If there are several recommendations, state each one in a separate subparagraph.*

4. BACKGROUND AND DISCUSSION. *Explain the origin of the action, why the problem exists, and a summary of events in chronological form. It helps put the problem in perspective and provides an understanding of the alternatives and the recommendation. If the decision paper is the result of a tasking document, refer to that document in this paragraph and place it at enclosure 2.*

5. IMPACTS. *State the impact of the recommended decision. Address each affected area in a separate subparagraph, for example, personnel, equipment, funding, environment, and stationing. State parties affected by the recommendation and the extent to which they are affected.*

6. COORDINATION. *The coordination, approval line, point of contact, signature block, and enclosures follow the same directions as for a staff study. (See chapter 5.)*

ACOS, G-1   **CONCUR/NONCONCUR**_____**CMT**_____**DATE:** _____

ACOS, G-3   **CONCUR/NONCONCUR**_____**CMT**_____**DATE:** _____

7. APPROVAL/DISAPPROVAL.

a. That the *(state the approving authority and recommended solution).*

APPROVED_____**DISAPPROVED**_____**SEE ME**_____

b. That the *(approving authority)* sign the implementing directive(s) (TAB A).

APPROVED_____**DISAPPROVED**_____**SEE ME**_____

[Signature Block] *Prepare the signature block as specified in Chapter 2 of AR 25-50.*

[#] Encl   *(Tab the enclosures)*

1.   Implementing document - *Enclosure 1 contains implementing memorandums, directives, or letters submitted for signature or approval. Since a staff study requests a decision, enclosure 1 contains the documents required to implement the decision (Tab A).*

2.   Tasking document - *Enclosure 2 contains the document that directed the staff study or decision paper. If the requirement was given verbally, include the memorandum for record that documents the conversation. If no record exists, enter "Not used" in the annex list in the body (Tab B).*

3.   Coordination list - *Enclosure 3 contains the staffing list if the list is too long for paragraph 6. If paragraph 6 contains the entire staffing list, enter "Not used" in the enclosure list in the body (Tab C).*

4.   Nonconcurrences - *Enclosure 4 contains statements of nonconcurrence and considerations of nonconcurrence. These documents may be placed in separate enclosures. Place concurrences with comment in either enclosure 4 or a separate enclosure. If there are no statements of nonconcurrence, enter "Not used" in the enclosure list in the body (Tab D).*

5–[#].   Other supporting documents, listed as separate enclosures - *Other enclosures contain detailed data, lengthy discussions, and bibliographies. Number the pages of each enclosure separately, except when an enclosure contains several distinct documents (such as, concurrences) (Tabs E through Z, if necessary).*

Figure 6-1. Decision paper format example

Chapter 7

Military Briefings

This chapter describes the four types of military briefings presented to commanders, staffs, or other audiences and describes the steps of these military briefings. It also provides instructions for developing military briefings.

TYPES OF MILITARY BRIEFINGS

7-1.   The Army uses four types of briefings: information, decision, mission, and staff.

INFORMATION BRIEFING

7-2.   An information briefing presents facts in a form the audience can easily understand. It does not include conclusions or recommendations, nor does it result in decisions. The main parts of an information briefing are the introduction, main body, and conclusion. (See figure 7-1.)

---

1. Introduction

Greeting. *Address the audience. Identify yourself and your organization.*

Type and Classification of Briefing. *Identify the type and classification of the briefing. For example, "This is an information briefing. It is unclassified."*

Purpose and Scope. *Describe complex subjects from general to specific.*

Outline or Procedure. *Briefly summarize the key points and general approach. Explain any special procedures (such as demonstrations, displays, or tours). For example, "During my briefing, I will discuss the six phases of our plan. I will refer to maps of our area of operations. Then my assistant will bring out a sand table to show you the expected flow of battle." The key points may be placed on a chart that remains visible throughout the briefing.*

2. Main Body

Arrange the main ideas in a logical sequence.

Use visual aids to emphasize main points.

Plan effective transitions from one main point to the next.

Be prepared to answer questions at any time.

3. Closing

Ask for questions.

Briefly recap main ideas and make a concluding statement.

Figure 7-1. Information briefing format example

7-3.   Examples of appropriate topics for information briefings include, but are not limited to—
- High-priority information requiring immediate attention.
- Information such as complicated plans, systems, statistics or charts, or other items that require detailed explanations.
- Information requiring elaboration and explanation.

DECISION BRIEFING

7-4.   A decision briefing obtains the answer to a question or a decision on a course of action. The briefer presents recommended solutions from the analysis or study of a problem. Decision briefings vary in formality and level of detail depending on the commander's or decisionmaker's knowledge of the subject.

7-5.   If the decisionmaker is unfamiliar with the problem, the briefing format adheres to the decision briefing format. (See figure 7-2.) Decision briefings include all facts and assumptions relevant to the problem, a discussion of alternatives, analysis-based conclusions, and any coordination required.

7-6.   When the decisionmaker is familiar with the subject or problem, the briefing format often resembles that of a decision paper: problem statement, essential background information, impacts, and recommended solution. In addition to this format, briefers must be prepared to present assumptions, facts, alternative solutions, reasons for recommendations, and any additional coordination required.

---

1. Introduction

Greeting. *Address the decisionmaker. Identify yourself and your organization. "This is a decision briefing."*

Type and Classification of Briefing. *Identify the type and classification of the briefing. For example, "This is a decision briefing. It is unclassified."*

Problem Statement. *State the problem.*

Recommendation. *State the recommendation.*

2. Main Body

Facts. *Provide an objective presentation of both positive and negative facts bearing upon the problem.*

Assumptions. *Identify necessary assumptions made to bridge any gaps in factual data.*

Solutions. *Discuss the various options that can solve the problem.*

Analysis. *List the screening and evaluation criteria by which the briefer will evaluate how to solve the problem. Discuss relative advantages and disadvantages for each course of action.*

Comparison. *Show how the courses of action compare against each other.*

Conclusion. *Describe why the recommended solution is best.*

3. Closing

Ask for questions.

Briefly recap main ideas and restate the recommendation.

If no decision is provided upon conclusion of the decision briefing, request a decision. "Sir/Ma'am, what is your decision?" The briefer ensures all participants clearly understand the decision and asks for clarification if necessary.

Figure 7-2. +Decision briefing format example

7-7.   The briefer clearly states and precisely words a recommendation presented during decision briefings to prevent ambiguity and to translate it easily into a decision statement. If the decision requires an implementation document, briefers present that document at the time of the briefing for the decisionmaker to sign. If the chief of staff or executive officer is absent, the briefer informs the secretary of the general staff or designated authority of the decision upon conclusion of the briefing.

MISSION BRIEFING

7-8.   Mission briefings are information briefings that occur during operations or training. Briefers may be commanders, staffs, or special representatives.

7-9.   Mission briefings serve to convey critical mission information not provided in the plan or order to individuals or small units. Mission briefings—

- Issue or enforce an order.
- Provide more detailed instructions or requirements.
- Instill a general appreciation for the mission.
- Review key points for an operation.
- Ensure participants know the mission objective, their contribution to the operation, problems they may confront, and ways to overcome them.

7-10. The nature and content of the information provided determines the mission briefing format. Typically a briefer will use the operation plan or order as a format for a mission briefing.

STAFF BRIEFING

7-11. Staff briefings inform the commander and staff of the current situation in order to coordinate and synchronize efforts within the unit. The individual convening the staff briefing sets the briefing agenda. Each staff element presents relevant information from its functional area. Staff briefings facilitate information exchange, announce decisions, issue directives, or provide guidance. The staff briefing format may include characteristics of the information briefing, decision briefing, and mission briefing. (See figure 7-1 [on page 7-1] and figure 7-2 for briefing formats.)

7-12. The commander, deputies or assistants, chiefs of staff or executive officers, coordinating personnel, and special staff officers often attend staff briefings. Representatives from other commands may also attend. The chief of staff or executive officer often presides over the briefing. The commander may take an active role during the briefing and normally concludes the briefing.

STEPS OF MILITARY BRIEFINGS

7-13. Staffs normally follow four steps when preparing an effective briefing:

- Plan—analyze the situation and prepare a briefing outline.
- Prepare—collect information and construct the briefing.
- Execute—deliver the briefing.
- Assess—follow up as required.

PLAN

7-14. Upon receipt of the task to conduct a briefing, the briefer analyzes the situation and determines the—

- Audience.
- Purpose and type of briefing.
- Subject.
- Classification.
- Physical facilities and support needed.
- Preparation timeline and schedule.

7-15. Based on the analysis, the briefer assembles a briefing outline and timeline. The briefing outline is the plan for the preparation, execution, and follow-up for the briefing. The timeline is a time management tool to manage briefing preparations and budget time if there is a need to refine the briefing as new information becomes available.

7-16. Briefers consider many factors while planning a briefing (see figure 7-3 on page 7-4). This planning includes, but is not limited to—

- Audience preferences for information delivery, such as how the decisionmaker prefers to see information presented.
- Time available.
- Facilities and briefing aids available.

---

1. Audience.
What is the size and composition? Single Service or joint? Civilians? Foreign nationals?
Who are the ranking members and their official duty positions?
How well do they know the subject?
Are they generalists or specialists?
What are their interests?
What is the anticipated reaction?

2. Purpose and Type.
Information briefing (to inform)?
Decision briefing (to obtain decision)?
Mission briefing (to review important details)?
Staff briefing (to exchange information)?

3. Subject.
What is the specific subject?
What is the desired depth of coverage?
How much time is allocated?

4. Classification.
What is the security classification?
Do all attendees meet this classification?

5. Physical Facilities and Support Needed.
Where is the briefing to be presented?
What support is needed?
What are the security requirements, if needed?
What are the equipment requirements? Computer? Projector? Screen?

6. Preparation Timeline and Schedule.
Prepare preliminary outline.
Determine requirements for training aids, assistants, and recorders.
Schedule rehearsals, facilities, and critiques.
Arrange for final review by responsible authority.

Figure 7-3. Planning considerations for military briefings

7-17. The briefer then estimates deadlines for each task and schedules the preparation effort accordingly. The briefer alerts support personnel and any assistants as soon as possible.

PREPARE

7-18. The briefing construction varies with type and purpose. (See figure 7-4.) The analysis of the briefing determines the basis for this. Briefers follow these key steps to prepare a briefing:

- Collect materials needed.
- Prepare first draft.
- Revise first draft and edit.
- Plan use of visual aids.
- Check audiovisual delivery systems (computer and other technical aids) to ensure availability and functionality.
- Practice.

```
1. Collect Materials Needed.
      Use the seven-step Army problem-solving process. (See chapter 4.)
      Research.
      Become familiar with the subject.
      Collect authoritative opinions and facts.
2. Prepare First Draft.
      Prepare draft outline.
      Include visual aids.
      Review with appropriate authority.
3. Revise First Draft and Edit.
      Verify facts, including those that are important and necessary.
      Include answers to anticipated questions.
      Refine materials.
4. Plan Use of Visual Aids.
      Check for simplicity.
      Check for readability.
5. Check Audiovisual delivery systems.
      Ensure availability and functionality.
6. Practice.
      Rehearse (with assistants and visual aids).
      Refine.
      Isolate key points.
      Memorize outline.
      Develop transitions.
      Anticipate and prepare for possible questions.
```

Figure 7-4. Preparation considerations for military briefings

EXECUTE

7-19. The success of a briefing depends on a concise, objective, accurate, clearly enunciated, and forceful delivery. The briefer must also be confident and relaxed. The briefer should consider the following:

- The basic purpose is to present the subject as directed and ensure the audience understands it.
- Brevity precludes a lengthy introduction or summary.
- Conclusions and recommendations must flow logically from facts and assumptions.

7-20. Interruptions and questions may occur at any point. If they occur, briefers answer each question before continuing, or they indicate that they will answer the question later in the briefing. When briefers answer questions later in the briefing, they specifically reference the earlier question when they introduce material. They anticipate possible questions and are prepared to answer them.

ASSESS

7-21. When the briefing is over, the briefer conducts a follow-up, as required. To ensure understanding, the briefer prepares a memorandum for record. This memorandum records the subject, date, time, and location of the briefing as well as the ranks, names, and positions of audience members. The briefer concisely records the briefing's content to help ensure understanding. The briefer records recommendations and their approval, disapproval, or approval with modification as well as instructions or directed actions. Recommendations can include who is to take action. The briefer records the decision. When a decision is involved and any ambiguity exists about the commander's intent, the briefer submits a draft of the memorandum for record for correction before preparing the final document. Lastly, the briefer informs proper authorities. The briefer distributes the final memorandum for record to staff elements and agencies required to act on the decisions or instructions or whose plans or operations may be affected.

This page intentionally left blank.

Chapter 8

Running Estimates

This chapter defines running estimate and describes how the commander and staff build and maintain their running estimates throughout the operations process. This chapter provides a generic running estimate format that the commander and each staff element may modify to fit their functional area. (See JP 5-0 for information on joint estimates.)

TYPES OF RUNNING ESTIMATES

8-1. A *running estimate* is the continuous assessment of the current situation used to determine if the current operation is proceeding according to the commander's intent and if planned future operations are supportable (ADP 5-0). The commander and each staff element maintain a running estimate. In their running estimates, the commander and each staff element continuously consider the effects of new information and update the following:

* Facts.
* Assumptions.
* Friendly force status.
* Enemy activities and capabilities.
* Civil considerations.
* Conclusions and recommendations.

8-2. Commanders maintain their running estimates to consolidate their understanding and visualization of an operation. The commander's running estimate summarizes the problem and integrates information and knowledge of the staff's and subordinate commanders' running estimates.

8-3. Each staff element builds and maintains running estimates. The running estimate helps the staff to track and record pertinent information and provide recommendations to commanders. Running estimates represent the analysis and expert opinion of each staff element by functional area. Staffs maintain running estimates throughout the operations process to assist commanders in the exercise of mission command.

8-4. Each staff element and command post functional cell maintains a running estimate focused on how its specific areas of expertise are postured to support future operations. Because an estimate may be needed at any time, running estimates must be developed, revised, updated, and maintained continuously while in garrison and during operations. While in garrison, staffs must maintain a running estimate on friendly capabilities. Running estimates can be presented verbally or in writing.

ESSENTIAL QUALITIES OF RUNNING ESTIMATES

8-5. A comprehensive running estimate addresses all aspects of operations and contains both facts and assumptions based on the staff's experience within a specific area of expertise. Each staff element modifies it to account for its specific functional areas. All running estimates cover essential facts and assumptions, including a summary of the current situation by the mission variables, conclusions, and recommendations. (See appendix A for information on the mission variables.) Once they complete the plan, commanders and staff elements continuously update their estimates. (See figure 8-1 on page 8-2 for the base format for a running estimate that parallels the planning process.)

---

1. **SITUATION AND CONSIDERATIONS.**

   a. **Area of Interest.** Identify and describe those factors of the area of interest that affect functional area considerations.

   b. **Characteristics of the Area of Operations.**

   (1) Terrain. State how terrain affects a functional area's capabilities.

   (2) Weather. State how weather affects a functional area's capabilities.

   (3) Enemy Forces. Describe enemy disposition, composition, strength, and systems within a functional area. Describe enemy capabilities and possible courses of action (COAs) and their effects on a functional area.

   (4) Friendly Forces. List current functional area resources in terms of equipment, personnel, and systems. Identify additional resources available for the functional area located at higher, adjacent, or other units. List those capabilities from other military and civilian partners that may be available to provide support within the functional area. Compare requirements to current capabilities and suggest solutions for satisfying discrepancies.

   (5) Civilian Considerations. Describe civil considerations that may affect the functional area, including possible support needed by civil authorities from the functional area as well as possible interference from civil aspects.

   c. **Facts/Assumptions.** List all facts and assumptions that affect the functional area.

2. **MISSION.** Show the restated mission resulting from mission analysis.

3. **COURSES OF ACTION.**

   a. List friendly COAs that were war-gamed.

   b. List enemy actions or COAs that were templated that impact the functional area.

   c. List the evaluation criteria identified during COA analysis. All staffs use the same criteria.

4. **ANALYSIS.** Analyze each COA using the evaluation criteria from COA analysis. Review enemy actions that impact the functional area as they relate to COAs. Identify issues, risks, and deficiencies these enemy actions may create with respect to the functional area.

5. **COMPARISON.** Compare COAs. Rank order COAs for each key consideration. Use a decision matrix to aid the comparison process.

6. **RECOMMENDATIONS AND CONCLUSIONS.**

   a. Recommend the most supportable COAs from the perspective of the functional area.

   b. Prioritize and list issues, deficiencies, and risks and make recommendations on how to mitigate them.

Figure 8-1. Generic base running estimate format

8-6. The base running estimate addresses information unique to each functional area. It serves as the staff element's initial assessment of the current readiness of equipment and personnel and of how the factors considered in the running estimate affect the staff's ability to accomplish the mission. Each staff element identifies functional area friendly and enemy strengths, systems, training, morale, leadership, and weather and terrain effects, and how all these factors impact the operational environment, including the area of operations. Because the running estimate is a picture relative to time, facts, and assumptions, each staff element constantly updates the estimate as new information arises, as assumptions become facts or are invalidated, when the mission changes, or when the commander requires additional input.

RUNNING ESTIMATES IN THE OPERATIONS PROCESS

8-7. Commanders and staff elements immediately begin updating their running estimates upon receipt of a mission. They continue to build and maintain their running estimates throughout the operations process in planning, preparation, execution, and assessment.

RUNNING ESTIMATES IN PLANNING

8-8. During planning, running estimates are key sources of information during mission analysis. Following mission analysis, commanders and staff elements update their running estimates throughout the rest of the military decisionmaking process. Based on the mission and the initial commander's intent, the staff develops one or more proposed courses of action (COAs) and continually refines its running estimates to account for the mission variables. The updated running estimates then support COA analysis (war-gaming) in which the staff identifies the strengths and weaknesses of each COA. The staff relies on its updated running estimate to provide input to the war game. Following COA analysis, the staff compares the proposed COAs against each other and recommends one of them to the commander for approval. During all these activities, each staff element continues to update and refine its running estimate to give commanders the best possible information available at the time to support their decisions. The selected COA provides each staff element an additional focus for its estimates and the key information it will need during orders production. Key information recorded in the running estimate may be included in orders, particularly in the functional annexes.

RUNNING ESTIMATES IN PREPARATION

8-9. The commander and staff transition from planning to execution. As they transition, they use running estimates to identify the current readiness of the unit in relationship to its mission. The commander and staff also use running estimates to develop, then track, mission readiness goals and additional requirements.

RUNNING ESTIMATES IN EXECUTION

8-10. During execution, the commander and staff incorporate information included in running estimates into the common operational picture. This enables the commander and staff to depict key information from each functional area or warfighting function as it impacts current and future operations. This information directly supports the commander's visualization and rapid decisionmaking during operations.

RUNNING ESTIMATES IN ASSESSMENT

8-11. Each staff element continuously analyzes new information during operations to create knowledge and to understand if operations are progressing according to plan. During planning, staffs develop measures of effectiveness and measures of performance to support assessment, including analysis of anticipated decisions during preparation and execution. The assessment of current operations also supports validation or rejection of additional information that will help update the estimates and support further planning. At a minimum, a staff element's running estimate assesses the following:

- Friendly force capabilities with respect to ongoing and planned operations.
- Enemy capabilities as they affect the staff element's area of expertise for current operations and plans for future operations.
- Civil considerations as they affect the staff element's area of expertise for current operations and plans for future operations.

This page intentionally left blank.

Chapter 9

The Military Decisionmaking Process

The military decisionmaking process is one of the Army's three planning methodologies. Before beginning an iteration of the military decisionmaking process, readers should review chapter 2 of ADRP 5-0 to understand the fundamentals of planning. This chapter defines and describes the characteristics of the military decisionmaking process. Next, it provides a detailed discussion of each step of the military decisionmaking process. The chapter concludes by providing guidance for conducting the military decisionmaking process in a time-constrained environment. Effectively conducting the military decisionmaking process requires leaders who understand the fundamentals of planning.

CHARACTERISTICS OF THE MILITARY DECISIONMAKING PROCESS

9-1.   The *military decisionmaking process* is an iterative planning methodology to understand the situation and mission, develop a course of action, and produce an operation plan or order (ADP 5-0). The military decisionmaking process (MDMP) helps leaders apply thoroughness, clarity, sound judgment, logic, and professional knowledge to understand situations, develop options to solve problems, and reach decisions. This process helps commanders, staffs, and others think critically and creatively while planning.

9-2.   The MDMP facilitates collaborative planning. The higher headquarters solicits input and continuously shares information concerning future operations through planning meetings, warning orders, and other means. It shares information with subordinate and adjacent units, supporting and supported units, and unified action partners. Commanders encourage active collaboration among all organizations affected by pending operations to build a shared understanding of the situation, participate in course of action development and decisionmaking, and resolve conflicts before publishing the plan or order.

9-3.   During planning, assessment focuses on developing an understanding of the current situation and determining what to assess and how to assess progress using measures of effectiveness and measures of performance. Developing the unit's assessment plan occurs during the MDMP—not after developing the plan or order. (See chapter 15 for details on assessment plans.)

9-4.   The MDMP also drives preparation. Since time is a factor in all operations, commanders and staffs conduct a time analysis early in the planning process. This analysis helps them determine when to begin certain actions to ensure forces are ready and in position before execution. This may require the commander to direct subordinates to start necessary movements, conduct task organization changes, begin information collection, and execute other preparation activities before completing the plan. As the commander and staff conduct the MDMP, they direct preparation tasks in a series of warning orders (WARNORDs).

9-5.   Depending on the situation's complexity, commanders can initiate the Army design methodology before or in parallel with the MDMP. If the problem is hard to identify or the operation's end state is unclear, commanders may initiate Army design methodology before engaging in detailed planning. Army design methodology can assist the commander and staff in understanding the operational environment, framing the problem, and considering an operational approach to solve or manage the problem. The understanding and products resulting from Army design methodology guide more detailed planning during the MDMP. When used in parallel, the commander may direct some staff members to conduct mission analysis while engaging others in Army design methodology activities prior to course of action development. Results of both mission analysis and Army design methodology inform commanders in development of their commander's intent and planning guidance. In time-constrained conditions, or when the problem is not complex, commanders may conduct the MDMP without incorporating formal Army design methodology efforts. During execution, the commander can use Army design methodology to help refine understanding and visualization as well as assessing and adjusting the plan as required.

THE SEVEN STEPS OF THE MILITARY DECISIONMAKING PROCESS

9-6.   The MDMP consists of seven steps, as shown in figure 9-1. Each step of the MDMP has various inputs, a step to conduct, and outputs. Each step also has a series of processes that commanders and staffs conduct to produce the outputs. The outputs lead to an increased understanding of the situation, facilitating the next step of the MDMP. Commanders and staffs generally perform these steps sequentially; however, they may revisit several steps in an iterative fashion as they learn more about the situation before producing the plan or order.

9-7.   Commanders initiate the MDMP upon receipt of, or in anticipation of, a mission. Commanders and staffs often begin planning in the absence of a complete and approved higher headquarters' operation plan (OPLAN) or operation order (OPORD). In these instances, the headquarters begins a new planning effort based on a WARNORD and other directives, such as a planning order or an alert order from its higher headquarters. This requires active collaboration with the higher headquarters and parallel planning among echelons as the plan or order is developed.

THE ROLE OF COMMANDERS AND STAFFS IN THE MILITARY DECISIONMAKING PROCESS

9-8.   The commander is the most important participant in the MDMP. More than simply decisionmakers in this process, commanders use their experience, knowledge, and judgment to guide staff planning efforts. While unable to devote all their time to the MDMP, commanders follow the status of the planning effort, participate during critical periods of the process, and make decisions based on the detailed work of the staff. During the MDMP, commanders focus their activities on understanding, visualizing, and describing.

9-9.   The MDMP stipulates several formal meetings and briefings between the commander and staff to discuss, assess, and approve or disapprove planning efforts as they progress. However, experience has shown that optimal planning results when the commander meets informally at frequent intervals with the staff throughout the MDMP. Such informal interaction between the commander and staff can improve the staff's understanding of the situation and ensure their planning efforts adequately reflect the commander's visualization of the operation.

9-10. The chief of staff (COS) (executive officer [XO]) is a key participant in the MDMP. The COS (XO) manages and coordinates the staff's work and provides quality control during the MDMP. To effectively supervise the entire process, this officer has to clearly understand the commander's intent and guidance. The COS (XO) provides timelines to the staff, establishes briefing times and locations, and provides any instructions necessary to complete the plan.

9-11. The staff's effort during the MDMP focuses on helping the commander understand the situation, make decisions, and synchronize those decisions into a fully developed plan or order. Staff activities during planning initially focus on mission analysis. The products the staff develops during mission analysis help commanders understand the situation and develop the commander's visualization. During course of action (COA) development and COA comparison, the staff provides recommendations to support the commander in selecting a COA. After the commander makes a decision, the staff prepares the plan or order that reflects the commander's intent, coordinating all necessary details. (See figure 9-1.)

Key inputs	Steps	Key outputs
• Higher headquarters' plan or order or a new mission anticipated by the commander	Step 1: Receipt of Mission	• Commander's initial guidance • Initial allocation of time
	Warning order	
• Commander's initial guidance • Higher headquarters' plan or order • Higher headquarters' knowledge and intelligence products • Knowledge products from other organizations • Army design methodology products	Step 2: Mission Analysis	• Problem statement • Mission statement • Initial commander's intent • Initial planning guidance • Initial CCIRs and EEFIs • Updated IPB and running estimates • Assumptions • Evaluation criteria for COAs
	Warning order	

• Mission statement
• Initial commander's intent, planning guidance, CCIRs, and EEFIs
• Updated IPB and running estimates
• Assumptions
• Evaluation criteria for COAs
	Step 3: Course of Action (COA) Development	• COA statements and sketches
		  - Tentative task organization
		  - Broad concept of operations
		• Revised planning guidance
		• Updated assumptions

• Updated running estimates • Revised planning guidance • COA statements and sketches • Updated assumptions	Step 4: COA Analysis (War Game)	• Refined COAs • Potential decision points • War-game results • Initial assessment measures • Updated assumptions
• Updated running estimates • Refined COAs • Evaluation criteria • War-game results • Updated assumptions	Step 5: COA Comparison	• Evaluated COAs • Recommended COAs • Updated running estimates • Updated assumptions
• Updated running estimates • Evaluated COAs • Recommended COAs • Updated assumptions	Step 6: COA Approval	• Commander approved COA and any modifications • Refined commander's intent, CCIRs, and EEFIs • Updated assumptions
	Warning order	
• Commander approved COA and any modifications • Refined commander's intent, CCIRs, and EEFIs • Updated assumptions	Step 7: Orders Production, Dissemination, and Transition	• Approved operation plan or order • Subordinates understand the plan or order

CCIR	commander's critical information requirement	EEFI	essential element of friendly information
COA	course of action	IPB	intelligence preparation of the battlefield

Figure 9-1. The seven steps of the military decisionmaking process

MODIFYING THE MILITARY DECISIONMAKING PROCESS

9-12. The MDMP can be as detailed as time, resources, experience, and the situation permit. Performing all steps of the MDMP is detailed, deliberate, and time-consuming. Commanders use the full MDMP when they have enough planning time and staff support to thoroughly examine two or more COAs and develop a fully synchronized plan or order. This typically occurs when planning for an entirely new mission.

9-13. Commanders may alter the steps of the MDMP to fit time-constrained circumstances and produce a satisfactory plan. In time-constrained conditions, commanders assess the situation, update the commander's visualization, and direct the staff to perform the MDMP activities that support the required decisions. In extremely compressed situations, commanders rely on more intuitive decisionmaking techniques, such as the rapid decisionmaking and synchronization process. (See paragraphs 9-205 through 9-210 for information on planning in a time-constrained environment.)

STEPS OF THE MILITARY DECISIONMAKING PROCESS

9-14. The remainder of this chapter describes the methods for conducting each step of the MDMP. It describes the key inputs and expected key outputs for each step. It also describes how the staff integrates intelligence preparation of the battlefield (IPB), targeting, risk management, and information collection throughout the MDMP.

STEP 1–RECEIPT OF MISSION

9-15. Commanders initiate the MDMP upon receipt or in anticipation of a mission. This step alerts all participants of the pending planning requirements, enabling them to determine the amount of time available for planning and preparation and decide on a planning approach, including guidance on using Army design methodology and how to abbreviate the MDMP, if required. When commanders identify a new mission, commanders and staffs perform the actions and produce the expected key outputs. (See figure 9-2.)

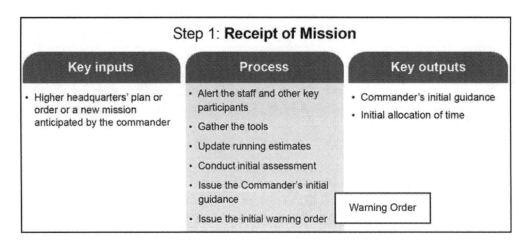

Figure 9-2. Step 1–receipt of the mission

Alert the Staff and Other Key Participants

9-16. As soon as a unit receives a new mission (or when the commander directs), the current operations integration cell alerts the staff of the pending planning requirement. Unit standard operating procedures (SOPs) should identify members of the planning staff who participate in mission analysis. In addition, the current operations integration cell also notifies other military, civilian, and host-nation organizations of pending planning events as required.

Gather the Tools

9-17. Once notified of the new planning requirement, the staff prepares for mission analysis by gathering the needed tools. These tools include, but are not limited to—

- Appropriate publications, including ADRP 1-02.
- All documents related to the mission and area of operations, including the higher headquarters' OPLAN and OPORD, maps and terrain products, and operational graphics.
- Higher headquarters' and other organizations' intelligence and assessment products.
- Estimates and products of other military and civilian agencies and organizations.
- Both their own and the higher headquarters' SOPs.
- Current running estimates.
- Any Army design methodology products.

9-18. The gathering of knowledge products continues throughout the MDMP. Staff officers carefully review the reference sections (located before paragraph **1. <u>Situation</u>**) of the higher headquarters' OPLANs and OPORDs to identify documents (such as theater policies and memoranda) related to the upcoming operation. If the MDMP occurs while in the process of replacing another unit, the staff begins collecting relevant documents—such as the current OPORD, branch plans, current assessments, operations and intelligence summaries, and SOPs—from that unit.

Update Running Estimates

9-19. While gathering the necessary tools for planning, each staff section begins updating its running estimate—especially the status of friendly units and resources and key civil considerations that affect each functional area. Running estimates not only compile critical facts and assumptions from the perspective of each staff section, but also include information from other staff sections and other military and civilian organizations. While listed at the beginning of the MDMP, this task of developing and updating running estimates continues throughout the MDMP and the operations process. (See chapter 8 for more information on running estimates.)

Conduct Initial Assessment

9-20. During receipt of mission, the commander and staff conduct an initial assessment of time and resources available to plan, prepare, and begin execution of an operation. This initial assessment helps commanders determine—

- The time needed to plan and prepare for the mission for both headquarters and subordinate units.
- Guidance on conducting the Army design methodology and abbreviating the MDMP, if required.
- Which outside agencies and organizations to contact and incorporate into the planning process.
- The staff's experience, cohesiveness, and level of rest or stress.

9-21. This assessment primarily identifies an initial allocation of available time. The commander and staff balance the desire for detailed planning against the need for immediate action. The commander provides guidance to subordinate units as early as possible to allow subordinates the maximum time for their own planning and preparation of operations. As a rule, commanders allocate a minimum of two-thirds of available time for subordinate units to conduct their planning and preparation. This leaves one-third of the time for commanders and their staffs to do their planning. They use the other two-thirds for their own preparation. Time, more than any other factor, determines the detail to which the commander and staff can plan.

9-22. Based on the commander's initial allocation of time, the COS (XO) develops a staff planning timeline that outlines how long the headquarters can spend on each step of the MDMP. The staff planning timeline indicates what products are due, who is responsible for them, and who receives them. It includes times and locations for meetings and briefings. It serves as a benchmark for the commander and staff throughout the MDMP.

Issue the Commander's Initial Guidance

9-23. Once time is allocated, the commander determines whether to initiate Army design methodology, perform Army design methodology in parallel with the MDMP, or proceed directly into the MDMP without the benefits of formal Army design methodology activities. In time-sensitive situations where commanders decide to proceed directly into the MDMP, they may also issue guidance on how to abbreviate the process. Having determined the time available together with the scope and scale of the planning effort, commanders issue initial planning guidance. Although brief, the initial guidance includes, but is not limited to—

- Initial time allocations.
- A decision to initiate Army design methodology or go straight into the MDMP.
- How to abbreviate the MDMP, if required.
- Necessary coordination to exchange liaison officers.
- Authorized movements and initiation of information collection.
- Collaborative planning times and locations.
- Initial information requirements.
- Additional staff tasks.

Issue the Initial Warning Order

9-24. The last task in receipt of mission is to issue a WARNORD to subordinate and supporting units. This order includes at a minimum the type of operation, the general location of the operation, the initial timeline, and any movement or information collection to initiate.

STEP 2–MISSION ANALYSIS

9-25. The MDMP continues with an assessment of the situation called mission analysis. Commanders (supported by their staffs and informed by subordinate and adjacent commanders and by other partners) gather, analyze, and synthesize information to orient themselves on the current conditions of the operational environment. The commander and staff conduct mission analysis to better understand the situation and problem, and identify *what* the command must accomplish, *when* and *where* it must be done, and most importantly *why*—the purpose of the operation.

9-26. Since no amount of subsequent planning can solve an insufficiently understood problem, mission analysis is the most important step in the MDMP. This understanding of the situation and the problem allows commanders to visualize and describe how the operation may unfold in their initial commander's intent and planning guidance. During mission analysis, the commander and staff perform the process actions and produce the outputs shown in figure 9-3.

9-27. Commanders and staffs may also begin the development of evaluation criteria during this step. These evaluation criteria are continually developed and refined throughout the MDMP and become a key input during Step 5—Course of Action Comparison.

Analyze the Higher Headquarters' Plan or Order

9-28. Commanders and staffs thoroughly analyze the higher headquarters' plan or order. They determine how their unit—by task and purpose—contributes to the mission, commander's intent, and concept of operations of the higher headquarters. The commander and staff seek to completely understand—

- The higher headquarters'—
  - Commander's intent.
  - Mission.
  - Concept of operations.
  - Available assets.
  - Timeline.

- The missions of adjacent, supporting, and supported units and their relationships to the higher headquarters' plan.
- The missions or goals of unified action partners that work in the operational areas.
- Their assigned area of operations.

Step 2: Mission Analysis

Key inputs	Process	Key outputs
• Commander's initial guidance • Higher headquarters' plan or order • Higher headquarters' intelligence and knowledge products • Knowledge products from other organizations • Army design methodology products	• Analyze the higher headquarters' plan or order • Perform initial IPB • Determine specified, implied, and essential tasks • Review available assets and identify resource shortfalls • Determine constraints • Identify critical facts and develop assumptions • Begin risk management • Develop initial CCIRs and EEFIs • Develop the initial information collection plan • Update plan for the use of available time • Develop initial themes and messages • Develop a proposed problem statement • Develop a proposed mission statement • Present the mission analysis briefing • Develop and issue initial commander's intent • Develop and issue initial planning guidance • Develop COA evaluation criteria • Issue a warning order	• Problem statement • Mission statement • Initial commander's intent • Initial planning guidance • Initial CCIRs and EEFIs • Updated IPB and running estimates • Assumptions • Evaluation criteria for COAs Warning Order

CCIR	commander's critical information requirement	IPB	intelligence preparation of the battlefield
COA	course of action	EEFI	essential element of friendly information

Figure 9-3. Step 2–mission analysis

9-29. If the commander misinterprets the higher headquarters' plan, time is wasted. Additionally, when analyzing the higher order, the commander and staff may identify difficulties and contradictions in the higher order. Therefore, if confused by the higher headquarters' order or guidance, commanders must seek immediate clarification. Liaison officers familiar with the higher headquarters' plan can help clarify issues. Collaborative planning with the higher headquarters also facilitates this task. Staffs use requests for information to clarify or obtain additional information from the higher headquarters.

Perform Initial Intelligence Preparation of the Battlefield

9-30. IPB is the systematic process of analyzing the mission variables of enemy, terrain, weather, and civil considerations in an area of interest to determine their effect on operations. The IPB process identifies critical gaps in the commander's knowledge of an operational environment. As a part of the initial planning guidance, commanders use these gaps as a guide to establish their initial intelligence requirements. IPB products enable the commander to assess facts about the operational environment and make assumptions about how friendly and threat forces will interact in the operational environment. The description of the operational environment's effects identifies constraints on potential friendly COAs. It also identifies key aspects of the operational environment, such as avenues of approach, engagement areas, and landing zones, which the staff integrates into potential friendly COAs and their running estimates. For mission analysis, the intelligence staff, along with the other staff elements, will use IPB to develop detailed threat COA models, which depict a COA available to the threat. The threat COA models provide a basis for formulating friendly COAs and completing the intelligence estimate.

9-31. The intelligence staff, in collaboration with other staffs, develops other IPB products during mission analysis. That collaboration should result in the drafting of initial priority intelligence requirements (PIRs), the production of a complete modified combined obstacles overlay, a list of high value targets, and unrefined event templates and matrices. IPB should provide an understanding of the threat's center of gravity, which then can be exploited by friendly forces.

Determine Specified, Implied, and Essential Tasks

9-32. The staff analyzes the higher headquarters' order and the higher commander's guidance to determine their specified and implied tasks. In the context of operations, a task is a clearly defined and measurable activity accomplished by Soldiers, units, and organizations that may support or be supported by other tasks. The "what" of a mission statement is always a task. From the list of specified and implied tasks, the staff determines essential tasks for inclusion in the recommended mission statement.

9-33. **A *specified task* is a task specifically assigned to a unit by its higher headquarters**. Paragraphs 2 and 3 of the higher headquarters' order or plan state specified tasks. Some tasks may be in paragraphs 4 and 5. Specified tasks may be listed in annexes and overlays. They may also be assigned verbally during collaborative planning sessions or in directives from the higher commander.

9-34. **An *implied task* is a task that must be performed to accomplish a specified task or mission but is not stated in the higher headquarters' order.** Implied tasks are derived from a detailed analysis of the higher headquarters' order, the enemy situation, the terrain, and civil considerations. Additionally, analysis of doctrinal requirements for each specified task might disclose implied tasks.

9-35. When analyzing the higher order for specified and implied tasks, the staff also identifies any be-prepared or on-order missions. **A *be-prepared mission* is a mission assigned to a unit that might be executed.** Generally a contingency mission, commanders execute it because something planned has or has not been successful. In planning priorities, commanders plan a be-prepared mission after any on-order mission. **An *on-order mission* is a mission to be executed at an unspecified time.** A unit with an on-order mission is a committed force. Commanders envision task execution in the concept of operations; however, they may not know the exact time or place of execution. Subordinate commanders develop plans and orders and allocate resources, task-organize, and position forces for execution.

9-36. Once staff members have identified specified and implied tasks, they ensure they understand each task's requirements and purpose. The staff then identifies essential tasks. **An *essential task* is a specified or implied task that must be executed to accomplish the mission.** Essential tasks are always included in the unit's mission statement.

Review Available Assets and Identify Resource Shortfalls

9-37. The commander and staff examine additions to and deletions from the current task organization, command and support relationships, and status (current capabilities and limitations) of all units. This analysis also includes capabilities of civilian and military organizations (joint, special operations, and multinational) that operate within their unit's area of operations. They consider relationships among specified, implied, and essential tasks, and between them and available assets. From this analysis, staffs determine if they have the assets needed to complete all tasks. If shortages occur, they identify additional resources needed for mission success to the higher headquarters. Staffs also identify any deviations from the normal task organization and provide them to the commander to consider when developing the planning guidance. A more detailed analysis of available assets occurs during COA development.

Determine Constraints

9-38. The commander and staff identify any constraints placed on their command. **A *constraint* is a restriction placed on the command by a higher command. A constraint dictates an action or inaction, thus restricting the freedom of action of a subordinate commander.** Constraints are found in paragraph 3 of the OPLAN or OPORD. Annexes to the order may also include constraints. The operation overlay, for example, may contain a restrictive fire line or a no fire area. Constraints may also be issued verbally, in WARNORDs, or in policy memoranda.

9-39. Constraints may also be based on resource limitations within the command, such as organic fuel transport capacity, or physical characteristics of the operational environment, such as the number of vehicles that can cross a bridge in a specified time.

9-40. The commander and staff should coordinate with the staff judge advocate for a legal review of perceived or obvious constraints, restraints, or limitations in the OPLAN, OPORD, or related documents.

Identify Critical Facts and Develop Assumptions

9-41. Plans and orders are based on facts and assumptions. Commanders and staffs gather facts and develop assumptions as they build their plan. A fact is a statement of truth or a statement thought to be true at the time. Facts concerning the operational and mission variables serve as the basis for developing situational understanding, for continued planning, and when assessing progress during preparation and execution.

9-42. An assumption is a supposition on the current situation or a presupposition on the future course of events, either or both assumed to be true in the absence of positive proof, necessary to enable the commander in the process of planning to complete an estimate of the situation and make a decision on the course of action. In the absence of facts, the commander and staff consider assumptions from their higher headquarters. They then develop their own assumptions necessary for continued planning.

9-43. Having assumptions requires commanders and staffs to continually attempt to replace those assumptions with facts. The commander and staff should list and review the key assumptions on which fundamental judgments rest throughout the MDMP. Rechecking assumptions is valuable at any time during the operations process prior to rendering judgments and making decisions.

Begin Risk Management

9-44. *Risk management* is the process of identifying, assessing, and controlling risks arising from operational factors and making decisions that balance risk cost with mission benefits (JP 3-0). During mission analysis, the commander and staff focus on identifying and assessing hazards. Developing specific control measures to mitigate those hazards occurs during course of action development.

9-45. The chief of protection (or operations staff officer [S-3] in units without a protection cell) in coordination with the safety officer integrates risk management into the MDMP. All staff sections integrate risk management for hazards within their functional areas. Units conduct the first four steps of risk management in the MDMP. FM 5-19 addresses the details for conducting risk management, including products of each step.

Develop Initial Commander's Critical Information Requirements and Essential Elements of Friendly Information

9-46. The mission analysis process identifies gaps in information required for further planning and decisionmaking during preparation and execution. During mission analysis, the staff develops information requirements. Some information requirements are of such importance to the commander that staffs nominate them to the commander to become a commander's critical information requirement (CCIR).

9-47. A commander's critical information requirement is an information requirement identified by the commander as being critical to facilitating timely decisionmaking. The two key elements are friendly force information requirements and priority intelligence requirements. A CCIR directly influences decisionmaking and facilitates the successful execution of military operations. A CCIR is—

- Specified by a commander for a specific operation.
- Applicable only to the commander who specifies it.
- Situation dependent—directly linked to a current or future mission.
- Time-sensitive.

9-48. Commanders consider staff input when determining their CCIRs. CCIRs are situation-dependent and specified by the commander for each operation. Commanders continuously review CCIRs during the planning process and adjust them as situations change. The initial CCIRs developed during mission analysis normally focus on decisions the commander needs to make to focus planning. Once the commander selects a COA, the CCIRs shift to information the commander needs in order to make decisions during preparation and execution. Commanders designate CCIRs to inform the staff and subordinates what they deem essential for making decisions. Typically, commanders identify ten or fewer CCIRs; minimizing the number of CCIRs assists in prioritizing the allocation of limited resources. CCIR fall into one of two categories: PIRs and friendly force information requirements (FFIRs).

9-49. A PIR is an intelligence requirement, stated as a priority for intelligence support, that the commander and staff need to understand the adversary or the operational environment. PIRs identify the information about the enemy and other aspects of the operational environment that the commander considers most important. Lessons from recent operations show that intelligence about civil considerations may be as critical as intelligence about the enemy. Thus, all staff sections may recommend information about civil considerations as PIRs. The intelligence officer manages PIRs for the commander through planning requirements and assessing collection.

9-50. An FFIR is information the commander and staff need to understand the status of friendly force and supporting capabilities. FFIRs identify the information about the mission, troops and support available, and time available for friendly forces that the commander considers most important. In coordination with the staff, the operations officer manages FFIRs for the commander.

9-51. In addition to nominating CCIRs to the commander, the staff also identifies and nominates essential elements of friendly information (EEFIs). An EEFI establishes an element of information to protect rather than one to collect. EEFIs identify those elements of friendly force information that, if compromised, would jeopardize mission success. Although EEFIs are not CCIRs, they have the same priority as CCIRs and require approval by the commander. Like CCIRs, EEFIs change as an operation progresses.

9-52. Depending on the situation, the commander and selected staff members meet prior to the mission analysis brief to approve the initial CCIRs and EEFIs. This is especially important if the commander intends to conduct information collection early in the planning process. The approval of the initial CCIRs early during planning assists the staff in developing the initial information collection plan. Approval of an EEFI allows the staff to begin planning and implementing measures to protect friendly force information, such as military deception and operations security.

Develop the Initial Information Collection Plan

9-53. The initial information collection plan is crucial to begin or adjust the information collection effort to help answer information requirements necessary in developing effective plans. The initial information collection plan sets reconnaissance, surveillance, and intelligence operations in motion. It may be issued as part of a WARNORD, a fragmentary order (FRAGORD), or an OPORD. As more information becomes available, it is incorporated into a complete information collection plan (Annex L) to the OPORD.

9-54. The intelligence staff creates the requirements management tools for the information collection plan. The operations staff is responsible for the information collection plan. During this step, the operations and intelligence staff work closely to ensure they fully synchronize and integrate information collection activities into the overall plan.

9-55. The operations officer considers several factors when developing the initial information collection plan, including:

- Requirements for collection assets in subsequent missions.
- The time available to develop and refine the initial information collection plan.
- The risk the commander is willing to accept if information collection missions are begun before the information collection plan is fully integrated into the scheme of maneuver.
- Insertion and extraction methods for reconnaissance, security, surveillance, and intelligence collection assets.
- Contingencies for inclement weather to ensure coverage of key named areas of interest or target areas of interest.
- The communications plan for transmission of reports from assets to command posts.
- The inclusion of collection asset locations and movements into the fire support plan.
- The reconnaissance handover with higher or subordinate echelons.
- The sustainment support.
- Legal support requirements.

FM 3-55 contains additional information on information collection, planning requirements, and assessing collection.

Update Plan for the Use of Available Time

9-56. As more information becomes available, the commander and staff refine their initial plan for the use of available time. They compare the time needed to accomplish tasks to the higher headquarters' timeline to ensure mission accomplishment is possible in the allotted time. They compare the timeline to the assumed enemy timeline with how they anticipate conditions will unfold. From this, they determine windows of opportunity for exploitation, times when the unit will be at risk for enemy activity, or when action to arrest deterioration in the local civilian population may be required.

9-57. The commander and COS (XO) also refine the staff planning timeline. The refined timeline includes the—

- Subject, time, and location of briefings the commander requires.
- Times of collaborative planning sessions and the medium over which they will take place.
- Times, locations, and forms of rehearsals.

Develop Initial Themes and Messages

9-58. Gaining and maintaining the trust of key actors is an important aspect of operations. Faced with the many different actors (individuals, organizations, and the public) connected with the operation, commanders identify and engage those actors who matter to operational success. These actors' behaviors can help solve or complicate the friendly forces' challenges as commanders strive to accomplish missions.

9-59. Themes and messages support operations and military actions. Commanders and their units coordinate what they do, say, and portray through themes and messages. A theme is a unifying or dominant idea or image that expresses the purpose for military action. Themes tie to objectives, lines of effort, and end state conditions. They are overarching and apply to capabilities of public affairs, military information support operations, and Soldier and leader engagements. A message is a verbal, written, or electronic communication that supports a theme focused on a specific actor or the public and in support of a specific action (task). Units transmit themes and messages to those actors or the public whose perceptions, attitudes, beliefs, and behaviors matter to the success of an operation.

9-60. The public affairs officer adjusts and refines themes and messages received from higher headquarters for use by the command. These themes and messages are designed to inform specific domestic and foreign audiences about current or planned military operations. The military information support operations element receives approved themes and messages. This element adjusts or refines depending on the situation. It employs themes and messages as part of planned activities designed to influence specific foreign audiences for various purposes that support current or planned operations. The commander and the chief of staff approve all themes and messages used to support operations. The information operations officer assists the G-3 (S-3) and the commander to de-conflict and synchronize the use of information-related capabilities used specifically to disseminate approved themes and messages during operations.

Develop a Proposed Problem Statement

9-61. A problem is an issue or obstacle that makes it difficult to achieve a desired goal or objective. The problem statement is the description of the primary issue or issues that may impede commanders from achieving their desired end states.

> *Note:* The commander, staff, and other partners develop the problem statement as part of Army design methodology. During mission analysis, the commander and staff review the problem statement and revise it as necessary based on the increased understanding of the situation. If Army design methodology activities do not precede mission analysis, then the commander and staff develop a problem statement prior to moving to Step 3—COA Development.

9-62. How the problem is formulated leads to particular solutions. It is important that commanders dedicate the time to identify the right problem to solve and describe it clearly in a problem statement. Ideally, the commander and staff meet to share their analysis of the situation. They talk with each other, synthesize the results of the current mission analysis, and determine the problem. If the commander is not available, the staff members talk among themselves.

9-63. As part of the discussion to help identify and understand the problem, the staff—
- Compares the current situation to the desired end state.
- Brainstorms and lists issues that impede the commander from achieving the desired end state.

9-64. Based on this analysis, the staff develops a proposed problem statement—a statement of the problem or set of problems to be solved—for the commander's approval.

Develop a Proposed Mission Statement

9-65. The COS (XO) or operations officer prepares a proposed mission statement for the unit based on the mission analysis. The commander receives and approves the unit's mission statement normally during the mission analysis brief. A *mission statement* is a short sentence or paragraph that describes the organization's essential task(s), purpose, and action containing the elements of who, what, when, where, and why (JP 5-0). The five elements of a mission statement answer these questions:
- <u>Who</u> will execute the operation (unit or organization)?
- <u>What</u> is the unit's essential task (tactical mission task)?
- <u>When</u> will the operation begin (by time or event) or what is the duration of the operation?
- <u>Where</u> will the operation occur (area of operations, objective, grid coordinates)?
- <u>Why</u> will the force conduct the operations (for what purpose)?

> ***Example 1.*** Not later than 220400 Aug 09 (**when**), 1st Brigade (**who**) secures ROUTE SOUTH DAKOTA (**what/task**) in AREA OF OPERATIONS JACKRABBIT (**where**) to enable the movement of humanitarian assistance materials (**why/purpose**).

> ***Example 2.*** 1-505th Parachute Infantry Regiment (**who**) seizes (**what/task**) JACKSON INTERNATIONAL AIRPORT (**where**) not later than D-day, H+3 (**when**) to allow follow-on forces to air-land into AREA OF OPERATIONS SPARTAN (**why/purpose**).

9-66. The mission statement may have more than one essential task. The following example shows a mission statement for a phased operation with a different essential task for each phase.

Example. 1-509th Parachute Infantry Regiment (**who**) seizes (**what/task**) JACKSON INTERNATIONAL AIRPORT (**where**) not later than D-day, H+3 (**when**) to allow follow-on forces to air-land into AREA OF OPERATIONS SPARTAN (**why/purpose**). On order (**when**), secures (**what/task**) OBJECTIVE GOLD (**where**) to prevent the 2nd Pandor Guards Brigade from crossing the BLUE RIVER and disrupting operations in AREA OF OPERATIONS SPARTAN (**why/purpose**).

9-67. The *who, where,* and *when* of a mission statement are straightforward. The *what* and *why* are more challenging to write and can confuse subordinates if not stated clearly. The *what* is a *task* and is expressed in terms of action verbs. These tasks are measurable and can be grouped as "actions by friendly forces" or "effects on enemy forces." The *why* puts the task into context by describing the reason for performing it. The *why* provides the mission's purpose—the reason the unit is to perform the task. It is extremely important to mission command and mission orders.

9-68. Commanders should use tactical mission tasks or other doctrinally approved tasks contained in combined arms field manuals or mission training plans in mission statements. These tasks have specific military definitions that differ from standard dictionary definitions. A *tactical mission task* is a specific activity performed by a unit while executing a form of tactical operation or form of maneuver. It may be expressed as either an action by a friendly force or effects on an enemy force (FM 7-15). FM 3-90-1 describes each tactical task. FM 3-07 provides a list of primary stability tasks which military forces must be prepared to execute. Commanders and planners should carefully choose the task that best describes the commander's intent and planning guidance.

Present the Mission Analysis Briefing

9-69. The mission analysis briefing informs the commander of the results of the staff's analysis of the situation. It helps the commander understand, visualize, and describe the operation. Throughout the mission analysis briefing, the commander, staff, and other partners discuss the various facts and assumptions about the situation. Staff officers present a summary of their running estimates from their specific functional area and how their findings impact or are impacted by other areas. This helps the commander and staff as a whole to focus on the interrelationships among the mission variables and to develop a deeper understanding of the situation. The commander issues guidance to the staff for continued planning based on situational understanding gained from the mission analysis briefing.

9-70. Ideally, the commander holds several informal meetings with key staff members before the mission analysis briefing, including meetings to assist the commander in developing CCIRs, the mission statement, and themes and messages. These meetings enable commanders to issue guidance for activities (such as reconnaissance, surveillance, security, and intelligence operations) and develop their initial commander's intent and planning guidance.

9-71. A comprehensive mission analysis briefing helps the commander, staff, subordinates, and other partners develop a shared understanding of the requirements of the upcoming operation. Time permitting, the staff briefs the commander on its mission analysis using the following outline:

- Mission and commander's intent of the headquarters two echelons up.
- Mission, commander's intent, and concept of operations of the headquarters one echelon up.
- A proposed problem statement.
- A proposed mission statement.
- Review of the commander's initial guidance.
- Initial IPB products, including civil considerations that impact the conduct of operations.
- Specified, implied, and essential tasks.
- Pertinent facts and assumptions.
- Constraints.
- Forces available and resource shortfalls.
- Initial risk assessment.
- Proposed themes and messages.

- Proposed CCIRs and EEFIs.
- Initial information collection plan.
- Recommended timeline.
- Recommended collaborative planning sessions.

9-72. During the mission analysis briefing or shortly thereafter, commanders approve the mission statement and CCIRs. They then develop and issue their initial commander's intent and planning guidance.

Develop and Issue Initial Commander's Intent

9-73. The *commander's intent* is a clear and concise expression of the purpose of the operation and the desired military end state that supports mission command, provides focus to the staff, and helps subordinate and supporting commanders act to achieve the commander's desired results without further orders, even when the operation does not unfold as planned (JP 3-0). The initial commander's intent describes the purpose of the operation, initial key tasks, and the desired end state (See ADRP 5-0 for more details on commander's intent).

9-74. The higher commander's intent provides the basis for unity of effort throughout the force. Each commander's intent nests within the higher commander's intent. The commander's intent explains the broader purpose of the operation beyond that of the mission statement. This explanation allows subordinate commanders and Soldiers to gain insight into what is expected of them, what constraints apply, and most importantly, why the mission is being conducted.

9-75. Based on their situational understanding, commanders summarize their visualization in their initial commander's intent statement. The initial commander's intent links the operation's purpose with conditions that define the desired end state. Commanders may change their intent statement as planning progresses and more information becomes available. The commander's intent must be easy to remember and clearly understood by leaders two echelons lower in the chain of command. The shorter the commander's intent, the better it serves these purposes. Typically, the commander's intent statement is three to five sentences long and contains the purpose, key tasks, and end state.

Develop and Issue Initial Planning Guidance

9-76. Commanders provide planning guidance along with their initial commander's intent. Planning guidance conveys the essence of the commander's visualization. This guidance may be broad or detailed, depending on the situation. The initial planning guidance outlines an *operational approach*—a description of the broad actions the force must take to transform current conditions into those desired at end state (JP 5-0). The initial planning guidance outlines specific COAs the commander desires the staff to look at as well as rules out any COAs the commander will not accept. That clear guidance allows the staff to develop several COAs without wasting effort on things that the commander will not consider. It reflects how the commander sees the operation unfolding. It broadly describes when, where, and how the commander intends to employ combat power to accomplish the mission within the higher commander's intent.

9-77. Commanders use their experience and judgment to add depth and clarity to their planning guidance. They ensure staffs understand the broad outline of their visualization while allowing the latitude necessary to explore different options. This guidance provides the basis for a detailed concept of operations without dictating the specifics of the final plan. As with their intent, commanders may modify planning guidance based on staff and subordinate input and changing conditions.

9-78. Commanders issue planning guidance initially after mission analysis. They continue to consider additional guidance throughout the MDMP including, but not limited, to the following:

- Upon receipt of or in anticipation of a mission (initial planning guidance).
- Following mission analysis (planning guidance for COA development).
- Following COA development (revised planning guidance for COA improvements).
- COA approval (revised planning guidance to complete the plan).

9-79. Table 9-1 lists commander's planning guidance by warfighting function. This list is not intended to meet the needs of all situations nor be all-inclusive, and providing guidance by warfighting function is not the only method. Commanders tailor planning guidance to meet specific needs based on the situation rather than address each item. Each item does not always fit neatly in a particular warfighting function, as it may be shared by more than one warfighting function. For example, although rules of engagement fall under the protection warfighting function, each other warfighting function chief has a vested interest in gaining guidance on rules of engagement. (See table 9-1.)

Table 9-1. Examples of commander's planning guidance by warfighting function

Mission Command	Commander's critical information requirements Rules of engagement Command post positioning Commander's location Initial themes and messages Succession of command	Liaison officer guidance Planning and operational guidance timeline Type of order and rehearsal Communications guidance Civil affairs operations Cyber electromagnetic considerations
Intelligence	Information collection guidance Information gaps Most likely and most dangerous enemy courses of action Priority intelligence requirements Most critical terrain and weather factors	Most critical local environment and civil considerations Intelligence requests for information Intelligence focus during phased operations Desired enemy perception of friendly forces
Movement and Maneuver	Commander's intent Course of action development guidance Number of courses of action to consider or not consider Critical events Task organization Task and purpose of subordinate units Forms of maneuver Reserve composition, mission, priorities, and control measures	Security and counterreconnaissance Friendly decision points Branches and sequels Task and direct collection Military deception Risk to friendly forces Collateral damage or civilian casualties Any condition that affects achievement of end state Information operations
Fires	Synchronization and focus of fires with maneuver Priority of fires High priority targets Special munitions Target acquisition zones Observer plan Air and missile defense positioning High-value targets	Task and purpose of fires Scheme of fires Suppression of enemy air defenses Fire support coordination measures Attack guidance Branches and sequels No strike list Restricted target list
Protection	Protection priorities Priorities for survivability assets Terrain and weather factors Intelligence focus and limitations for security Acceptable risk Protected targets and areas	Vehicle and equipment safety or security constraints Environmental considerations Unexploded ordnance Operations security risk tolerance Rules of engagement Escalation of force and nonlethal weapons Counterintelligence
Sustainment	Sustainment priorities—manning, fueling, fixing, arming, moving the force, and sustaining Soldiers and systems Health system support Sustainment of detainee and resettlement operations	Construction and provision of facilities and installations Detainee movement Anticipated requirements of Classes III, IV, and V Controlled supply rates

Develop Course of Action Evaluation Criteria

9-80. Evaluation criteria are standards the commander and staff will later use to measure the relative effectiveness and efficiency of one COA relative to other COAs. Developing these criteria during mission analysis or as part of commander's planning guidance helps to eliminate a source of bias prior to COA analysis and comparison. Evaluation criteria address factors that affect success and those that can cause failure. Criteria change from mission to mission and must be clearly defined and understood by all staff members before starting the war game to test the proposed COAs. Normally, the COS (XO) initially determines each proposed criterion with weights based on the assessment of its relative importance and the commander's guidance. Commanders adjust criterion selection and weighting according to their own experience and vision. The staff member responsible for a functional area scores each COA using those criteria. The staff presents the proposed evaluation criteria to the commander at the mission analysis brief for approval.

Issue a Warning Order

9-81. Immediately after the commander gives the planning guidance, the staff sends subordinate and supporting units a WARNORD. (See appendix C for sample WARNORD.) It contains, at a minimum—

- The approved mission statement.
- The commander's intent.
- Changes to task organization.
- The unit area of operations (sketch, overlay, or some other description).
- CCIRs and EEFIs.
- Risk guidance.
- Priorities by warfighting functions.
- Military deception guidance.
- Essential stability tasks.
- Initial information collection plan.
- Specific priorities.
- Updated operational timeline.
- Movements.

Step 3–Course of Action Development

9-82. A COA is a broad potential solution to an identified problem. The COA development step generates options for subsequent analysis and comparison that satisfy the commander's intent and planning guidance. During COA development, planners use the problem statement, mission statement, commander's intent, planning guidance, and various knowledge products developed during mission analysis. (See figure 9-4.)

Step 3: Course of Action (COA) Development

Key inputs	Process	Key outputs
• Mission statement • Initial commander's intent, planning guidance, CCIRs, and EEFIs • Updated IPB and running estimates • Assumptions • Evaluation criteria for COAs	• Assess relative combat power • Generate options • Array forces • Develop a broad concept • Assign headquarters • Develop COA statements and sketches • Conduct COA briefing • Select or modify COAs for continued analysis	• COA statements and sketches - Tentative task organization - Broad concept of operations • Revised planning guidance • Updated assumptions

CCIR commander's critical information requirement
COA course of action
EEFI essential element of friendly information
IPB intelligence preparation of the battlefield

Figure 9-4. Step 3–course of action development

9-83. Embedded in COA development is the application of operational and tactical art. Planners develop different COAs by varying combinations of the elements of operational art, such as phasing, lines of effort, and tempo. (See ADRP 3-0 for more information on operational art.) Planners convert the approved COA into the concept of operations.

9-84. The commander's direct involvement in COA development greatly aids in producing comprehensive and flexible COAs within the time available. To save time, the commander may also limit the number of COAs staffs develop or specify particular COAs not to explore. Planners examine each prospective COA for validity using the following screening criteria:

- Feasible. The COA can accomplish the mission within the established time, space, and resource limitations.
- Acceptable. The COA must balance cost and risk with the advantage gained.
- Suitable. The COA can accomplish the mission within the commander's intent and planning guidance.
- Distinguishable. Each COA must differ significantly from the others (such as scheme of maneuver, lines of effort, phasing, use of the reserve, and task organization).
- Complete. A COA must incorporate—
  - How the decisive operation leads to mission accomplishment.
  - How shaping operations create and preserve conditions for success of the decisive operation or effort.
  - How sustaining operations enable shaping and decisive operations or efforts.
  - How to account for offensive, defensive, and stability or defense support of civil authorities tasks.
  - Tasks to be performed and conditions to be achieved.

9-85. It is important in COA development that commanders and staffs appreciate the unpredictable and uncertain nature of the operational environment, and understand how to cope with ambiguity. Some problems that commanders face are straightforward, as when clearly defined guidance is provided from higher headquarters, or when resources required for a mission are available and can easily be allocated. In such cases, the COA is often self evident. However, for problems that are unfamiliar or ambiguous, Army design methodology may assist commanders in better understanding the nature of the problem, and afford both the commander and staff a level of comfort necessary to effectively advance through COA development. Commanders and staffs that are comfortable with ambiguity will often find that the Army design methodology provides flexibility in developing COAs that contain multiple options for dealing with changing circumstances. Staffs tend to focus on specific COAs for specific sets of circumstances, when it is usually best to focus on flexible COAs that provide the greatest options to account for the widest range of circumstances.

9-86. Commanders and staffs must be cautious not to attempt to identify and resolve every possible outcome to military operations. The interaction of multiple variables within an operational environment can lead to countless possible options and outcomes. Commanders and staffs should focus their efforts around known variables and analyze COAs that provide flexible options to the commander during execution. If commanders and staffs focus on what is known about a situation, it often becomes clear that the known information provides sufficient guidance to develop flexible COAs. It is important to clearly identify which variables the unit can control, which it does not control, and the implications of those that it does not control. Even when there are few facts available, it is often possible to reduce key issues to either an ability to do "X", or an inability to do "X" as a starting point. Such a reduction is preferred over trying to derive a wide range of possibilities. It is just as important not to see facts as constraining flexibility, but seek to use them to generate flexibility. Staffs work to confirm or deny facts before developing options. Staffs must also determine what risks are associated with various COAs.

9-87. As an example, a commander may know with reasonable certainty that an enemy force is positioned on the outskirts of a town. The commander may not be certain of the exact size of the enemy force, all the resources available to the enemy force, or actions the enemy may take over time. Such unknowns are a reality in an ambiguous operational environment. But, by focusing on the known information, that is, the position of the enemy at a point in time, the staff can develop COAs that provide maximum flexibility for the commander. Known information can also apply to friendly actions, such as an established time for crossing a line of departure, or transition to a subsequent phase of an operation. COAs should allow for variances in timelines and resources as additional information on the enemy, as well as friendly forces, becomes available. Variances may also occur as changes in guidance from high headquarters arrives, or significant national policy decisions are made. Staffs identify risks associated with both friendly and enemy actions, as well as who is accepting the risk, and what resources should be allocated to mitigate the risks.

9-88. COA development should also identify decision points, the person responsible for making the decision, and what measures may be taken to provide the commander with additional time before making a decision. (See paragraph 9-127 for a discussion of decision points.) Good COAs provide commanders with options they can take based on anticipated and unanticipated changes in the situation. (See Chapter 14 for further discussion on decisionmaking in execution.) Staffs should highlight to the commander options that may be critical to mission success. Staffs should also identify points in time when options may no longer be viable, while working to keep options open to the commander as long as possible. In all cases, staffs provide commanders with options that are flexible, while clearly identifying risks associated with committing to options. Staffs also assess how possible options may impact on a commander's options at a higher echelon.

9-89. The unpredictable and uncertain nature of the operational environment should not in itself result in paralysis or hesitancy in military operations. By focusing COA development around information that is known to the staff, staffs can better steer their efforts toward developing COAs that provide maximum flexibility and viable options for the commander in the execution of military operations.

Assess Relative Combat Power

9-90. *Combat power* is the total means of destructive, constructive, and information capabilities that a military unit or formation can apply at a given time (ADRP 3-0). Combat power is the effect created by combining the elements of intelligence, movement and maneuver, fires, sustainment, protection, mission command, information, and leadership. The goal is to generate overwhelming combat power to accomplish the mission at minimal cost.

9-91. To assess relative combat power, planners initially make a rough estimate of force ratios of maneuver units two levels below their echelon. For example, at division level, planners compare all types of maneuver battalions with enemy maneuver battalion equivalents. Planners then compare friendly strengths against enemy weaknesses, and vice versa, for each element of combat power. From these comparisons, they may deduce particular vulnerabilities for each force that may be exploited or may need protection. These comparisons provide planners insight into effective force employment.

9-92. In troop-to-task analysis for stability and defense support of civil authorities, staffs determine relative combat power by comparing available resources to specified or implied stability or defense support of civil authorities tasks. This analysis provides insight as available options and needed resources. In such operations, the elements of sustainment, movement and maneuver, nonlethal effects, and information may dominate.

9-93. By analyzing force ratios and determining and comparing each force's strengths and weaknesses as a function of combat power, planners can gain insight into—
- Friendly capabilities that pertain to the operation.
- The types of operations possible from both friendly and enemy perspectives.
- How and where the enemy may be vulnerable.
- How and where friendly forces are vulnerable.
- Additional resources needed to execute the mission.
- How to allocate existing resources.

9-94. Planners must not develop and recommend COAs based solely on mathematical analysis of force ratios. Although the process uses some numerical relationships, the estimate is largely subjective. Assessing combat power requires assessing both tangible and intangible factors, such as morale and levels of training. A relative combat power assessment identifies exploitable enemy weaknesses, identifies unprotected friendly weaknesses, and determines the combat power necessary to conduct essential stability or defense support of civil authorities tasks.

Generate Options

9-95. Based on the commander's guidance and the initial results of the relative combat power assessment, the staff generates options. A good COA can defeat all feasible enemy COAs while accounting for essential stability tasks. In an unconstrained environment, planners aim to develop several possible COAs. Depending on available time, commanders may limit the options in the commander's guidance. Options focus on enemy COAs arranged in order of their probable adoption or on those stability tasks that are most essential to prevent the situation from deteriorating further.

9-96. Brainstorming can be used for generating options. It requires time, imagination, and creativity, but it produces the widest range of choices. The staff (and members of organizations outside the headquarters) remains unbiased and open-minded when developing proposed options.

9-97. In developing COAs, staff members determine the doctrinal requirements for each proposed operation, including doctrinal tasks for subordinate units. For example, a deliberate breach requires a breach force, a support force, and an assault force. Essential stability tasks require the ability to provide a level of civil security, civil control, and certain essential services. In addition, the staff considers the potential capabilities of attachments and other organizations and agencies outside military channels.

9-98. Army leaders are responsible for clearly articulating their visualization of operations in time, space, purpose, and resources in order to generate options. ADRP 3-0 describes in detail three established operational frameworks. Army leaders are not bound by any specific framework in organizing operations, but three operational frameworks, mentioned below, have proven valuable in the past. The higher headquarters will direct the specific framework or frameworks to be used by subordinate headquarters; the frameworks should be consistent throughout all echelons. The three operational frameworks are—

- Deep-close-security.
- Main and supporting effort.
- Decisive-shaping-sustaining.

9-99. For example, when generating options for a decisive-shaping-sustaining operation, the staff starts with the decisive operation identified in the commander's planning guidance. The staff checks that the decisive operation nests within the higher headquarters' concept of operations. The staff clarifies the decisive operation's purpose and considers ways to mass the effects (lethal and nonlethal) of overwhelming combat power to achieve it.

9-100. Next, the staff considers shaping operations. The staff establishes a purpose for each shaping operation tied to creating or preserving a condition for the decisive operation's success. Shaping operations may occur before, concurrently with, or after the decisive operation. A shaping operation may be designated as the main effort if executed before or after the decisive operation.

9-101. The staff then determines sustaining operations necessary to create and maintain the combat power required for the decisive operation and shaping operation. After developing the basic operational organization for a given COA, the staff then determines the essential tasks for each decisive, shaping, and sustaining operation.

9-102. Once staff members have explored possibilities for each COA, they examine each COA to determine if it satisfies the screening criteria stated in paragraph 9-81. In doing so, they change, add, or eliminate COAs as appropriate. During this process, staffs avoid focusing on the development of one good COA among several throwaway COAs.

Array Forces

9-103. After determining the decisive and shaping operations and their related tasks and purposes, planners determine the relative combat power required to accomplish each task. Often, planners use minimum historical planning ratios as a starting point. For example, historically, defenders have over a 50 percent probability of defeating an attacking force approximately three times their equivalent strength.

Therefore, as a starting point, commanders may defend on each avenue of approach with roughly a 1:3 force ratio. (See table 9-2.)

Table 9-2. Historical minimum planning ratios

Friendly Mission	Position	Friendly : Enemy
Delay		1:6
Defend	Prepared or fortified	1:3
Defend	Hasty	1:2.5
Attack	Prepared or fortified	3:1
Attack	Hasty	2.5:1
Counterattack	Flank	1:1

9-104.   Planners determine whether these and other intangibles increase the relative combat power of the unit assigned the task to the point that it exceeds the historical planning ratio for that task. If it does not, planners determine how to reinforce the unit. Combat power comparisons are provisional at best. Arraying forces is tricky, inexact work, affected by factors that are difficult to gauge, such as impact of past engagements, quality of leaders, morale, maintenance of equipment, and time in position. Levels of electronic warfare support, fire support, close air support, civilian support, and many other factors also affect arraying forces.

9-105.   In counterinsurgency operations, planners can develop force requirements by gauging troop density—the ratio of security forces (including host-nation military and police forces as well as foreign counterinsurgents) to inhabitants. Most density recommendations fall within a range of 20 to 25 counterinsurgents for every 1,000 residents in an area of operations. A ratio of twenty counterinsurgents per 1,000 residents is often considered the minimum troop density required for effective counterinsurgency operations; however, as with any fixed ratio, such calculations strongly depend on the situation. (See FM 3-24 for more information on counterinsurgency planning.)

9-106.   Planners also determine relative combat power with regard to civilian requirements and conditions that require attention, and then they array forces and capabilities for stability tasks. For example, a COA may require a follow-on force to establish civil security, maintain civil control, and restore essential services in a densely populated urban area over an extended period. Planners conduct a troop-to-task analysis to determine the type of units and capabilities needed to accomplish these tasks.

9-107.   Planners then proceed to initially array friendly forces starting with the decisive operation and continuing with all shaping and sustaining operations. Planners normally array ground forces two levels below their echelon. The initial array focuses on generic ground maneuver units without regard to specific type or task organization and then considers all appropriate intangible factors. For example, at corps level, planners array generic brigades. During this step, planners do not assign missions to specific units; they only consider which forces are necessary to accomplish their task. In this step, planners also array assets to accomplish essential stability tasks.

9-108.   The initial array identifies the total number of units needed and identifies possible methods of dealing with the enemy and stability tasks. If the number arrayed is less than the number available, planners place additional units in a pool for use when they develop the initial concept of the operation. (See paragraph 9-106.) If the number of units arrayed exceeds the number available and the difference cannot be compensated for with intangible factors, the staff determines whether the COA is feasible. Ways to make up the shortfall include requesting additional resources, accepting risk in that portion of the area of operations, or executing tasks required for the COA sequentially rather than simultaneously. Commanders should also consider requirements to minimize and relieve civilian suffering. Establishing civil security and providing essential services such as medical care, water, food, and shelter are implied tasks for commanders during any combat operation. (See FM 3-07 for a full discussion on stability tasks.)

Develop a Broad Concept

9-109. In developing the broad concept of the operation, the commander describes how arrayed forces will accomplish the mission within the commander's intent. The broad concept concisely expresses the *how* of the commander's visualization and will eventually provide the framework for the concept of operations and summarizes the contributions of all warfighting functions. The staff develops the initial concept of the operation for each COA expressed in both narrative and graphic forms. A sound COA is more than the arraying of forces. It presents an overall combined arms idea that will accomplish the mission. The initial concept of the operation includes, but is not limited to, the following:

- The purpose of the operation.
- A statement of where the commander will accept risk.
- Identification of critical friendly events and transitions between phases (if the operation is phased).
- Designation of the reserve, including its location and composition.
- Information collection activities.
- Essential stability tasks.
- Identification of maneuver options that may develop during an operation.
- Assignment of subordinate areas of operations.
- Scheme of fires.
- Themes, messages, and means of delivery.
- Military deception operations (on a need to know basis).
- Key control measures.
- Designate the operational framework for this operation: deep-close-security, main and supporting effort, or decisive-shaping-sustaining.
- Designation of the decisive operation, along with its task and purpose, linked to how it supports the higher headquarters' concept.

> *NOTE*: For the purpose of this section, the decisive-shaping-sustaining operational framework is an example. Planners use the same process when analyzing the other two operational frameworks—deep-close-security and main and supporting effort—to develop initial concepts of the operation.

9-110. Planners select control measures, including graphics, to control subordinate units during an operation. These establish responsibilities and limits that prevent subordinate units' actions from impeding one another. These measures also foster coordination and cooperation between forces without unnecessarily restricting freedom of action. Good control measures foster decisionmaking and individual initiative. (See FM 3-90-1 for a discussion of control measures associated with offensive and defensive tasks. See ADRP 1-02 for doctrinally correct unit symbols, control measures, and rules for drawing control measures on overlays and maps.)

9-111. Planners may use both lines of operations and lines of effort to build their broad concept. Lines of operations portray the more traditional links among objectives, decisive points, and centers of gravity. A line of effort, however, helps planners link multiple tasks with goals, objectives, and end state conditions. Combining lines of operations with lines of effort allows planners to include nonmilitary activities in their broad concept. This combination helps commanders incorporate stability or defense support of civil authorities tasks that, when accomplished, help set end state conditions of an operation.

9-112. Based on the commander's planning guidance (informed by the Army design methodology concept if this preceded the MDMP), planners develop lines of effort by—

- Confirming end state conditions from the initial commander's intent and planning guidance.
- Determining and describing each line of effort.
- Identifying objectives (intermediate goals) and determining tasks along each line of effort.

9-113. During COA development, lines of effort are general and lack specifics, such as tasks to subordinate units associated to objectives along each line of effort. Units develop and refine lines of effort, including specific tasks to subordinate units, during war-gaming. (See ADRP 5-0 and FM 3-07 for examples of operations depicted along lines of effort.)

9-114. As planning progresses, commanders may modify lines of effort and add details while war-gaming. Operations with other instruments of national power support a broader, comprehensive approach to stability tasks. Each operation, however, differs. Commanders develop and modify lines of effort to focus operations on achieving an end state, even as the situation evolves.

Assign Headquarters

9-115. After determining the broad concept, planners create a task organization by assigning headquarters to groupings of forces. They consider the types of units to be assigned to a headquarters and the ability of that headquarters to control those units. Generally, a headquarters controls at least two subordinate maneuver units (but not more than five) for fast-paced offensive or defensive tasks. The number and type of units assigned to a headquarters for stability tasks vary based on factors of the mission variables: mission, enemy, terrain and weather, troops and support available, time available, and civil considerations (METT-TC). If planners need additional headquarters, they note the shortage and resolve it later. Task organization takes into account the entire operational organization. It also accounts for the special command requirements for operations, such as a passage of lines, or air assault.

Develop Course of Action Statements and Sketches

9-116. The G-3 (S-3) prepares a COA statement and supporting sketch for each COA. The COA statement clearly portrays how the unit will accomplish the mission. The COA statement briefly expresses how the unit will conduct the combined arms concept. The sketch provides a picture of the movement and maneuver aspects of the concept, including the positioning of forces. Together, the statement and sketch cover the *who* (generic task organization), *what* (tasks), *when*, *where*, and *why* (purpose) for each subordinate unit.

9-117. The COA sketch includes the array of generic forces and control measures, such as—
- The unit and subordinate unit boundaries.
- Unit movement formations (but not subordinate unit formations).
- The line of departure or line of contact and phase lines, if used.
- Information collection graphics.
- Ground and air axes of advance.
- Assembly areas, battle positions, strong points, engagement areas, and objectives.
- Obstacle control measures and tactical mission graphics.
- Fire support coordination and airspace coordinating measures.
- Main effort.
- Location of command posts and critical communications nodes.
- Known or templated enemy locations.
- Population concentrations.

9-118. Planners can include identifying features (such as cities, rivers, and roads) to help orient users. The sketch may be on any medium. What it portrays is more important than its form. (See figure 9-5 on page 9-24 for a sample COA sketch and COA statement for a brigade combat team using the operational framework of decisive-shaping-sustaining.)

Conduct a Course of Action Briefing

9-119. After developing COAs, the staff briefs them to the commander. A collaborative session may facilitate subordinate planning. The COA briefing includes—

- An updated IPB (if there are significant changes).
- As many threat COAs as necessary (or specified by the commander). At a minimum the most likely and most dangerous threat COAs must be developed.
- The approved problem statement and mission statement.
- The commander's and higher commander's intents.
- COA statements and sketches, including lines of effort if used.
- The rationale for each COA, including—
  - Considerations that might affect enemy COAs.
  - Critical events for each COA.
  - Deductions resulting from the relative combat power analysis.
  - The reason units are arrayed as shown on the sketch. (See ADRP 1-02 for doctrine on COA sketches.)
  - The reason the staff used the selected control measures.
  - The impact on civilians.
  - How the COA accounts for minimum essential stability tasks.
  - New facts and new or updated assumptions.
  - Refined COA evaluation criteria.

Select or Modify Courses of Action for Continued Analysis

9-120.   After the COA briefing, the commander selects or modifies those COAs for continued analysis. The commander also issues planning guidance. If commanders reject all COAs, the staff begins again. If commanders accept one or more of the COAs, staff members begin COA analysis. The commander may create a new COA by incorporating elements of one or more COAs developed by the staff. The staff then prepares to war-game this new COA. The staff incorporates those modifications and ensures all staff members understand the changed COA.

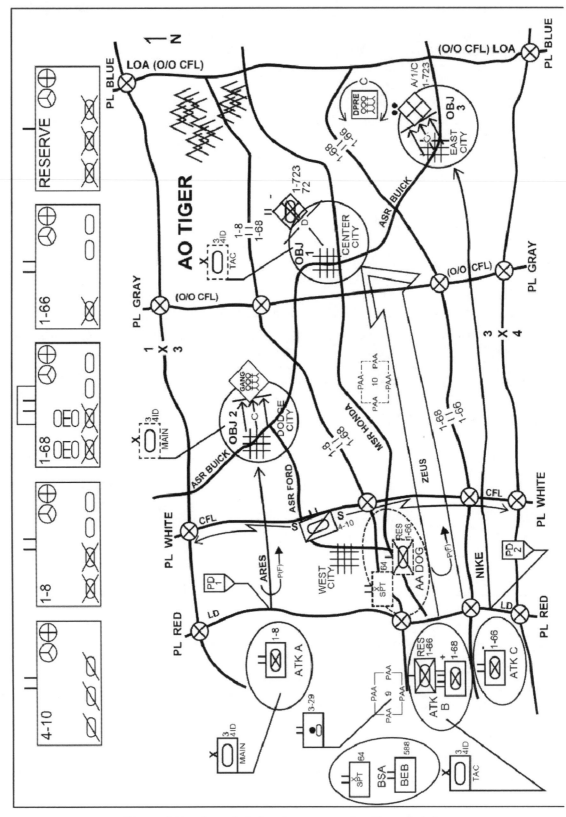

Figure 9-5. +Sample brigade course of action sketch

MISSION: On order, 3rd ABCT destroys remnants of the 72nd BDE in AO TIGER to establish security and enable the host-nation in reestablishing civil control in the region.

COMMANDER'S INTENT: The purpose of this operation is to provide a safe and secure environment in AO TIGER to enable the host-nation and other civilian organizations to reestablish civil control, restore essential services, and reestablish local governance within the area. The key tasks are: 1) destroy remnants of the 72nd BDE; 2) secure population centers vic OBJs 1, 2, and 3; 3) transition authority to the host nation. At end state, the BCT has destroyed remnant enemy forces in AO TIGER, secured population centers, and is prepared to transition responsibility for security to host nation authority.

INFORMATION COLLECTION: Priority of reconnaissance initially to locate enemy forces between PL RED (LD) and PL WHITE. Information collection operations subsequently focus on: 1) identifying the location and disposition of enemy forces vic OBJ 1; 2) observation of MSR HONDA between PL WHITE and PL BLUE; 3) observation of dislocated civilian traffic from CENTER CITY to EAST CITY.

SHAPING OPERATIONS:
4-10 CAV (ME) initially screens along PL WHITE IOT deny enemy reconnaissance and provide freedom of maneuver for follow on operations. On order, conducts FPOL at PL WHITE IOT move 1-8 CAB and 1-66 CAB(-) forward to conduct operations while maintaining contact with enemy.
O/O, 1-8 CAB (SE) in the north moves from ATK A, crosses LD at PD1 on DIRECTION OF ATTACK ARES, conducts FPOL, and clears hostile gang vic OBJ 2 IOT enable NGO delivery of humanitarian assistance to WEST CITY and DODGE CITY.
TF 1-68 (SE) in the center occupies ATK B IOT prepare for follow on operations.
On order, 1-66 CAB(-) (SE) in the south moves from ATK C, crosses LD at PD 2, attacks along DIRECTION OF ATTACK NIKE, and clears enemy vic OBJ 3 IOT prevent disruption of DO vic OBJ 1.
588 BEB (SE) occupies BSA IOT set conditions for follow on operations.
RESERVE initially establishes vic ATK B. On order, displace to AA DOG (east). Priority of commitment to DO vic OBJ 1.

DECISIVE OPERATION:
4-10 CAV (SE) conducts FPOL vic PL WHITE IOT move 1-68 CAB (ME) forward to conduct operations while maintaining enemy contact. On order, occupy AA DOG (south) IOT prepare for future operations. BPT conduct security operations in northeastern portion of AO TIGER IOT provide early and accurate warning of enemy or hostile threats to the security of population centers.
1-8 CAB (SE) controls ASRs BUICK and FORD in assigned AO IOT facilitate sustaining operations and prevent civilians interference with DO vic OBJ 1.

O/O, TF 1-68 (ME) moves from ATK B along AXIS ZEUS, conducts FPOL, and attacks to destroy elements of 72nd BDE vic OBJ 1 IOT provide a secure environment for the CENTER CITY population. Bypass criteria is platoon-size or smaller.
1-66 CAB(-) (SE) controls DPRE camp vic EAST CITY IOT provide a secure environment and controls ASR BUICK in assigned AO IOT facilitate sustaining operations and prevent civilian interference with DO vic OBJ 1.
588 BEB (SE) conducts operations as required IOT support DO.
RESERVE establishes in AA DOG (east). Priority of commitment is to reinforce DO vic OBJ 1.

FIRES:
(Shaping Operations): Priority of fires to 4-10 CAV, 1-8 CAB, 1-66 CAB, and TF 1-68 initially from PAA 9. O/O displace to PAA 10. HPTs are enemy reconnaissance forces, indirect fire systems, and mechanized infantry forces.
(Decisive Operations): Priority of fires to TF 1-68 (ME), 1-66 CAB, 1-8 CAB, and 4-10 CAV from PAA 10. HPTs are enemy armor, mechanized infantry forces, and indirect fire systems.
FSCM: CFL initially PL WHITE, O/O PL GRAY, O/O PL BLUE (LOA).

SUSTAINING OPERATIONS:
(Shaping Operations): 64 BSB will initially establish operations in BSA. O/O, establish BSA in AA DOG vic WEST CITY using MSR HONDA, ASR FORD, and ASR BUICK as primary routes IOT sustain operations. Establish FLEs as required to support operations. Priority of support to 4-10 CAV (ME) will be class III, V, maintenance, and medical.
(Decisive Operations): Priority of support to TF 1-68 (ME) will be class III, V, maintenance, and medical. Coordinate with humanitarian relief agencies IOT facilitate rapid restoration of essential services in AO TIGER.

MISSION COMMAND:
(Command): 3rd ABCT commander located with TAC CP and executive officer located with MAIN CP throughout mission.
(Control/Signal): 3rd ABCT MAIN CP initially located vic ATK A. O/O, displaces vic OBJ 2. 3rd ABCT TAC CP initially located vic ATK B. O/O, displaces vic OBJ 1.

RISK: Based on intelligence reports of negative enemy activity in the northeast mountainous portion of AO TIGER, risk is assumed with no ground maneuver forces initially allocated to conduct reconnaissance or surveillance operations. Mitigation will be accomplished by assigning a BPT mission to 4-10 CAV to conduct security operations IOT provide early and accurate warning of enemy or hostile threats to the security of population centers.

AA	assembly area	BSA	brigade support area	FSCM	fire support coordination measures	ME	main effort	SE	supporting effort
ABCT	armored brigade combat team	BSB	brigade support battalion	FLE	forward logistics element	MSR	main supply route	SPT	support
AO	area of operations	CAB	combat aviation brigade	FPOL	forward passage of lines	NGO	nongovernmental organization	TAC	tactical
ASR	alternate supply route	CAV	cavalry	HPT	high payoff targets	O/O	on order	TF	task force
ATK	attack position	CFL	coordinated fire line	ID	infantry division	OBJ	objective	vic	vicinity
BCT	brigade combat team	CP	command post	IOT	in order to	PAA	position area of artillery		
BDE	brigade	DO	decisive operations	LD	line of departure	PD	point of departure		
BEB	brigade engineer battalion	DPRE	displaced persons, refugees, and evacuees	LOA	limit of advance	PL	phase line		
BPT	be prepared to					RES	reserve		

Figure 9-5. +Sample brigade course of action sketch (continued)

STEP 4–COURSE OF ACTION ANALYSIS AND WAR-GAMING

9-121.   COA analysis enables commanders and staffs to identify difficulties or coordination problems as well as probable consequences of planned actions for each COA being considered. It helps them think through the tentative plan. COA analysis may require commanders and staffs to revisit parts of a COA as discrepancies arise. COA analysis not only appraises the quality of each COA, but it also uncovers potential execution problems, decisions, and contingencies. In addition, COA analysis influences how commanders and staffs understand a problem and may require the planning process to restart. (See figure 9-6.)

Step 4: COA Analysis (War Game)

Key inputs	Process	Key outputs
• Updated running estimates • Revised planning guidance • COA statements and sketches • Updated assumptions	• Gather the tools • List all friendly forces • List assumptions • List known critical events and decision points • Select the war-gaming method • Select a technique to record and display results • War-game the operation and assess the results • Conduct a war-game briefing (optional)	• Refined COAs • Potential decision points • War-game results • Initial assessment measures • Updated assumptions

COA   course of action

Figure 9-6. Step 4–course of action analysis and war-gaming

9-122.   War-gaming is a disciplined process, with rules and steps that attempt to visualize the flow of the operation, given the force's strengths and dispositions, the enemy's capabilities, and possible COAs; the impact and requirements of civilians in the area of operations; and other aspects of the situation. The simplest form of war-gaming is the manual method, often using a tabletop approach with blowups of matrixes and templates. The most sophisticated form of war-gaming is computer-aided modeling and simulation. Regardless of the form used, each critical event within a proposed COA should be war-gamed using the action, reaction, and counteraction methods of friendly and enemy forces interaction. This basic war-gaming method (modified to fit the specific mission and environment) applies to offensive, defensive, and stability or defense support of civil authorities operations. When conducting COA analysis, commanders and staffs perform the process actions and produce the outputs shown in figure 9-6.

9-123.   War-gaming results in refined COAs, a completed synchronization matrix, and decision support templates and matrixes for each COA. A synchronization matrix records the results of a war game. It depicts how friendly forces for a particular COA are synchronized in time, space, and purpose in relation to an enemy COA or other events in stability or defense support of civil authorities operations. The decision support template and matrix portray key decisions and potential actions that are likely to arise during the execution of each COA.

9-124.   COA analysis allows the staff to synchronize the six warfighting functions for each COA. It also helps the commander and staff to—

- Determine how to maximize the effects of combat power while protecting friendly forces and minimizing collateral damage.
- Further develop a visualization of the operation.
- Anticipate operational events.

- Determine conditions and resources required for success.
- Determine when and where to apply force capabilities.
- Identify coordination needed to produce synchronized results.
- Determine the most flexible COA.

9-125.   During the war game, the staff takes each COA and begins to develop a detailed plan while determining its strengths or weaknesses. War-gaming tests and improves COAs. The commander, staff, and other available partners (and subordinate commanders and staffs if the war game is conducted collaboratively) may change an existing COA or develop a new COA after identifying unforeseen events, tasks, requirements, or problems.

Gather the Tools

9-126.   The first task for COA analysis is to gather the necessary tools to conduct the war game. The COS (XO) directs the staff to gather tools, materials, and data for the war game. Units war-game with maps, sand tables, computer simulations, or other tools that accurately reflect the terrain. The staff posts the COA on a map displaying the area of operations. Tools required include, but are not limited to—

- Running estimates.
- Threat templates and models.
- Civil considerations overlays, databases, and data files.
- Modified combined obstacle overlays and terrain effects matrices.
- A recording method.
- Completed COAs, including graphics.
- A means to post or display enemy and friendly unit symbols and other organizations.
- A map of the area of operations.

List All Friendly Forces

9-127.   The commander and staff consider all units that can be committed to the operation, paying special attention to support relationships and constraints. This list includes assets from all participants operating in the area of operations. The friendly forces list remains constant for all COAs.

List Assumptions

9-128.   The commander and staff review previous assumptions for continued validity and necessity. Any changes resulting from this review are noted for record.

List Known Critical Events and Decision Points

9-129.   **A *critical event* is an event that directly influences mission accomplishment**. Critical events include events that trigger significant actions or decisions (such as commitment of an enemy reserve), complicated actions requiring detailed study (such as a passage of lines), and essential tasks. The list of critical events includes major events from the unit's current position through mission accomplishment. It includes reactions by civilians that potentially affect operations or require allocation of significant assets to account for essential stability tasks.

9-130.   A *decision point* is a point in space and time when the commander or staff anticipates making a key decision concerning a specific course of action (JP 5-0). Decision points may be associated with the friendly force, the status of ongoing operations, and with CCIRs that describe what information the commander needs to make the anticipated decision. A decision point requires a decision by the commander. It does not dictate what the decision is, only that the commander must make one, and when and where it should be made to maximally impact friendly or enemy COAs or the accomplishment of stability tasks.

Select the War-Gaming Method

9-131.   Three recommended war-gaming methods exist: belt, avenue-in-depth, and box. Each considers the area of interest and all enemy forces that can affect the outcome of the operation. Planners can use the methods separately or in combination and modified for long-term operations dominated by stability.

9-132.   The belt method divides the area of operations into belts (areas) running the width of the area of operations. The shape of each belt is based on the factors of METT-TC. The belt method works best when conducting offensive and defensive tasks on terrain divided into well-defined cross-compartments, during phased operations (such as gap crossings, air assaults, or airborne operations), or when the enemy is deployed in clearly defined belts or echelons. Belts can be adjacent to or overlap each other.

9-133.   This war-gaming method is based on a sequential analysis of events in each belt. Commanders prefer it because it focuses simultaneously on all forces affecting a particular event. A belt might include more than one critical event. Under time-constrained conditions, the commander can use a modified belt method. The modified belt method divides the area of operations into not more than three sequential belts. These belts are not necessarily adjacent or overlapping but focus on the critical actions throughout the depth of the area of operations. (See figure 9-7.)

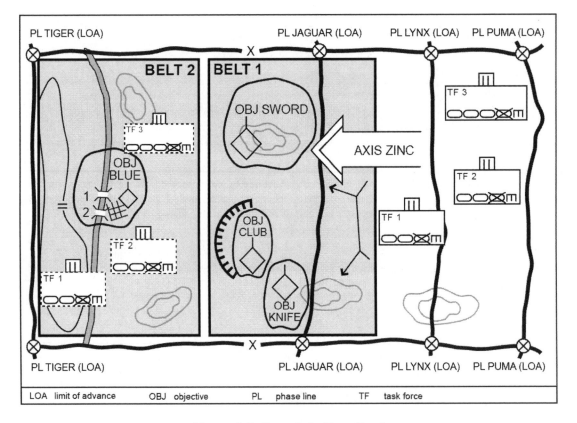

Figure 9-7. Sample belt method

9-134.   In stability tasks, the belt method can divide the COA by events, objectives (goals not geographic locations), or events and objectives in a selected slice across all lines of effort. The belt method consists of war-gaming relationships among events or objectives on all lines of effort in the belt. (See figure 9-8 on page 9-29.)

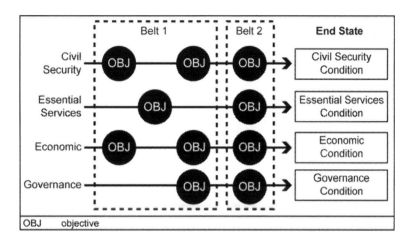

Figure 9-8. Sample modified belt method using lines of effort

9-135. The avenue-in-depth method focuses on one avenue of approach at a time, beginning with the decisive operation. This method is good for offensive COAs or in the defense when canalizing terrain inhibits mutual support. (See figure 9-9.)

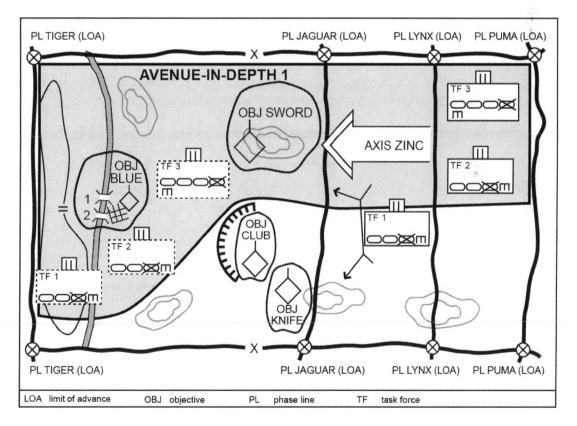

Figure 9-9. Sample avenue-in-depth method

9-136. In stability tasks, planners can modify the avenue-in-depth method. Instead of focusing on a geographic avenue, the staff war-games a line of effort. This method focuses on one line of effort at a time, beginning with the decisive line. The avenue-in-depth method includes not only war-gaming events and objectives in the selected line, but also war-gaming relationships among events or objectives on all lines of effort with respect to events in the selected line. (See figure 9-10 on page 9-30.)

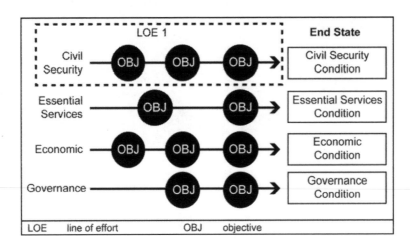

Figure 9-10. Sample modified avenue-in-depth method using lines of effort

9-137.   The box method is a detailed analysis of a critical area, such as an engagement area, a wet gap crossing site, or a landing zone. It works best in a time-constrained environment, such as a hasty attack. The box method is particularly useful when planning operations in noncontiguous areas of operation. When using this method, the staff isolates the area and focuses on critical events in it. Staff members assume that friendly units can handle most situations in the area of operations and focus their attention on essential tasks. (See figure 9-11).

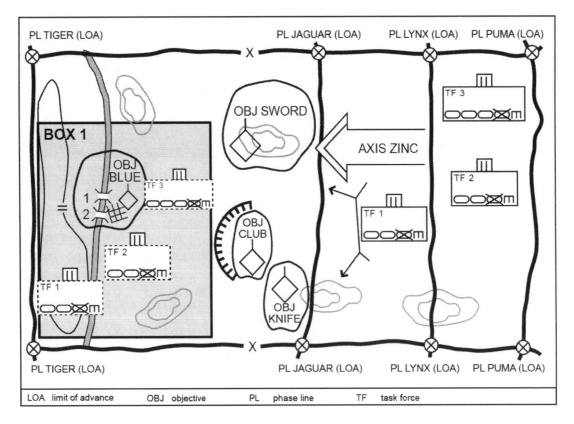

Figure 9-11. Sample box method

9-138.   In stability tasks, the box method may focus analysis on a specific objective along a line of effort, such as development of local security forces as part of improving civil security. (See figure 9-12.)

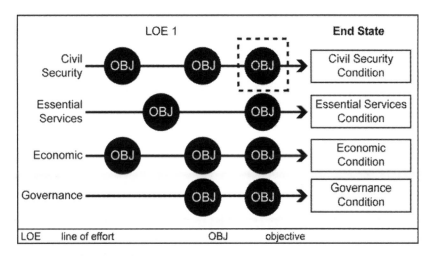

Figure 9-12. Sample modified box method using lines of effort

Select a Technique to Record and Display Results

9-139.   The war-game results provide a record from which to build task organizations, synchronize activities, develop decision support templates, confirm and refine event templates, prepare plans or orders, and compare COAs. Two techniques are commonly used to record and display results: the synchronization matrix technique and the sketch note technique. In both techniques, staff members record any remarks regarding the strengths and weaknesses they discover. The amount of detail depends on the time available. Unit SOPs address details and methods of recording and displaying war-gaming results.

9-140.   The synchronization matrix is a tool the staff uses to record the results of war-gaming that helps them synchronize a course of action across time, space, and purpose in relationship to potential enemy and civil actions. The first entry in the left column is the time, event, or phase of the operation. The second entry is the most likely enemy action. The third entry is the most likely civilian action. The fourth entry is the decision points for the friendly COA. The remainder of the matrix focuses on selected warfighting functions, their subordinate tasks, and the unit's major subordinate commands. (See table 9-3 on page 9-32.)

Table 9-3. Sample synchronization matrix tool

Time/Event/Phase		H - 24 hours (or event or phase)	H-hour (or event or phase)	H + 24 (or event or phase)
Enemy Action		Initiates threat activities and movements	Defends from security zone	Commits reserve
Population or Civilian Action		Orderly evacuation from area continues		
Decision Points		Conduct aviation attack of OBJ Irene		
Control Measures				
Movement and Maneuver	1st ABCT	Move on Route Irish	Cross LD	Seize on OBJ Irene
	2d ABCT	Move on Route Longstreet	Cross LD	Seize on OBJ Rose
	3d ABCT			FPOL with 1st BCT
	Avn Bde	Attack enemy reserve on OBJ Irene		
	BFSB			
Reserve				
Information Collection				
Fires		Prep fires initiated at H-5 Suppression of enemy air defense initiated		
Protection	Engineer			
	PMO			
	CBRN			
Sustainment				
Mission Command			Main CP with 1st BCT	
Close Air Support				
Electronic Warfare			Enemy command and control jammed	
Nonlethal Effects		Surrender broadcasts and leaflets		
Host Nation				
Interagency				
NGOs			Begins refugee relief	

Note: The first column is representative only and can be modified to fit formation needs.

AMD	air and missile defense		H	hour
Avn Bde	aviation brigade		LD	line of departure
ABCT	armored brigade combat team		NGO	nongovernmental organization
CBRN	chemical, biological, radiological, and nuclear		OBJ	objective
CP	command post		PMO	provost marshal office
FPOL	forward passage of lines			

9-141.   The sketch note method uses brief notes concerning critical locations or tasks and purposes. These notes refer to specific locations or relate to general considerations covering broad areas. The commander and staff mark locations on the map and on a separate war-game work sheet. Staff members use sequential numbers to link the notes to the corresponding locations on the map or overlay. Staff members also identify actions by placing them in sequential action groups, giving each subtask a separate number. They use the war-game work sheet to identify all pertinent data for a critical event. (See table 9-4.) They assign each event a number and title and use the columns on the work sheet to identify and list in sequence—

- Units and assigned tasks.
- Expected enemy actions and reactions.
- Friendly counteractions and assets.
- Total assets needed for the task.
- Estimated time to accomplish the task.
- The decision point tied to executing the task.
- CCIRs.
- Control measures.
- Remarks.

Table 9-4. Sample sketch note method

Critical Event	Seize OBJ Sword
Sequence number	1
Action	TF 3 attacks to destroy enemy company on OBJ Sword
Reaction	Enemy company on OBJ Club counterattacks
Counteraction	TF 1 suppresses enemy company on OBJ Club
Assets	TF 3, TF 1, and TF2
Time	H+1 to H+4
Decision point	DP 3a and 3b
Commander's critical information requirements	Location of enemy armor reserve west of PL Jaguar
Control measures	Axis Zinc and support by fire position 1
Remarks	none

DP	decision point	PL	phase line
OBJ	objective	TF	task force

War-Game the Operation and Assess the Results

9-142.   War-gaming is a conscious attempt to visualize the flow of operations given the friendly force's strengths and dispositions, the enemy's capabilities and possible COAs, and civilian locations and activities. During the war game, the commander and staff try to foresee the actions, reactions, and counteractions of all participants, including civilians. The staff analyzes each selected event. It identifies tasks that the force one echelon below it must accomplish, using assets two echelons below the staff. Identifying strengths and weaknesses of each COA allows the staff to adjust the COAs as necessary.

9-143.   The war game focuses not so much on the tools used but on the people who participate. Staff members who participate in war-gaming should be the individuals deeply involved in developing COAs. Red team members (who can provide alternative points of view) provide insight on each COA. In stability tasks, subject matter experts in areas such as economic or local governance can also help assess the probable results of planned actions, including identifying possible unintended effects.

9-144.   The war game follows an action-reaction-counteraction cycle. Actions are those events initiated by the side with the initiative. Reactions are the opposing side's actions in response. With regard to stability tasks, the war game tests the effects of actions, including intended and unintended effects, as they stimulate anticipated responses from civilians and civil institutions. Counteractions are the first side's responses to reactions. This sequence of action-reaction-counteraction continues until the critical event is completed or until the commander decides to use another COA to accomplish the mission.

9-145.   The staff considers all possible forces, including templated enemy forces outside the area of operations, that can influence the operation. The staff also considers the actions of civilians in the area of operations, the diverse kinds of coverage of unfolding events, and their consequences in the global media. The staff evaluates each friendly move to determine the assets and actions required to defeat the enemy at that point or to accomplish stability tasks. The staff continually considers branches to the plan that promote success against likely enemy counteractions or unexpected civilian reactions. Lastly, the staff lists assets used in the appropriate columns of the work sheet and lists the totals in the assets column (not considering any assets lower than two command levels below the staff).

9-146.   The commander and staff examine many areas during the war game. These include, but are not limited to—

- All friendly capabilities.
- All enemy capabilities and critical civil considerations that impact operations.
- Global media responses to proposed actions.
- Movement considerations.
- Closure rates.
- Lengths of columns.
- Formation depths.
- Ranges and capabilities of weapon systems.
- Desired effects of fires.

9-147.   The commander and staff consider how to create conditions for success, protect the force, and shape the operational environment. Experience, historical data, SOPs, and doctrinal literature provide much of the necessary information. During the war game, staff officers perform a risk assessment for their functional areas for each COA. They then propose appropriate control measures. They continually assess the risk of adverse reactions from population and media resulting from actions taken by all sides in the operation. Staff officers develop ways to mitigate those risks.

9-148.   The staff continually assesses the risk to friendly forces, balancing between mass and dispersion. When assessing the risk of weapons of mass destruction to friendly forces, planners view the target that the force presents through the eyes of an enemy target analyst. They consider ways to reduce vulnerability and determine the appropriate level of mission-oriented protective posture consistent with mission accomplishment.

9-149.   The staff identifies the required assets of the warfighting functions to support the concept of operations, including those needed to synchronize sustaining operations. If requirements exceed available assets, the staff recommends priorities based on the situation, commander's intent, and planning guidance. To maintain flexibility, the commander may decide to create a reserve to maintain assets for unforeseen tasks or opportunities.

9-150.   The commander can modify any COA based on how things develop during the war game. When doing this, the commander validates the composition and location of the decisive operation, shaping operations, and reserve forces. Control measures are adjusted as necessary. The commander may also identify situations, opportunities, or additional critical events that require more analysis. The staff performs this analysis quickly and incorporates the results into the war-gaming record.

9-151.   An effective war game results in the commander and staff refining, identifying, analyzing, developing, and determining several effects. (See table 9-5.)

Table 9-5. Effective war game results

The commander and staff refine (or modify)—
Each course of action, to include identifying branches and sequels that become on-order or be-prepared missions.
The locations and times of decisive points.
The enemy event template and matrix.
The task organization, including forces retained in general support.
Control requirements, including control measures and updated operational graphics.
Commander's critical information requirements and other information requirements—including the latest time information is of value—and incorporate them into the information collection plan.

The commander and staff identify—
Key or decisive terrain and determining how to use it.
Tasks the unit retains and tasks assigned to subordinates.
Likely times and areas for enemy use of weapons of mass destruction and friendly chemical, biological, radiological, and nuclear defense requirements.
Potential times or locations for committing the reserve.
The most dangerous enemy course of action.
The most likely enemy course of action.
The most dangerous civilian reaction.
Locations for the commander and command posts.
Critical events.
Requirements for support of each warfighting function.
Effects of friendly and enemy actions on civilians and infrastructure and on military operations.
Or confirming the locations of named areas of interest, target areas of interest, decision points, and intelligence requirements needed to support them.
Analyzing, and evaluating strengths and weaknesses of each course of action.
Hazards, assessing their risk, developing control measures for them, and determining residual risk.
The coordination required for integrating and synchronizing interagency, host-nation, and nongovernmental organization involvement.

The commander and staff analyze—
Potential civilian reactions to operations.
Potential media reaction to operations.
Potential impacts on civil security, civil control, and essential services in the area of operations.

The commander and staff develop—
Decision points.
A synchronization matrix.
A decision support template and matrix.
Solutions to achieving minimum essential stability tasks in the area of operations.
The information collection plan and graphics.
Themes and messages.
Fires, protection, and sustainment plans and graphic control measures.

The commander and staff determine—
The requirements for military deception and surprise.
The timing for concentrating forces and starting the attack or counterattack.
The movement times and tables for critical assets, including information systems nodes.
The estimated the duration of the entire operation and each critical event.
The projected the percentage of enemy forces defeated in each critical event and overall.
The percentage of minimum essential tasks that the unit can or must accomplish.
The media coverage and impact on key audiences.
The targeting requirements in the operation, to include identifying or confirming high-payoff targets and establishing attack guidance.
The allocation of assets to subordinate commanders to accomplish their missions.

Conduct a War-Game Briefing (Optional)

9-152.   Time permitting, the staff delivers a briefing to all affected elements to ensure everyone understands the results of the war game. The staff uses the briefing for review and ensures that it captures all relevant points of the war game for presentation to the commander, COS (XO), or deputy or assistant commander. In a collaborative environment, the briefing may include selected subordinate staffs. A war-game briefing format includes the following:

- Higher headquarters' mission, commander's intent, and military deception plan.
- Updated IPB.
- Assumptions.
- Friendly and enemy COAs that were war-gamed, including—
    - Critical events.
    - Possible enemy actions and reactions.
    - Possible impact on civilians.
    - Possible media impacts.
    - Modifications to the COAs.
    - Strengths and weaknesses.
    - Results of the war game.
- War-gaming technique used.

General War-Gaming Rules and Responsibilities

9-153.   War gamers need to—

- Remain objective, not allowing personality or their sense of "what the commander wants" to influence them.
- Avoid defending a COA just because they personally developed it.
- Record advantages and disadvantages of each COA accurately as they emerge.
- Continually assess feasibility, acceptability, and suitability of each COA. If a COA fails any of these tests, reject it.
- Avoid drawing premature conclusions and gathering facts to support such conclusions.
- Avoid comparing one COA with another during the war game. This occurs during Step 5—COA Comparison.

Mission Command Responsibilities

9-154.   The commander has overall responsibility for the war-gaming process, and the commander can determine the staff members who are involved in war-gaming. Traditionally, certain staff members have key and specific roles.

9-155.   The COS (XO) coordinates actions of the staff during the war game. This officer is the unbiased controller of the process, ensuring the staff stays on a timeline and achieves the goals of the war-gaming session. In a time-constrained environment, this officer ensures that, at a minimum, the decisive operation is war-gamed.

9-156.   The G-3 (S-3) assists the commander with the rehearsal. The G-3 (S-3)—

- Portrays the friendly scheme of maneuver, including the employment of information-related capabilities.
- Ensures subordinate unit actions comply with the commander's intent.
- Normally provides the recorder.

9-157.   The assistant chief of staff, signal (G-6 [S-6]) assesses network operations, spectrum management operations, network defense, and information protection feasibility of each war-gamed COA. The G-6 (S-6) determines communications systems requirements and compares them to available assets, identifies potential shortfalls, and recommends actions to eliminate or reduce their effects.

9-158.   The information operations officer assesses the information operations concept of support against the ability of information-related capabilities to execute tasks in support of each war-gamed COA and the effectiveness of integrated information-related capabilities to impact various audiences and populations in and outside the area of operations. The information operations officer, in coordination with the electronic warfare officer, also integrates information operations with cyber electromagnetic activities.

9-159.   The assistant chief of staff, civil affairs operations (G-9 [S-9]) ensures each war-gamed COA effectively integrates civil considerations (the "C" of METT-TC). The civil affairs operations officer considers not only tactical issues but also sustainment issues. This officer assesses how operations affect civilians and estimates the requirements for essential stability tasks commanders might have to undertake based on the ability of the unified action partners. Host-nation support and care of dislocated civilians are of particular concern. The civil affairs operations officer's analysis considers how operations affect public order and safety, the potential for disaster relief requirements, noncombatant evacuation operations, emergency services, and the protection of culturally significant sites. This officer provides feedback on how the culture in the area of operations affects each COA. If the unit lacks an assigned civil affairs officer, the commander assigns these responsibilities to another staff member.

9-160.   The red team staff section provides the commander and assistant chief of staff, intelligence (G-2) with an independent capability to fully explore alternatives. The staff looks at plans, operations, concepts, organizations, and capabilities of the operational environment from the perspectives of enemies, unified action partners, and others.

9-161.   The electronic warfare officer provides information on the electronic warfare target list, electronic attack taskings, electronic attack requests, and the electronic warfare portion of the collection matrix and the attack guidance matrix. Additionally, the electronic warfare officer assesses threat vulnerabilities, friendly electronic warfare capabilities, and friendly actions relative to electronic warfare activities and other cyber electromagnetic activities not covered by the G-6 or G-2.

9-162.   The staff judge advocate advises the commander on all matters pertaining to law, policy, regulation, good order, and discipline for each war-gamed COA. This officer provides legal advice across the range of military operations on law of war, rules of engagement, international agreements, Geneva Conventions, treatment and disposition of noncombatants, and the legal aspects of targeting.

9-163.   The operations research and systems analysis staff section provides analytic support to the commander for planning and assessment of operations. Specific responsibilities include—

- Providing quantitative analytic support, including regression and trend analysis, to planning and assessment activities.
- Assisting other staff members in developing customized analytical tools for specific requirements, providing a quality control capability, and conducting assessments to measure the effectiveness of operations.

9-164.   The safety officer provides input to influence accident and incident reductions by implementing risk management procedures throughout the mission planning and execution process.

9-165.   The knowledge management officer assesses the effectiveness of the knowledge management plan for each course of action.

9-166.   The space operations officer provides and represents friendly, threat, and non-aligned space capabilities.

Intelligence Responsibilities

9-167.   During the war game the G-2 (S-2) role-plays the enemy commander, other threat organizations in the area of operations, and critical civil considerations in the area of operations. This officer develops critical enemy decision points in relation to the friendly COAs, projects enemy reactions to friendly actions, and projects enemy losses. The intelligence officer assigns different responsibilities to available staff members within the section (such as the enemy commander, friendly intelligence officer, and enemy recorder) for war-gaming. The intelligence officer captures the results of each enemy, threat group, and civil considerations action and counteraction as well as the corresponding friendly and enemy strengths and vulnerabilities. By trying to realistically win the war game for the enemy, the intelligence officer ensures that the staff fully addresses friendly responses for each enemy COA. For the friendly force, the intelligence officer—

- Refines intelligence and information requirements and the planning requirements tools.
- Refines the situation and event templates, including named areas of interest that support decision points.
- Refines the event template with corresponding decision points, target areas of interest, and high-value targets.
- Participates in targeting to select high-payoff targets from high-value targets identified during IPB.
- Recommends priority intelligence requirements that correspond to the decision points.
- Refines civil considerations overlays, databases, and data files.
- Refines the modified combined obstacle overlays and terrain effects matrices.
- Refines weather products that outline the critical weather impacts on operations.

Movemenent and Maneuver Responsibilities

9-168.   During the war game, the G-3 (S-3) and assistant chief of staff, plans (G-5 [S-5]) are responsible for movement and maneuver. The G-3 (S-3) normally selects the technique for the war game and role-plays the friendly maneuver commander. Various staff officers assist the G-3 (S-3), such as the aviation officer and engineer officer. The G-3 (S-3) executes friendly maneuver as outlined in the COA sketch and COA statement. The G-5 (S-5) assesses warfighting function requirements, solutions, and concepts for each COA; develops plans and orders; and determines potential branches and sequels arising from various war-gamed COAs. The G-5 (S-5) also coordinates and synchronizes warfighting functions in all plans and orders. The planning staff ensures that the war game of each COA covers every operational aspect of the mission. The members of the staff record each event's strengths and weaknesses and the rationale for each action. They complete the decision support template and matrix for each COA. They annotate the rationale for actions during the war game and use it later with the commander's guidance to compare COAs.

Fires Responsibilities

9-169.   The chief of fires (fire support officer) assesses the fire support feasibility of each war-gamed COA. This officer develops a proposed high-payoff target list, target selection standards, and attack guidance matrix. The chief of fires works with the intelligence officer to identify named and target areas of interest for enemy indirect fire weapon systems, and identifies high-payoff targets and additional events that may influence the positioning of field artillery and air defense artillery assets. The chief of fires should also offer a list of possible defended assets for air defense artillery forces and assist the commander in making a final determination about asset priority.

Protection Responsibilities

9-170.   The chief of protection assesses protection element requirements, refines EEFIs, and develops a scheme of protection for each war-gamed COA. The chief of protection—

- Refines the critical asset list and the defended asset list.
- Assesses hazards.
- Develops risk control measures and mitigation measures of threats and hazards.
- Establishes personnel recovery coordination measures.
- Implements operational area security to include security of lines of communications, antiterrorism measures, and law enforcement operations.
- Ensures survivability measures reduce vulnerabilities.
- Refines chemical, biological, radiological, and nuclear operations.

Sustainment Responsibilities

9-171.   During the war game, the assistant chief of staff, personnel (G-1 [S-1]) assesses the personnel aspect of building and maintaining the combat power of units. This officer identifies potential shortfalls and recommends COAs to ensure units maintain adequate manning to accomplish their mission. As the primary staff officer assessing the human resources planning considerations to support sustainment operations, the G-1 (S-1) provides human resources support for the operation.

9-172.   The assistant chief of staff, logistics (G-4 [S-4]) assesses the logistics feasibility of each war-gamed COA. This officer determines critical requirements for each logistics function (classes I through VII, IX, and X) and identifies potential problems and deficiencies. The G-4 (S-4) assesses the status of all logistics functions required to support the COA, including potential support required to provide essential services to the civilians, and compares it to available assets. This officer identifies potential shortfalls and recommends actions to eliminate or reduce their effects. While improvising can contribute to responsiveness, only accurately predicting requirements for each logistics function can ensure continuous sustainment. The logistics officer ensures that available movement times and assets support each COA.

9-173.   During the war game, the assistant chief of staff, financial management (G-8) assesses the commander's area of operations to determine the best COA for use of resources. This assessment includes both core functions of financial management: resource management and finance operations. This officer determines partner relationships (joint, interagency, intergovernmental, and multinational), requirements for special funding, and support to the procurement process.

9-174.   The surgeon section coordinates, monitors, and synchronizes the execution of the health system activities for the command for each war-gamed COA to ensure a fit and healthy force.

Recorders

9-175.   The use of recorders is particularly important. Recorders capture coordinating instructions, subunit tasks and purposes, and information required to synchronize the operation. Recorders allow the staff to write part of the order before they complete the planning. Automated information systems enable recorders to enter information into preformatted forms that represent either briefing charts or appendixes to orders. Each staff section keeps formats available to facilitate networked orders production.

STEP 5–COURSE OF ACTION COMPARISON

9-176.   COA comparison is an objective process to evaluate COAs independently and against set evaluation criteria approved by the commander and staff. The goal is to identify the strengths and weaknesses of COAs, enable selecting a COA with the highest probability of success, and further developing it in an OPLAN or OPORD. The commander and staff perform certain actions and processes that lead to key outputs. (See figure 9-13.)

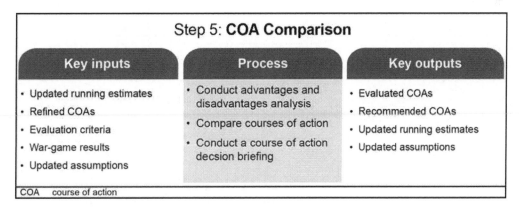

Figure 9-13. Step 5–course of action comparison

Conduct Advantages and Disadvantages Analysis

9-177.   The COA comparison starts with all staff members analyzing and evaluating the advantages and disadvantages of each COA from their perspectives. Staff members each present their findings for the others' consideration. Using the evaluation criteria developed before the war game, the staff outlines each

COA, highlighting its advantages and disadvantages. Comparing the strengths and weaknesses of the COAs identifies their advantages and disadvantages with respect to each other. (See table 9-6.)

Table 9-6. Sample advantages and disadvantages

Course of Action	Advantages	Disadvantages
Course of action 1	Decisive operation avoids major terrain obstacles. Adequate maneuver space available for units conducting the decisive operation and the reserve.	Units conducting the decisive operation face stronger resistance at the start of the operation. Limited resources available to establishing civil control to town X.
Course of action 2	Shaping operations provide excellent flank protection of the decisive operations. Upon completion of decisive operations, units conducting shaping operations can quickly transition to establish civil control and provide civil security to the population in town X.	Operation may require the early employment of the division's reserve.

Compare Courses of Action

9-178. Comparison of COAs is critical. The staff uses any technique that helps develop those key outputs and recommendations and assists the commander to make the best decision. A common technique is the decision matrix. This matrix uses evaluation criteria developed during mission analysis and refined during COA development to help assess the effectiveness and efficiency of each COA. (See table 9-7.)

Table 9-7. Sample decision matrix

Weight[1]	1	2	1	1	2	
Criteria[2] Course of Action	Simplicity	Maneuver	Fires	Civil control	Mass	Total
COA 1[3]	2	2 (4)	2	1	1 (2)	8 (11)
COA 2[3]	1	1 (2)	1	2	2 (4)	7 (10)

Notes:
[1] The COS (XO) may emphasize one or more criteria by assigning weights to them based on a determination of their relative importance. Lower weights are preferred.
[2] Criteria are those assigned in step 5 of COA analysis.
[3] COAs are those selected for war-gaming with rankings assigned to them based on comparison between them with regard to relative advantages and disadvantages of each, such as when compared for relative simplicity COA 2 is by comparison to COA 1 simpler and therefore is ranked as 1 with COA 1 ranked as 2.

9-179. The decision matrix is a tool to compare and evaluate COAs thoroughly and logically. However, the process may be based on highly subjective judgments that can change dramatically during the course of evaluation. In table 9-7, the numerical rankings reflect the relative advantages or disadvantages of each criterion for each COA as initially estimated by a COS (XO) during mission analysis. Rankings are assigned from 1 to however many COAs exist. Lower rankings are more preferred. At the same time, the COS (XO) determines weights for each criterion based on a subjective determination of their relative value. The lower weights signify a more favorable advantage, such as the lower the number, the more favorable the weight. After assigning ranks to COAs and weights to criteria, the staff adds the unweighted ranks in each row horizontally and records the sum in the Total column on the far right of each COA. The staff then multiplies the same ranks by the weights associated with each criterion and notes the product in parenthesis underneath the unweighted rank. No notation is required if the weight is 1. The staff adds these weighted products horizontally and records the sum in parenthesis underneath the unweighted total in the Total column to the right of each COA. The staff then compares the totals to determine the most preferred (lowest number) COA based on both unweighted and weighted ranks. Upon review and consideration, the commander—based on personal judgment—may elect to change either the weight or ranks for any criterion. Although the lowest total denotes a most preferred solution, the process for estimating relative ranks assigned to criterion and weighting may be highly subjective.

9-180.   Commanders and staffs cannot solely rely on the outcome of a decision matrix, as it only provides a partial basis for a solution. During the decision matrix process, planners carefully avoid reaching conclusions from a quantitative analysis of subjective weights. Comparing and evaluating COAs by criterion is probably more useful than merely comparing totaled ranks. Judgments often change with regard to the relative weighting of criteria during close analysis of COAs, which will change weighted rank totals and possibly the most preferred COA.

9-181.   The staff compares feasible COAs to identify the one with the highest probability of success against the most likely enemy COA, the most dangerous enemy COA, the most important stability task, or the most damaging environmental impact. The selected COA should also—

- Pose the minimum risk to the force and mission accomplishment.
- Place the force in the best posture for future operations.
- Provide maximum latitude for initiative by subordinates.
- Provide the most flexibility to meet unexpected threats and opportunities.
- Provide the most secure and stable environment for civilians in the area of operations.
- Best facilitate information themes and messages.

9-182.   Staff officers often use their own matrix to compare COAs with respect to their functional areas. Matrixes use the evaluation criteria developed before the war game. Their greatest value is providing a method to compare COAs against criteria that, when met, produce operational success. Staff officers use these analytical tools to prepare recommendations. Commanders provide the solution by applying their judgment to staff recommendations and making a decision.

Conduct a Course of Action Decision Briefing

9-183.   After completing its analysis and comparison, the staff identifies its preferred COA and makes a recommendation. If the staff cannot reach a decision, the COS (XO) decides which COA to recommend. The staff then delivers a decision briefing to the commander. The COS (XO) highlights any changes to each COA resulting from the war game. The decision briefing includes—

- The commander's intent of the higher and next higher commanders.
- The status of the force and its components.
- The current IPB.
- The COAs considered, including—
    - Assumptions used.
    - Results of running estimates.
    - A summary of the war game for each COA, including critical events, modifications to any COA, and war-game results.
    - Advantages and disadvantages (including risks) of each COA.
    - The recommended COA. If a significant disagreement exists, then the staff should inform the commander and, if necessary, discuss the disagreement.

STEP 6–COURSE OF ACTION APPROVAL

9-184.   After the decision briefing, the commander selects the COA to best accomplish the mission. If the commander rejects all COAs, the staff starts COA development again. If the commander modifies a proposed COA or gives the staff an entirely different one, the staff war-games the new COA and presents the results to the commander with a recommendation. (See figure 9-14.)

Step 6: **COA Approval**		
Key inputs	**Process**	**Key outputs**
• Updated running estimates	• Commander approves a COA	• Commander approved COA and any modifications
• Evaluated COAs		• Refined commander's intent, CCIRs, and EEFIs
• Recommended COA		
• Updated assumptions		• Updated assumptions

CCIR	commander's critical information requirement	EEFI	essential element of friendly information
COA	course of action		

Figure 9-14. Step 6–course of action approval

9-185.   After approving a COA, the commander issues the final planning guidance. The final planning guidance includes a refined commander's intent (if necessary) and new CCIRs to support execution. It also includes any additional guidance on priorities for the warfighting functions, orders preparation, rehearsal, and preparation. This guidance includes priorities for resources needed to preserve freedom of action and ensure continuous sustainment.

9-186.   Commanders include the risk they are willing to accept in the final planning guidance. If there is time, commanders use a video teleconference to discuss acceptable risk with adjacent, subordinate, and senior commanders. However, commanders still obtain the higher commander's approval to accept any risk that might imperil accomplishing the higher commander's mission.

9-187.   Based on the commander's decision and final planning guidance, the staff issues a WARNORD to subordinate headquarters. This WARNORD contains the information subordinate units need to refine their plans. It confirms guidance issued in person or by video teleconference and expands on details not covered by the commander personally. The WARNORD issued after COA approval normally contains—

- The area of operations.
- Mission.
- Commander's intent.
- Updated CCIRs and EEFIs.
- Concept of operations.
- Principal tasks assigned to subordinate units.
- Preparation and rehearsal instructions not included in the SOPs.
- A final timeline for the operations.

STEP 7–ORDERS PRODUCTION, DISSEMINATION, AND TRANSITION

9-188.   The staff prepares the order or plan by turning the selected COA into a clear, concise concept of operations and the required supporting information. The COA statement becomes the concept of operations for the plan. The COA sketch becomes the basis for the operation overlay. If time permits, the staff may conduct a more detailed war game of the selected COA to more fully synchronize the operation and complete the plan. (See figure 9-15 on page 9-42.) The staff writes the OPORD or OPLAN using the Army's operation order format. (See appendix C.)

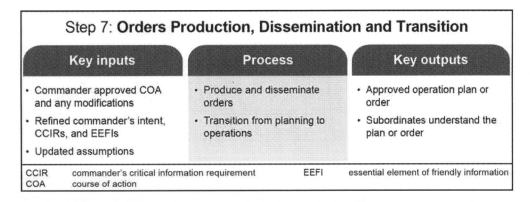

Figure 9-15. Step 7–orders production, dissemination, and transition

9-189.   Normally, the COS (XO) coordinates with staff principals to assist the G-3 (S-3) in developing the plan or order. Based on the commander's planning guidance, the COS (XO) dictates the type of order, sets and enforces the time limits and development sequence, and determines which staff section publishes which attachments.

9-190.   Prior to the commander approving the plan or order, the staff ensures the plan or order is internally consistent and is nested with the higher commander's intent. They do this through—

- Plans and orders reconciliation.
- Plans and orders crosswalk.

Plans and Orders Reconciliation

9-191.   Plans and orders reconciliation occurs internally as the staff conducts a detailed review of the entire plan or order. This reconciliation ensures that the base plan or order and all attachments are complete and in agreement. It identifies discrepancies or gaps in planning. If staff members find discrepancies or gaps, they take corrective actions. Specifically, the staff compares the commander's intent, mission, and commander's CCIRs against the concept of operations and the different schemes of support (such as scheme of fires or scheme of sustainment). The staff ensures attachments are consistent with the information in the base plan or order.

Plans and Orders Crosswalk

9-192.   During the plans and orders crosswalk, the staff compares the plan or order with that of the higher and adjacent commanders to achieve unity of effort and ensure the plan meets the superior commander's intent. The crosswalk identifies discrepancies or gaps in planning. If staff members find discrepancies or gaps, they take corrective action.

Approving the Plan or Order

9-193.   The final action in plan and order development is the approval of the plan or order by the commander. Commanders normally do not sign attachments; however, they should review them before signing the base plan or order.

9-194.   Step 7 bridges the transition between planning and preparations. The plans-to-operations transition is a preparation activity that occurs within the headquarters. It ensures members of the current operations cell fully understand the plan before execution. During preparation, the responsibility for developing and maintaining the plan shifts from the plans (or future operations) cell to the current operations cell. This transition is the point at which the current operations cell becomes responsible for controlling execution of the operation order. This responsibility includes answering requests for information concerning the order and maintaining the order through fragmentary orders. This transition enables the plans cell to focus its planning efforts on sequels, branches, and other planning requirements directed by the commander. (See

ADRP 5-0 for information on the plans to operations handover and chapter 12 of this manual for information on rehearsals.)

9-195.   Commanders review and approve orders before the staff reproduces and disseminates them, unless commanders have delegated that authority. Subordinates immediately acknowledge receipt of the higher order. If possible, the higher commander and staff brief the order to subordinate commanders in person. The commander and staff conduct confirmation briefings with subordinates immediately afterwards. Confirmation briefings can be conducted collaboratively with several commanders at the same time or with single commanders. These briefings may be conducted in person or by video teleconference.

PLANNING IN A TIME-CONSTRAINED ENVIRONMENT

9-196.   Any planning process aims to quickly develop a flexible, sound, and fully integrated and synchronized plan. However, any operation may "outrun" the initial plan. The most detailed estimates cannot anticipate every possible branch or sequel, enemy action, threat action, or reaction from the local population, unexpected opportunity, or change in mission directed from higher headquarters. Fleeting opportunities or unexpected enemy action may require a quick decision to implement a new or modified plan. When this occurs, units often find themselves pressed for time in developing a new plan.

9-197.   Before a unit can effectively conduct planning in a time-constrained environment, it must master the steps in the full MDMP. A unit can only shorten the process if it fully understands the role of each and every step of the process and the requirements to produce the necessary products. Training on these steps must be thorough and result in a series of staff battle drills that can be tailored to the time available.

9-198.   Quality staffs produce simple, flexible, and tactically sound plans in time-constrained environments. Any METT-TC factor, but especially limited time, may make it difficult to complete every step of the MDMP in detail. Applying an inflexible process to all situations does not work. Anticipation, organization, and prior preparation are the keys to successful planning under time-constrained conditions.

9-199.   Staffs can use the time saved on any step of the MDMP to—
- Refine the plan more thoroughly.
- Conduct a more deliberate and detailed war game.
- Consider potential branches and sequels in detail.
- Focus more on rehearsing and preparing the plan.
- Allow subordinate units more planning and preparation time.

THE COMMANDER'S RESPONSIBILITY

9-200.   The commander decides how to adjust the MDMP, giving specific guidance to the staff to focus on the process and save time. Commanders shorten the MDMP when they lack time to perform each step in detail. The most significant factor to consider is time. It is the only nonrenewable, and often the most critical, resource. Commanders (who have access to only a small portion of the staff or none at all) rely even more than normal on their own expertise, intuition, and creativity as well as on their understanding of the environment and of the art and science of war. They may have to select a COA, mentally war-game it, and confirm their decision to the staff in a short time. If so, they base their decision more on experience than on a formal, integrated staff process.

9-201.   Effective commanders avoid changing their guidance unless a significantly changed situation requires major revisions. Making frequent, minor changes to the guidance can easily result in lost time as the staff constantly adjusts the plan with an adverse ripple effect throughout overall planning.

9-202.   Commanders consult with subordinate commanders before making a decision, if possible. Subordinate commanders are closer to the operation and can more accurately describe enemy, friendly, and civilian situations. Additionally, consulting with subordinates gives commanders insights into the upcoming operation and allows parallel planning. White boards and collaborative digital means of communicating greatly enhance parallel planning.

9-203.   In situations where commanders must decide quickly, they advise their higher headquarters of the selected COA, if time is available. However, commanders do not let an opportunity pass just because they cannot report their actions.

THE STAFF'S RESPONSIBILITY

9-204.   Staff members keep their running estimates current. When time constraints exist, they can provide accurate, up-to-date assessments quickly and move directly into COA development. Under time-constrained conditions, commanders and staffs use as much of the previously analyzed information and as many of the previously created products as possible. The importance of running estimates increases as time decreases. Decisionmaking in a time-constrained environment usually occurs after a unit has entered the area of operations and begun operations. This means that the IPB, an updated common operational picture, and some portions of the running estimates should already exist. Civilian and military joint and multinational organizations operating in the area of operations should have well-developed plans and information to add insights to the operational environment. Detailed planning provides the basis for information that the commander and staff need to make decisions during execution.

TIME-SAVING TECHNIQUES

9-205.   Paragraphs 9-206 through 9-210 discuss time-saving techniques to speed the planning process.

Increase Commander's Involvement

9-206.   While commanders cannot spend all their time with their planning staffs, the greater the commander's involvement in planning, the faster the staff can plan. In time-constrained conditions, commanders who participate in the planning process can make decisions (such as COA selection) without waiting for a detailed briefing from the staff.

Limit the Number of Courses of Action to Develop

9-207.   Limiting the number of COAs developed and war-gamed can save planning time. If time is extremely short, the commander can direct development of only one COA. In this case, the goal is an acceptable COA that meets mission requirements in the time available. This technique saves the most time. The fastest way to develop a plan has the commander directing development of one COA with branches against the most likely enemy COA or most damaging civil situation or condition. However, this technique should be used only when time is severely limited. In such cases, this choice of COA is often intuitive, relying on the commander's experience and judgment. The commander determines which staff officers are essential to assist in COA development. Normally commanders require the intelligence officer, operations officer, plans officer, chief of fires (fire support officer), engineer officer, civil affairs operations officer, information operations officer, military information support operations officer, electronic warfare officer, and COS (XO). They may also include subordinate commanders, if available, either in person or by video teleconference. This team quickly develops a flexible COA that it feels will accomplish the mission. The commander mentally war-games this COA and gives it to the staff to refine.

Maximize Parallel Planning

9-208.   Although parallel planning is the norm, maximizing its use in time-constrained environments is critical. In a time-constrained environment, the importance of WARNORDs increases as available time decreases. A verbal WARNORD now, followed by a written order later, saves more time than a written order one hour from now. The staff issues the same WARNORDs used in the full MDMP when abbreviating the process. In addition to WARNORDs, units must share all available information with subordinates, especially IPB products, as early as possible. The staff uses every opportunity to perform parallel planning with the higher headquarters and to share information with subordinates.

Increase Collaborative Planning

9-209.   Planning in real time with higher headquarters and subordinates improves the overall planning effort of the organization. Modern information systems and a common operational picture shared electronically allow collaboration with subordinates from distant locations, can increase information sharing, and can improve the commander's visualization. Additionally, taking advantage of subordinates' input and knowledge of the situation in their areas of operations often results in developing better COAs quickly.

Use Liaison Officers

9-210. Liaison officers posted to higher headquarters and unified action partners' headquarters allow commanders to have representation in their higher headquarters' planning session. These officers assist in passing timely information to their parent headquarters and directly to the commander. Effective liaison officers have the commander's full confidence and the necessary rank and experience for the mission. Commanders may elect to use a single individual or a liaison team. As representatives, liaison officers must—

- Understand how their commander thinks and interpret verbal and written guidance.
- Convey their commander's intent, planning guidance, mission, and concept of operations.
- Represent their commander's position.
- Know the unit's mission; tactics, techniques, and procedures; organization; capabilities; and communications equipment.
- Observe the established channels of command and staff functions.
- Be trained in their functional responsibilities.
- Be tactful.
- Possess the necessary language expertise.

Chapter 10

Troop Leading Procedures

Troop leading procedures provide small-unit leaders with a framework for planning and preparing for operations. Leaders of company and smaller units use troop leading procedures to develop plans and orders. This chapter describes the eight steps of troop leading procedures and their relationship to the military decisionmaking process (MDMP). While this chapter explains troop leading procedures from a ground-maneuver perspective, it applies to all types of small units.

BACKGROUND AND COMPARISON TO THE MDMP

10-1. Troop leading procedures (TLP) extend the MDMP to the small-unit level. The MDMP and TLP are similar but not identical. They are both linked by the basic Army problem-solving process (see chapter 4). Commanders with a coordinating staff use the MDMP as their primary planning process. Company-level and smaller units lack formal staffs and use TLP to plan and prepare for operations. This places the responsibility for planning primarily on the commander or small-unit leader.

10-2. *Troop leading procedures* are a dynamic process used by small-unit leaders to analyze a mission, develop a plan, and prepare for an operation (ADP 5-0). These procedures enable leaders to maximize available planning time while developing effective plans and preparing their units for an operation. (See paragraphs 10-10 to 10-41 for a discussion on the eight steps of TLP.)

10-3. Leaders use TLP when working alone or with a small group to solve problems. For example, a company commander may use the executive officer, first sergeant, fire support officer, supply sergeant, and communications sergeant to assist during TLP.

10-4. The type, amount, and timeliness of information passed from higher to lower headquarters directly impact the lower unit leader's TLP. Figure 10-1 on page 10-2 illustrates the parallel sequences of the MDMP of a battalion with the TLP of a company and a platoon. The solid arrows depict when a higher headquarters' planning event could start the TLP of a subordinate unit. However, events do not always occur in the order shown. For example, TLP may start with receipt of a warning order (WARNORD), or they may not start until the higher headquarters has completed the MDMP and issued an operation order (OPORD). WARNORDs from higher headquarters may arrive at any time during TLP. Leaders remain flexible. They adapt TLP to fit the situation rather than try to alter the situation to fit a preconceived idea of how events should flow.

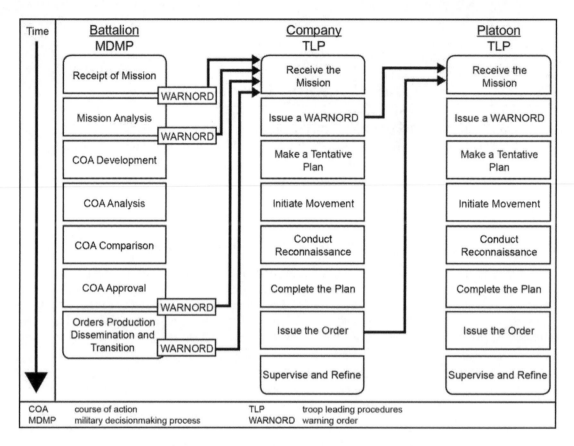

Figure 10-1. Parallel sequences of the MDMP and troop leading procedures

10-5. Normally, the first three steps (receive the mission, issue a WARNORD, and make a tentative plan) of TLP occur in order. However, the sequence of subsequent steps is based on the situation. The tasks involved in some steps (for example, initiate movement and conduct reconnaissance) may occur several times. The last step, supervise and refine, occurs throughout.

10-6. A tension exists between executing current operations and planning for future operations. The small-unit leader must balance both. If engaged in a current operation, leaders have less time for TLP. If in a lull, transition, or an assembly area, leaders have more time to perform TLP thoroughly. In some situations, time constraints or other factors may prevent leaders from performing each step of TLP as thoroughly as they would like. For example, during the step make a tentative plan, small-unit leaders often develop only one acceptable course of action (COA) instead of multiple COAs. If time permits, leaders develop, compare, and analyze several COAs before deciding which one to execute.

10-7. Ideally, a battalion headquarters issues at least three WARNORDs to subordinates when conducting the MDMP as depicted in figure 10-1. WARNORDs are issued upon receipt of mission, completion of mission analysis, and when the commander approves a COA. However, the number of WARNORDs is not fixed. WARNORDs serve a function in planning similar to that of fragmentary orders (FRAGORDs) during execution. Commanders may issue a WARNORD whenever they need to disseminate additional planning information or initiate necessary preparatory action, such as movement or reconnaissance.

10-8. Leaders begin TLP when they receive the initial WARNORD or receive a new mission. As each subsequent order arrives, leaders modify their assessments, update tentative plans, and continue to supervise and assess preparations. In some situations, the higher headquarters may not issue the full sequence of WARNORDs; security considerations or tempo may make it impractical. Commanders carefully consider decisions to eliminate WARNORDs. Subordinate units always need to have enough information to plan and prepare for the operation. In other cases, leaders may initiate TLP before receiving a WARNORD based on existing plans and orders and on their understanding of the situation.

10-9. Parallel planning hinges on distributing information as it is received or developed. Leaders cannot complete their plans until they receive their unit mission. If each successive WARNORD contains enough information, the higher headquarters' final order will confirm what subordinate leaders have already analyzed and put into their tentative plans. In other cases, the higher headquarters' order may change or modify the subordinate's tasks enough that additional planning and reconnaissance are required.

STEPS OF TROOP LEADING PROCEDURES

10-10. TLP provide small-unit leaders a framework for planning and preparing for operations. TLP begin when the leader receives a mission and continues throughout the operations process (plan, prepare, execute and assess). TLP are a sequence of actions that assist leaders to effectively and efficiently use available time to issue orders and execute tactical operations.

10-11. TLP consist of eight steps. TLP are also supported by risk management. (See FM 5-19 for more information on risk management.) The sequence of the steps of TLP is not rigid. Leaders modify the sequence to meet the mission, situation, and available time. Some steps are done concurrently, while others may go on continuously throughout the operation:

- Step 1–Receive the mission.
- Step 2–Issue a warning order.
- Step 3–Make a tentative plan.
- Step 4–Initiate movement.
- Step 5–Conduct reconnaissance.
- Step 6–Complete the plan.
- Step 7–Issue the order.
- Step 8–Supervise and refine.

STEP 1–RECEIVE THE MISSION

10-12. Receive the mission may occur in several ways. It may begin when the initial WARNORD or OPORD arrives from higher headquarters or when a leader anticipates a new mission. Frequently, leaders receive a mission in a FRAGORD over the radio. Ideally, they receive a series of WARNORDs, the OPORD, and a briefing from their commander. Normally, after receiving an OPORD leaders give a confirmation brief to their higher commander to ensure they understand the higher commander's intent and concept of operations. The leader obtains clarification on any portions of the higher headquarters plan as required.

10-13. When they receive the mission, leaders perform an initial assessment of the situation (mission analysis) and allocate the time available for planning and preparation. (Preparation includes rehearsals and movement.) When a higher headquarters assigns a mission, it provides small-unit leaders an analysis of its operational environment. Often, higher headquarters will provide this assessment using the operational variables (political, military, economic, social, information, infrastructure, physical environment, and time [PMESII-PT]). From this higher level assessment, they can draw information relevant to their own operational environments and supplement it with their own knowledge of local conditions. During mission analysis, they filter relevant information into the categories of the mission variables (mission, enemy, terrain and weather, troops and support available, time available, and civil considerations [METT-TC]). (See appendix A for a more detailed description of PMESII-PT and METT-TC). This initial assessment and time allocation form the basis of their initial WARNORDs and addresses the factors of METT-TC. The order and detail in which leaders analyze the factors of METT-TC is flexible. It depends on the amount of information available and the relative importance of each factor. For example, leaders may concentrate on the mission, enemy, and terrain, leaving weather and civil considerations until they receive more detailed information.

10-14. Often, leaders do not receive their final unit mission until the WARNORD is disseminated after COA approval or after the OPORD. Effective leaders do not wait until their higher headquarters completes planning to begin their planning. Using all information available, leaders develop their unit mission as completely as they can. They focus on the mission, commander's intent, and concept of operations of their higher and next higher headquarters. They pick major tasks their unit will probably be assigned and develop a mission statement based on information they have received. At this stage, the mission may be incomplete. For example, an initial mission statement could be, "First platoon conducts an ambush in the next 24 hours." While not complete, this information allows subordinates to start preparations. Leaders complete a formal mission statement during TLP step 3 (make a tentative plan) and step 6 (complete the plan).

10-15.   Based on what they know, leaders estimate the time available to plan and prepare for the mission. They begin by identifying the times they must complete major planning and preparation events, including rehearsals. Reverse planning helps them do this. Leaders identify critical times specified by higher headquarters and work back from them, estimating how much time each event will consume. Critical times might include times to load aircraft, the line of departure, or the start point for movement.

10-16.   Leaders ensure that all subordinate echelons have sufficient time for their own planning and preparation needs. Generally, leaders at all levels use no more than one-third of the available time for planning and issuing the OPORD. Leaders allocate the remaining two-thirds of it to subordinates. Figure 10-2 illustrates a possible time schedule for an infantry company. The company adjusts the tentative schedule as necessary.

0600–Execute mission.

0530–Finalize or adjust the plan based on leader's reconnaissance.

0400–Establish the objective rallying point; begin leader reconnaissance.

0200–Begin movement.

2100–Conduct platoon inspections.

1900–Conduct rehearsals.

1800–Eat meals.

1745–Hold backbriefs (squad leaders to platoon leaders).

1630–Issue platoon OPORDs.

1500–Hold backbriefs (platoon leaders to company commander).

1330–Issue company OPORD.

1045–Conduct reconnaissance.

1030–Update company WARNORD.

1000–Receive battalion OPORD.

0900–Receive battalion WARNORD; issue company WARNORD.

| OPORD | operation order | WARNORD | warning order |

Figure 10-2. Sample schedule

STEP 2–ISSUE A WARNING ORDER

10-17.   As soon as leaders finish their initial assessment of the situation and available time, they issue a WARNORD. Leaders do not wait for more information. They issue the best WARNORD possible with the information at hand and update it as needed with additional WARNORDs.

10-18.   The WARNORD contains as much detail as possible. It informs subordinates of the unit mission and gives them the leader's timeline. Leaders may also pass on any other instructions or information they think will help subordinates prepare for the new mission. This includes information on the enemy, the nature of the higher headquarters' plan, and any specific instructions for preparing their units. The most important thing is that leaders not delay in issuing the initial WARNORD. As more information becomes available, leaders can—and should—issue additional WARNORDs. By issuing the initial WARNORD as quickly as possible, leaders enable their subordinates to begin their own planning and preparation.

10-19.   WARNORDs follow the five-paragraph OPORD format. (See appendix C for the WARNORD format.) Normally an initial WARNORD issued below battalion level includes—

- The mission or nature of the operation.
- The time and place for issuing the OPORD.
- Units or elements participating in the operation.
- Specific tasks not addressed by unit standard operating procedures (SOPs).
- The timeline for the operation.

STEP 3–MAKE A TENTATIVE PLAN

10-20. Once they have issued the initial WARNORD, leaders develop a tentative plan. This step combines the MDMP steps 2 through 6: mission analysis, COA development, COA analysis, COA comparison, and COA approval. At levels below battalion, these steps are less structured than for units with staffs. Often, leaders perform them mentally. They may include their principal subordinates—especially during COA development, analysis, and comparison. However, leaders—not their subordinates—select the COA on which to base the tentative plan.

Mission Analysis

10-21. To frame the tentative plan, leaders perform mission analysis. This mission analysis follows the METT-TC format, continuing the initial assessment performed in TLP step 1. (See table 10-1 for a brief description of the mission variables and appendix A for a more detailed description.)

Table 10-1. Mission variables

Variable	Description
Mission	Commanders and staffs view all of the mission variables in terms of their impact on mission accomplishment. The mission is the task, together with the purpose, that clearly indicates the action to be taken and the reason therefore. It is always the first variable commanders consider during decisionmaking. A mission statement contains the "who, what, when, where, and why" of the operation.
Enemy	The second variable to consider is the enemy—dispositions (including organization, strength, location, and tactical mobility), doctrine, equipment, capabilities, vulnerabilities, and probable courses of action.
Terrain and weather	Terrain and weather analysis are inseparable and directly influence each other's impact on military operations. Terrain includes natural features (such as rivers and mountains) and man-made features (such as cities, airfields, and bridges). Commanders analyze terrain using the five military aspects of terrain expressed in the memory aid OAKOC: observation and fields of fire, avenues of approach, key and decisive terrain, obstacles, cover and concealment. The military aspects of weather include visibility, wind, precipitation, cloud cover, temperature, and humidity.
Troops and support available	This variable includes the number, type, capabilities, and condition of available friendly troops and support. These include supplies, services, and support available from joint, host nation, and unified action partners. They also include support from civilians and contractors employed by military organizations, such as the Defense Logistics Agency and the Army Materiel Command.
Time available	Commanders assess the time available for planning, preparing, and executing tasks and operations. This includes the time required to assemble, deploy, and maneuver units in relationship to the enemy and conditions.
Civil considerations	*Civil considerations* are the influence of man-made infrastructure, civilian institutions, and activities of the civilian leaders, populations, and organizations within an area of operations on the conduct of military operations (ADRP 5-0). Civil considerations comprise six characteristics, expressed in the memory aid ASCOPE: areas, structures, capabilities, organizations, people, and events.

Course of Action Development

10-22. Mission analysis provides information needed to develop COAs. COA development aims to determine one or more ways to accomplish the mission. At lower echelons, the mission may be a single task. Most missions and tasks can be accomplished in more than one way. Normally, leaders develop two or more COAs. However, in a time-constrained environment, they may develop only one. Leaders do not wait for a complete order before beginning COA development. Usable COAs are suitable, feasible, acceptable, distinguishable, and complete. Leaders develop COAs as soon as they have enough information to do so. To develop COAs, leaders focus on the actions the unit takes at the objective and conduct a reverse plan to the starting point.

Analyze Relative Combat Power

10-23. During COA development, leaders determine whether the unit has enough combat power to defeat the force (or accomplish a task in stability or defense support of civil authorities tasks) against which it is arrayed by comparing the combat power of friendly and enemy forces. Leaders seek to determine where, when, and how friendly combat power (the elements of intelligence, movement and maneuver, fires, sustainment, protection, mission command, leadership, and information) can overwhelm the enemy. It is a particularly difficult process if the unit is fighting a dissimilar unit, for example, if an infantry unit is attacking or defending against an enemy mechanized force. Below battalion level, relative combat power comparisons are rough and generally rely on professional judgment instead of numerical analysis. When an enemy is not the object of a particular mission or tasks, leaders conduct a troop-to-task analysis to determine if they have enough combat power to accomplish the tasks. For example, a company commander assigned the task "establish civil control in town X" would need to determine if there were enough Soldiers and equipment (including vehicles and barrier materials) to establish the necessary check points and security stations within the town to control the population in town X.

Generate Options

10-24. Leaders brainstorm different ways to accomplish the mission. They determine the doctrinal requirements for the operation, including the tactical tasks normally assigned to subordinates. Doctrinal requirements give leaders a framework from which to develop COAs.

10-25. Next, leaders identify where and when the unit can mass overwhelming combat power to achieve specific results (with respect to enemy, terrain, time, or civil considerations) that accomplish the mission. Offensive and defensive tasks focus on the destructive effects of combat power. Stability tasks, on the other hand, emphasize constructive effects. Leaders identify any decisive points and determine what result they must achieve at the decisive points to accomplish the mission. This helps leaders determine the required tasks and the amount of combat power to apply at a decisive point.

10-26. After identifying tasks, leaders next determine the purpose for each task. There is normally one primary task for each mission. The unit assigned this task is the main effort. The other tasks should support the accomplishment of the primary task.

Develop an Initial Concept of Operations

10-27. The concept of operations describes how the leader envisions the operation unfolding from its start to its conclusion or end state. It determines how accomplishing each task leads to executing the next. It identifies the best ways to use available terrain and to employ unit strengths against enemy weaknesses. Fire support considerations make up an important part of the concept of operations. Planners identify essential stability tasks. Leaders develop the graphic control measures necessary to convey and enhance the understanding of the concept of operations, prevent fratricide, and clarify the task and purpose of the main effort.

Assign Responsibilities

10-28. Leaders assign responsibility for each task to a subordinate. Whenever possible, they depend on the existing chain of command. They avoid fracturing unit integrity unless the number of simultaneous tasks exceeds the number of available elements. Different command and support arrangements may be the distinguishing feature among COAs.

Prepare a Course of Action Statement and Sketch

10-29. Leaders base the COA statement on the concept of operations for that COA. The COA statement focuses on all significant actions, from the start of the COA to its finish. Whenever possible, leaders prepare a sketch showing each COA. It is useful to provide the amount of time it takes to achieve each movement and task in the COA sketch. This helps subordinate leaders gain an appreciation for how much time will pass as they execute each task of the COA. The COA contains the following information:

- Form of movement or defense to be used.
- Designation of the main effort.
- Tasks and purposes of subordinate units.
- Necessary sustaining operations.
- Desired end state.

10-30. Table 10-2 provides a sample mission statement and COA statement for an infantry company in the defense.

Table 10-2. Sample mission and course of action statements

Mission Statement:	B Co/1-31 IN defends NLT (not later than) 281700(Z) AUG 2005 from GL 375652 to GL 389650 to GL 394660 to GL 373665 to prevent the envelopment of A Co, the battalion main effort.
COA Statement:	The company defends with two platoons (PLTs) forward and one PLT in depth from PLT battle positions. The northern PLT (2 squads) destroys enemy forces to prevent enemy bypass of the main effort PLT on Hill 657. The southern PLT (3 squads, 2 Javelins) destroys enemy forces to prevent an organized company attack against the Co main effort on Hill 657. The main effort PLT (3 squads, 2 TOWS) retains Hill 657 (vicinity GL378659) to prevent the envelopment of Co A (battalion main effort) from the south. The anti-armor section (1 squad, 4 Javelins) establishes ambush positions at the road junction (vicinity GL 377653) to destroy enemy recon to deny observation of friendly defensive position and to prevent a concentration of combat power against the main effort PLT. The company mortars establish a mortar firing point vicinity GL 377664 to suppress enemy forces to protect the main effort platoon.

Analyze Courses of Action (War-game)

10-31. For each COA, leaders think through the operation from start to finish. They compare each COA with the enemy's most probable COA. At the small-unit level, the enemy's most probable COA is what the enemy is most likely to do given what friendly forces are doing at that instant. The leader visualizes a set of actions and reactions. The object is to determine what can go wrong and what decision the leader will likely have to make as a result.

Course of Action Comparison and Selection

10-32. Leaders compare COAs by weighing the advantages, disadvantages, strengths, and weaknesses of each, as noted during the war game. They decide which COA to execute based on this comparison and on their professional judgment. They take into account—

- Mission accomplishment.
- Time available to execute the operation.
- Risks.
- Results from unit reconnaissance.
- Subordinate unit tasks and purposes.
- Casualties incurred.
- Posturing of the force for future operations.

STEP 4–INITIATE MOVEMENT

10-33.   Leaders conduct any movement directed by higher headquarters or deemed necessary to continue mission preparation or position the unit for execution. They do this as soon as they have enough information to do so or the unit is required to move to position itself for a task. This is also essential when time is short. Movements may be to an assembly area, a battle position, a new area of operations, or an attack position. They may include movement of reconnaissance elements, guides, or quartering parties.

STEP 5–CONDUCT RECONNAISSANCE

10-34.   Whenever time and circumstances allow, or as directed by higher headquarters, leaders personally observe the area of operations for the mission prior to execution. No amount of information from higher headquarters can substitute for firsthand assessment of the mission variables from within the area of operations. Unfortunately, many factors can keep leaders from performing a personal reconnaissance. The minimum action necessary is a thorough map reconnaissance supplemented by imagery and intelligence products. As directed, subordinates or other elements (such as scouts) may conduct reconnaissance while the leader completes other TLP steps.

10-35.   Leaders use results of the war game to identify information requirements. Reconnaissance tasks seek to confirm or deny information that supports the tentative plan. They focus first on information gaps identified during mission analysis. Leaders ensure their leader's reconnaissance complements the higher headquarters' information collection plan. The unit may conduct additional reconnaissance tasks as the situation allows. This step may also precede making a tentative plan if commanders lack enough information to begin planning. Reconnaissance may be the only way to develop the information required for planning.

STEP 6–COMPLETE THE PLAN

10-36.   During this step, leaders incorporate the results of reconnaissance into their selected COA to complete the plan or order. This includes preparing overlays, refining the indirect fire target list, coordinating sustainment with signal requirements, and updating the tentative plan because of reconnaissance. At lower levels, this step may entail only confirming or updating information contained in the tentative plan. If time allows, leaders make final coordination with adjacent units and higher headquarters before issuing the order.

STEP 7–ISSUE THE ORDER

10-37.   Small-unit orders are normally issued verbally and supplemented by graphics and other control measures. An order follows the standard five-paragraph OPORD format. (See appendix C for the OPORD format.) Typically, leaders below company level do not issue a commander's intent. They reiterate the intent of their higher and next higher commanders.

10-38.   The ideal location for issuing the order is a point in the area of operations with a view of the objective and other aspects of the terrain. The leader may perform a leader's reconnaissance, complete the order, and then summon subordinates to a specified location to receive it. Sometimes security or other constraints make it impractical to issue the order on the terrain. Then leaders use a sand table, a detailed sketch, maps, and other products to depict the area of operations and the situation.

STEP 8–SUPERVISE AND REFINE

10-39.   Throughout TLP, leaders monitor mission preparations, refine the plan, coordinate with adjacent units, and supervise and assess preparations. Normally, unit SOPs state individual responsibilities and the sequence of preparation activities. To ensure the unit is ready for the mission, leaders supervise subordinates and inspect their personnel and equipment.

10-40.   A crucial component of preparation is the rehearsal. Rehearsals allow leaders to assess their subordinates' preparations. They may identify areas that require more supervision. Leaders conduct rehearsals to—

- Practice essential tasks.
- Identify weaknesses or problems in the plan.
- Coordinate subordinate element actions.
- Improve Soldier understanding of the concept of operations.
- Foster confidence among Soldiers.

10-41.   Company and smaller sized units use four types of rehearsals, and they are discussed in chapter 12:

- Backbrief.
- Combined arms rehearsal.
- Support rehearsal.
- Battle drill or SOP rehearsal.

This page intentionally left blank.

Chapter 11

Military Deception

This chapter provides information on military deception. Initially this chapter addresses the principles of military deception. It then discusses how commanders use military deception to shape the area of operations in support of decisive action. The chapter concludes with a discussion of how to plan, prepare, execute, and assess military deception.

MILITARY DECEPTION PROCESS AND CAPABILITY

11-1. Modern military deception is both a process and a capability. As a process, military deception is a methodical, information-based strategy that systematically, deliberately, and cognitively targets individual decisionmakers. The objective is the purposeful manipulation of decisionmaking. As a capability, military deception is useful to a commander when integrated early in the planning process as a component of the operation focused on causing an enemy to act or react in a desired manner. (See JP 3-13 for a discussion in information operations and JP 3-13.4 for a more detailed discussion on military deception.)

PRINCIPLES OF MILITARY DECEPTION

11-2. Military deception is applicable during any phase of military operations in order to create conditions to accomplish the commander's intent. The Army echelon that plans a military deception often determines its type. The levels of war define and clarify the relationship between strategic and tactical actions. The levels have no finite limits or boundaries. They correlate to specific levels of responsibility and military deception planning. They help organize thought and approaches to a problem. Decisions at one level always affect other levels. Common to all levels of military deception is a set of guiding principles:

- Focus on the target.
- Motivating the target to act.
- Centralized planning and control.
- Security.
- Conforming to the time available.
- Integration.

FOCUS ON THE TARGET

11-3. Leaders determine which targeted decisionmaker has the authority to make the desired decision and then can act or fail to act upon that decision. Many times it is one, key individual, or it could be a network of decisionmakers who rely on each other for different aspects of their mission or operation.

MOTIVATING THE TARGET TO ACT

11-4. Leaders determine what motivates the targeted decisionmaker and which information-related capabilities are capable of inducing the targeted decisionmaker to think a certain way. The desired result is that the targeted decisionmaker acts or fails to act as intended. This result is favorable to friendly forces. Often, the military objective is to manipulate the targeted decisionmaker's thinking and subsequent actions. This can be accomplished in a variety of ways. Leaders—

- Exploit target biases. Leaders provide the targeted decisionmakers with information that fulfills their expectations. This reinforces the target's preexisting perceptions and can be exceptionally powerful.
- Employ variety. The target should receive information, true and false, through multiple means and methods, from many angles, throughout the information and operational environment.
- Avoid windfalls. Important military information that is too easy to obtain is usually suspect. Information that "falls" into the enemy's hands must appear to be the result of legitimate collection activities.
- Leverage the truth. Any deception must conform to the target's perception of reality. It is much simpler to have the deception adhere to the target's belief than to make the target accept an unexpected reality as truth.

CENTRALIZED PLANNING AND CONTROL

11-5. Centralized planning and control ensures continuity. The assistant chief of staff, plans (G-5 [S-5]) usually leads the planning. However, there may be times when the commander designates a military deception officer to assist the G-5 (S-5) throughout the planning, hand-off, and termination of the deception operation. Centralizing the planning and control is imperative. It keeps the deception operation on track and limits unintended leaks and compromises.

SECURITY

11-6. Successful military deception requires strict security and protection measures to prevent compromise of both the deception and the actual operation. This includes counterintelligence, computer network defense, operations security, camouflage, and concealment. (See FM 5-19 for a discussion on risk management. See AR 530-1 for more detailed information and regulations on operations security.)

CONFORMING TO THE TIME AVAILABLE

11-7. Planning, preparing, executing, and assessing military deception must conform to the time available for both sides to "play their parts" in the deception. The targeted decisionmaker requires time to see, interpret, decide, and act upon the deception. Equally important, friendly forces require time to detect and assess the targeted decisionmaker's reaction to the deception.

INTEGRATION

11-8. A military deception is an integral part of the concept of an operation. It is not an afterthought or a stand-alone operation. The military deception officer assists the staff in integrating the deception operation throughout all phases of the operation. This begins with planning, the hand-off to current operations, and eventually the termination of the deception. Integration involves the use of information-related capabilities and activities. Military information support operations can contribute to the deception plan by providing a means to disseminate both accurate and deceptive information to the targeted decisionmakers by discreetly conveying approved tailored deception messages to selected target audiences. Therefore, the individual assigned as the military deception officer is often well versed in the use and integration information-related capabilities and activities.

MILITARY DECEPTION IN SUPPORT OF OPERATIONS

11-9. Military deception often relies on the basic understanding that the complexities and uncertainties of combat make decisionmakers susceptible to deception. The basic mechanism for any deception is either to increase or decrease the level of uncertainty, or ambiguity, in the mind of the deception target (or targeted decisionmaker). Military deception and deception in support of operations security present false or misleading information to the targeted decisionmaker with the deliberate intent to manipulate uncertainty. The aim of deception is to either increase or decrease the targeted decisionmaker's ambiguity in order to manipulate the target to perceive friendly motives, intentions, capabilities, and vulnerabilities erroneously and thereby alter the target's perception of reality.

AMBIGUITY-DECREASING DECEPTION

11-10. Ambiguity-decreasing deception reduces uncertainty and normally confirms the enemy decisionmaker's preconceived beliefs, so the decisionmaker becomes very certain about the selected course of action (COA). This type of deception presents false information that shapes the enemy decisionmaker's thinking, so the enemy makes and executes a specific decision that can be exploited by friendly forces. By making the wrong decision, which is the deception objective, the enemy could misemploy forces and provide friendly forces an operational advantage. For example, ambiguity-decreasing deceptions can present supporting elements of information concerning a specific enemy's COA. These deceptions are complex to plan and execute, but the potential rewards are often worth the increased effort and resources.

AMBIGUITY-INCREASING DECEPTION

11-11. Ambiguity-increasing deception presents false information aimed to confuse the enemy decisionmaker, thereby increasing the decisionmaker's uncertainty. This confusion can produce different results. Ambiguity-increasing deceptions can challenge the enemy's preconceived beliefs. These deceptions draw enemy attention from one set of activities to another, create the illusion of strength where weakness exists, create the illusion of weakness where strength exists, and accustom the enemy to particular patterns of activity that are exploitable at a later time. For example, ambiguity-increasing deceptions can cause the target to delay a decision until it is too late to prevent friendly mission success. They can place the target in a dilemma for which there is no acceptable solution. They may even prevent the target from taking any action at all. Deceptions in support of operations security (OPSEC) are typically executed as this type of deception.

TACTICAL DECEPTION

11-12. Most often, Army commanders will be faced with deciding when and where to employ military deception in support of tactical operations. The intent of tactical deception is to induce the enemy decisionmakers to act in a manner prejudicial to their interests. This is accomplished by either increasing or decreasing the ambiguity of the enemy decisionmaker through the manipulation, distortion, or falsification of evidence. Military deception undertaken at the tactical level supports engagements, battles, and stability tasks. This focus is what differentiates tactical deception from other forms of military deception. (See JP 3-13.4 for more information on military deception.)

STRATEGIC AND OPERATIONAL MILITARY DECEPTION

11-13. Less frequently, Army commanders will employ strategic and operational military deception to influence enemy strategic decisionmakers' abilities to successfully oppose U.S. national interests and goals or to influence enemy decisionmakers' abilities to conduct operations. These deceptions are joint or multinational efforts. In these cases, Army commanders usually opt to form a military deception cell to plan, coordinate, integrate, assess, and terminate the deception.

11-14. On occasion, Army commanders will employ deception in support of OPSEC. This is a military deception that protects friendly operations, personal, programs, equipment, and other assets against foreign intelligence security services collection. The intent of deception in support of OPSEC is to create multiple false indicators to confuse or make friendly intentions harder to interpret by foreign intelligence security services and other enemy intelligence gathering apparatus. This deception limits the ability of foreign intelligence security services to collect accurate intelligence on friendly forces. Deceptions in support of OPSEC are general in nature, and are not specifically targeted against particular enemy decisionmakers. Deceptions in support of OPSEC are instead used to protect friendly operations and forces by obscuring friendly capabilities, intentions, or vulnerabilities. (See chapter 14 for information on risk management and AR 530-1 for information and regulations on OPSEC.)

MILITARY DECEPTION TACTICS

11-15. The selection of military deception tactics and their use depends on an understanding of the current situation as well as the desired military deception goal and objective. (See appendix A for a discussion of operational and mission variables.) As a rule, Army commanders should be familiar with planning and conducting feints, ruses, demonstrations, and displays.

- A *feint,* in military deception, is an offensive action involving contact with the adversary conducted for the purpose of deceiving the adversary as to the location and/or time of the actual main offensive action (JP 3-13.4).
- A *ruse,* in military deception, is a trick of war designed to deceive the adversary, usually involving the deliberate exposure of false information to the adversary's intelligence collection system (JP 3-13.4).
- A *demonstration,* in military deception, is a show of force in an area where a decision is not sought that is made to deceive an adversary. It is similar to a feint but no actual contact with the adversary is intended (JP 3-13.4).
- A *display,* in military deception, is a static portrayal of an activity, force, or equipment intended to deceive the adversary's visual observation (JP 3-13.4).

COMMON MILITARY DECEPTION MEANS

11-16. Army commanders should also be familiar with some of the more commonly available military deception means that can be employed to support a given military deception. They cover the full scope of units, forces, personnel, capabilities, and resources available to the commander for the conduct of decisive action. In most cases, Army commanders have at their disposal the use of the following six information-related capabilities and other activities to support a planned military deception:

- Military information support operations.
- OPSEC.
- Camouflage, concealment and decoys.
- Cyber electromagnetic activities.
- Physical attack and destruction capabilities.
- Presence, posture, and profile.

MILITARY INFORMATION SUPPORT OPERATIONS

11-17. Dedicated military information support operations (MISO) assets have the ability to discretely convey intended information to the targeted decisionmaker via selected target audiences and appropriate key communicators. MISO assets can add additional fidelity to ruses, demonstrations, and displays.

OPERATIONS SECURITY

11-18. Military deception and OPSEC are complementary. They both seek to control the information available to the targeted decisionmaker. The intent is to protect indicators and deny information which could reveal the true operation. OPSEC measures do not expose the military deception while promoting and exposing those indicators and information supportive of the military deception. A deception in support of OPSEC uses false information about friendly forces' intentions, capabilities, or vulnerabilities to shape the enemy's perceptions. It targets the enemy's intelligence, surveillance, and reconnaissance capabilities to distract the enemy's intelligence collection away from, or provide cover for, unit operations. A deception in support of OPSEC is a relatively easy form of deception to use and is very appropriate for use at battalion-level and below. To be successful, a balance must be achieved between OPSEC and military deception requirements.

CAMOUFLAGE, CONCEALMENT, AND DECOYS

11-19. Camouflage, concealment, and decoy activities are normally individual or unit responsibilities and governed by standard operating procedures (SOPs). They can also play a role in a larger military deception or deception in support of OPSEC where camouflage, concealment, and decoys comprise just a few of many elements that mislead the enemy's intelligence, surveillance, and reconnaissance capabilities. Merely hiding forces may not be adequate, as the enemy may need to "see" these forces elsewhere. In such cases, cover and concealment can hide the presence of friendly forces, but decoy placement should be coordinated as part of the deception in support of OPSEC.

CYBER ELECTROMAGNETIC ACTIVITIES

11-20.   Commanders exploit cyberspace and the electromagnetic spectrum for deception purposes. Cyber electromagnetic activities can be used to show friendly intentions and to shape perceptions of friendly actions. Cyber electromagnetic activities can add fidelity and believability to feints, ruses, demonstrations, and displays.

PHYSICAL ATTACK AND DESTRUCTION CAPABILITIES

11-21.   Nothing is perhaps more effective at shaping an enemy' perceptions than the attack and destruction of enemy assets, units, resources, and capabilities. When used to support a military deception, fires and physical attacks (feints, demonstrations, and displays) can exploit perceptions and biases as to where the enemy believes the friendly decisive operation will be committed.

SUSTAINMENT CAPABILITIES

11-22.   Many times sustainment operations are much more visible than combat preparations, and become a key indicator of when, where, and how combat operations will be conducted. When linked with fires and physical attacks, sustainment operations used in support of military deception seek to confirm the targeted decisionmaker's perceptions and biases as to where friendly forces will commit decisive operations.

MILITARY DECEPTION IN THE OPERATIONS PROCESS

11-23.   Military deception is considered in all activities of the operations process. Planning, preparing, executing, and continually assessing military deception does not take place in isolation. It occurs simultaneously with the operations process. If it does not, then the risk increases exponentially for the military deception to be under resourced and not integrated into the larger operation as the military deception evolves. It is unlikely that an under resourced and nonintegrated military deception will succeed. Because military deception supports a range of missions, and to prevent one unit's military deception from compromising another unit's operations, leaders coordinate military deceptions both laterally and vertically. Deception operations are approved by the headquarters two operational echelons higher than the originating command. Only two authorities can direct a military deception: a higher headquarters and the originating unit commander.

PLANNING

11-24.   Planning develops the information needed to prepare, execute, and assess a military deception. The output of the military deception mission analysis is the running estimate, prepared by the military deception officer. The running estimate identifies military deception opportunities, information and capability requirements, and recommends feasible deception goals and objectives. The military deception officer presents this estimate during the mission analysis briefing. The estimate considers current capabilities based on enemy susceptibilities, preconceptions, and biases; available time; and available military deception means. A key outcome of the running estimate is the determination of whether or not there is a viable military deception opportunity. (See chapter 8 for more information on running estimates.) Military deception may be a feasible option, if it is appropriate to the mission, and if there is a possibility of success. Issues to consider when determining if military deception is a viable course of action include:

- Availability of assets.
- Understanding the military deception target.
- Suitability.
- Time.

Chapter 11

Availability of Assets

11-25. The commander determines if sufficient assets exist to support both the operation and the military deception. There are few assets specifically designed and designated for military deception. This means the commander must shift assets from the operation to support the military deception. Commanders must be certain that shifting assets to support a military deception does not adversely affect the operation or prevent mission success.

Understanding the Military Deception Target

11-26. The commander determines if sufficient information exists on how the military deception target acquires information and makes decisions, what knowledge the target has of the situation, and how the target views the friendly force. The commander also determines if sufficient information exists to reveal the targeted decisionmaker's biases, beliefs, and fears. If necessary, the staff can make assumptions about the military deception target, but it must avoid mirror imaging its preconceptions onto the military deception's targeted decisionmaker.

Suitability

11-27. Some missions are better suited to military deception than others. When a unit has the initiative and has some control over the area of operations, then military deception is more suitable.

Time

11-28. The commander determines if sufficient time exists to execute a military deception. Execution of the military deception must provide sufficient time for the military deception target to observe the military deception activities, form the desired perceptions, and act in a manner consistent with the deception objective.

Military Deception Planning Steps

11-29. The basic steps of military deception planning come together during COA analysis, comparison, and approval and are overseen by the military deception officer. (These are MDMP steps 2, 3, and 4. See chapter 9 for a detailed discussion of the MDMP.) The G-5 (S-5)-developed COAs provide the basis for military deception COAs. The military deception officer develops military deception COAs in conjunction with the G-5 (S-5). Basing the military deception COAs on the operational COAs ensures deception COAs are feasible, practical, and nested and effectively support the operational COAs.

11-30. The military deception officer and G-5(S-5) planners consider the military deception COAs as the staff war-games the COAs. They analyze the strengths and weaknesses of each military deception COA and compare it against the criteria established by the military deception officer for evaluating the military deception COAs.

11-31. The military deception officer, working with the G-5 (S-5) planners, prepares the military deception plan after the commander approves the military deception COA. Once the G-5 (S-5) planner completes, coordinates, and reviews the military deception for consistency, it is presented to the commander for tentative approval. To ensure synchronization of military deception at all levels, approval authority for military deception resides two echelons above the originating command. After the approving authority has approved the military deception plan, it becomes a part of the operation plan (OPLAN) or operation order (OPORD). It is important that military deception plans are not widely distributed. In order to ensure every opportunity to succeed and to protect the military deception from compromise, access to the military deception operation is strictly limited to those with a need to know.

11-32. The military deception officer ensures that each military deception plan is properly constructed. There are ten steps in military deception planning:
- Step 1—Determine the military deception goal.
- Step 2—Determine the deception objective.
- Step 3—Identify the military deception target.

- Step 4—Identify required perceptions of the military deception target.
- Step 5—Develop the military deception story.
- Step 6—Identify the military deception means.
- Step 7—Develop military deception events.
- Step 8—Develop OPSEC and other protection measures.
- Step 9—Develop assessment criteria.
- Step 10—Develop a termination plan.

Step 1—Determine the Military Deception Goal

11-33.   The military deception goal is the desired contribution of the military deception to friendly mission success. The military deception goal is often expressed in terms of the desired optimal situation under which the commander wants to conduct the primary operation. The military deception goal is usually recommended in the running estimate and confirmed by the commander's planning guidance at the conclusion of mission analysis. Alternatively, the commander can identify the military deception role and leave it to the staff to identify desired military deception actions.

Step 2—Determine the Deception Objective

11-34.   The military deception objective is the purpose of the military deception expressed in terms of what the enemy is to do or not to do at the critical time and location. Like the military deception goal, the military deception objective is also recommended in the running estimate and confirmed by the commander in the commander's planning guidance at the conclusion of mission analysis.

Step 3—Identify the Military Deception Target

11-35.   The military deception target is the enemy decisionmaker or a select set of decisionmakers with the authority to make the decision that will achieve the deception objective.

Step 4—Identify Required Perceptions of the Military Deception Target

11-36.   The military deception target perceptions are what the military deception target must believe in order to make the decision that will achieve the deception objective. This perception of friendly force actions is based on the deception objective and exploits the military deception target's information processing cycle. This includes the supporting information and network enabled systems, decisionmaking processes, beliefs, biases, and preconceptions regarding friendly forces and the situation. It is often more effective to tell the military deception target what the target wants to believe than it is to convince the target of something different.

Step 5—Develop the Military Deception Story

11-37.   The military deception story is a plausible, but essentially false, view of the situation that leads the military deception target to act in a manner that accomplishes the military deception objective. It weaves military deception events together into a coherent whole that describes the situation that the commander wants the military deception target to perceive. If the military deception target is to develop the desired perceptions, the military deception story must be believable, verifiable, and consistent. The story must be doctrinally correct for the situation. Ideally, the military deception target should form the exact mental picture projected by the military deception story as the military deception unfolds. To develop the military deception story, the military deception officer thinks about how the target sees the situation and then writes the story from the target's perspective. An example would be to write the story similar to the military deception target's own intelligence estimate. The military deception story is based upon what the military deception target believes and understands already and the evidence or observables (friendly force actions, units, and real or fake resources) that reinforce the military deception target's beliefs and understanding.

Step 6—Identify the Military Deception Means

11-38.   The military deception means are the methods, resources, and techniques used to create required observables (things the military deception target needs to see in order to deduce the desired perceptions) and act out the military deception story. The nature of the desired perception, with the indicators needed to convey the perception to the deception target, determines the deception means employed. Physical means are observable physical activities of forces, systems, and individuals that present visual indicators. Technical means could include cyber-based messaging and information sharing venues, smart phone and mobile wireless communications, radio broadcasts, radar emissions, and electromagnetic deception. Administrative means are used to convey oral, pictorial, documentary, or other material evidence to the deception target. While there may be many means available, the means employed must be consistent.

Step 7—Develop Military Deception Events

11-39.   The military deception events are the activities conducted through military deception means at a specific time and location to convey the military deception story to the target. To convey the military deception story, the events must be observed and sensed by the enemy. To determine this, the military deception officer pairs up military deception means with the enemy's intelligence collection system capabilities. If the enemy intelligence system can "see" the military deception event, then it can collect the information it needs to piece together the military deception story. The systematic, yet seemingly random, projection of deception story elements by multiple means also makes the deception more believable. The military deception officer must also take care to ensure that information reaching the enemy appears as legitimately collected. Important military information that is too easy to obtain is usually suspect.

Step 8—Develop Operations Security and Other Protection Measures

11-40.   OPSEC and other protection measures are employed with military deception in order to ensure that only the desired military deception events reach the enemy and that actions in support of the supported operations are concealed. False indicators are wrapped in significant amounts of factual information to enhance their acceptance but not to compromise the supported operation. Without OPSEC, the deceptive activities may not convince the enemy to believe the military deception story if the preparations for the supported operation are also observable. Equally important is risk assessment. All military deception involves risk and cost. Commanders base the decision to conduct a military deception on a deliberate assessment that weighs costs against benefits. Risk can be mitigated by ensuring the success of the supported operation does not hinge upon the success of the military deception, anticipating conditions that could compromise the military deception, and developing responses in the event of unintended effects. (See chapter 14 for information on risk management and a discussion on risk management and AR 530-1 for information and regulations on operations security.)

Step 9—Develop Assessment Criteria

11-41.   Commanders and staffs focus assessment efforts by developing criteria and feedback mechanisms that they use to assess the progress of the military deception. In particular, early and frequent coordination with the assistant chief of staff, intelligence (G-2 [S-2]) is important. The commander and staff monitor feedback and compare it against the measures of effectiveness (MOEs) established for the operation. Feedback comes in the form of information that reveals how the military deception target is responding to the military deception story and if the plan is working. Assessment efforts focus on two types of military deception feedback:

- Target feedback—information, analytical determinations and evidence (MOEs) that the target is receiving and acting on the military deception.
- Conduit feedback—information and evidence (MOEs) that the conduits are receiving, processing, and transmitting elements of the military deception.

Ideally, there will be indicators of whether the target is receiving the military deception story as planned, and if the target is acting in accordance with the military deception objective. (See chapter 15 for more information on assessment.)

Step 10—Develop a Termination Plan

11-42. Military deception does not just simply end. It must be guided by a commander-approved termination plan that in essence represents a coherent, structured, and implementable exit strategy. This is important because the commander terminates a military deception after it meets its objective. Like the military deception story, the exit strategy must also be believable and consistent with friendly operational profiles. Additionally, the enemy should not know what deception means, techniques, and events were used. Otherwise, the next deception operation may not have the desired effect due to the enemy gaining insights into friendly tactics, techniques, and procedures.

PREPARING

11-43. During preparation, commanders take every opportunity to refine the military deception plan based on updated intelligence and friendly information. OPSEC activities also continue during preparation for the military deception. OPSEC is a dynamic effort that anticipates and reacts to enemy collection efforts.

11-44. Military deception plans are not static and are continually adjusted. The military deception officer normally moves with the military deception plan from the G-5 to the future operations integrating cell to oversee the refinement of the plan and ensure it is fully integrated with the operation. As assumptions prove true or false, enemy perceptions are confirmed, or the status of friendly units change, the military deception officer adjusts the military deception for the commander, or recommends aborting it if the military deception can no longer significantly influence the situation and achieve the military deception goal.

EXECUTING

11-45. Execution takes place in a dynamic environment and as part of the operation. The commander, assisted by the military deception officer, continually assesses and refines the military deception as it unfolds during execution. Consequently, the military deception officer must move from the future operations integrating cell to the current operations integrating cell with the military deception plan in order to direct the military deception operation and its termination.

11-46. Terminating a military deception is the final execution decision. When the decision to terminate is made, the appropriate termination branch or sequel becomes the basis for execution of a deliberate series of events designed to end the military deception while protecting its existence and the means and techniques employed to execute it.

ASSESSING

11-47. Assessment is the continuous monitoring—throughout planning, preparation, and execution—and evaluation of the current situation to measure the overall effectiveness of the operations (see ADRP 5-0 and chapter 15). This involves receiving information about the implementation of the military deception and evaluating it against established MOEs. It also includes continual reassessment of the military deception objective, target, story, and events to ensure they are still important to the accomplishment of the mission objectives. There are four types of assessments conducted during a military deception:

- Monitoring and evaluating the military deception to ensure it continues to support the supported operations.
- Evaluating how the target is acting or not acting in response to the military deception story.
- Monitoring for unintended consequences resulting from the military deception.
- Determining when termination criteria are met.

11-48. Commanders continually assess military deception events. A military deception's effectiveness is directly related to the validity of the projected situation when the supported operation starts. Validating this projection with updated information is essential to any assessment. Such assessment is necessary to determine when to commence, modify, or terminate the military deception. (See chapter 15 for more information on assessment.)

This page intentionally left blank.

Chapter 12

Rehearsals

Rehearsing key actions before execution allows Soldiers to become familiar with the operation and translate the abstract ideas of the written plan into concrete actions. This chapter describes types of rehearsals. It then lists the responsibilities of those involved. It also contains guidelines for conducting rehearsals.

REHEARSAL BASICS

12-1. Rehearsals allow leaders and their Soldiers to practice key aspects of the concept of operations. These actions help Soldiers orient themselves to their environment and other units before executing the operation. Rehearsals help Soldiers build a lasting mental picture of the sequence of key actions within the operation.

12-2. Rehearsals are the commander's tool to ensure staffs and subordinates understand the commander's intent and the concept of operations. They allow commanders and staffs to identify shortcomings in the plan not previously recognized. Rehearsals also contribute to external and internal coordination, as the staff identifies additional coordinating requirements.

12-3. Effective and efficient units habitually rehearse during training. Commanders at every level routinely train and practice various rehearsal types. Local standard operating procedures (SOPs) identify appropriate rehearsal types and standards for their execution. All leaders conduct periodic after action reviews to ensure their units conduct rehearsals to standard and correct substandard performances. After action reviews also enable leaders to incorporate lessons learned into existing plans and orders, or into subsequent rehearsals.

12-4. Adequate time is essential when conducting rehearsals. The time required varies with the complexity of the mission, the type and technique of rehearsal, and the level of participation. Units conduct rehearsals at the lowest possible level, using the most thorough technique possible, given the time available. Under time-constrained conditions, leaders conduct abbreviated rehearsals, focusing on critical events determined by reverse planning. Each unit will have different critical events based on the mission, unit readiness, and the commander's assessment.

12-5. The rehearsal is a coordination event, not an analysis. It does not replace war-gaming. Commanders war-game during the military decisionmaking process (MDMP) to analyze different courses of action to determine the optimal one. Rehearsals practice that selected course of action. Commanders avoid making major changes to operation orders (OPORDs) during rehearsals. They make only those changes essential to mission success and risk mitigation.

REHEARSAL TYPES

12-6. Each rehearsal type achieves a different result and has a specific place in the preparation timeline. The four types of rehearsals are the—
- Backbrief.
- Combined arms rehearsal.
- Support rehearsal.
- Battle drill or SOP rehearsal.

BACKBRIEF

12-7. A *backbrief* **is a briefing by subordinates to the commander to review how subordinates intend to accomplish their mission**. Normally, subordinates perform backbriefs throughout preparation. These briefs allow commanders to clarify the commander's intent early in subordinate planning. Commanders use the backbrief to identify any problems in the concept of operations.

12-8. The backbrief differs from the confirmation brief (a briefing subordinates give their higher commander immediately following receipt of an order) in that subordinate leaders are given time to complete their plan. Backbriefs require the fewest resources and are often the only option under time-constrained conditions. Subordinate leaders explain their actions from the start to the finish of the mission. Backbriefs are performed sequentially, with all leaders reviewing their tasks. When time is available, backbriefs can be combined with other types of rehearsals. Doing this lets all subordinate leaders coordinate their plans before performing more elaborate drills.

COMBINED ARMS REHEARSAL

12-9. A combined arms rehearsal is a rehearsal in which subordinate units synchronize their plans with each other. A maneuver unit headquarters normally executes a combined arms rehearsal after subordinate units issue their OPORD. This rehearsal type helps ensure that subordinate commanders' plans achieve the higher commander's intent.

SUPPORT REHEARSAL

12-10. The support rehearsal helps synchronize each warfighting function with the overall operation. This rehearsal supports the operation so units can accomplish their missions. Throughout preparation, units conduct support rehearsals within the framework of a single or limited number of warfighting functions. These rehearsals typically involve coordination and procedure drills for aviation, fires, engineer support, or casualty evacuation. Support rehearsals and combined arms rehearsals complement preparations for the operation. Units may conduct rehearsals separately and then combine them into full-dress rehearsals. Although these rehearsals differ slightly by warfighting function, they achieve the same result.

BATTLE DRILL OR STANDARD OPERATING PROCEDURE REHEARSAL

12-11. A battle drill is a collective action rapidly executed without applying a deliberate decisionmaking process. A battle drill or SOP rehearsal ensures that all participants understand a technique or a specific set of procedures. Throughout preparation, units and staffs rehearse battle drills and SOPs. These rehearsals do not need a completed order from higher headquarters. Leaders place priority on those drills or actions they anticipate occurring during the operation. For example, a transportation platoon may rehearse a battle drill on reacting to an ambush while waiting to begin movement.

12-12. All echelons use these rehearsal types; however, they are most common for platoons, squads, and sections. They are conducted throughout preparation and are not limited to published battle drills. All echelons can rehearse such actions as a command post shift change, an obstacle breach lane-marking SOP, or a refuel-on-the-move site operation.

METHODS OF REHEARSAL

12-13. Methods for conducting rehearsals are limited only by the commander's imagination and available resources. Several methods are illustrated in figure 12-1. Resources required for each method range from broad to narrow. As listed from left to right, each successive method takes more time and more resources. Each rehearsal method also imparts a different level of understanding to participants.

12-14. Paragraphs 12-15 through 12-51 address these implications for each method:
- Time—the amount of time required to conduct (plan, prepare, execute, and assess) the rehearsal.
- Echelons involved—the number of echelons that can participate in the rehearsal.
- Operations security (OPSEC) risks—the ease by which an enemy can exploit friendly actions from the rehearsal.
- Terrain—the amount of space needed for the rehearsal.

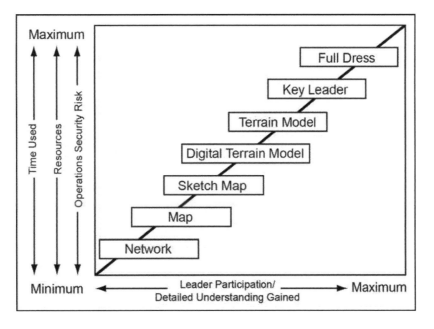

Figure 12-1. Types of rehearsals

FULL-DRESS REHEARSAL

12-15.   A full-dress rehearsal produces the most detailed understanding of the operation. It includes every participating Soldier and system. Leaders conduct the rehearsal on terrain similar to the area of operations, initially under good light conditions, and then in limited visibility. Leaders repeat small-unit actions until units execute them to standard. A full-dress rehearsal helps Soldiers clearly understand what commanders expect of them. It helps them gain confidence in their ability to accomplish the mission. Supporting elements, such as aviation crews, meet and rehearse with Soldiers to synchronize the operation.

12-16.   A unit may conduct full-dress rehearsals. The higher headquarters may conduct and support full-dress rehearsals. The full-dress rehearsal is most difficult to accomplish at higher echelons. At those levels, commanders may develop an alternate rehearsal plan that mirrors the actual plan but fits the terrain available for the rehearsal.

12-17.   Full-dress rehearsals consume more time than any other rehearsal type. For companies and smaller units, full-dress rehearsals most effectively ensure all units in the operation understand their roles. However, brigade and task force commanders consider how much time their subordinates need to plan and prepare when deciding whether to conduct a full-dress rehearsal.

12-18.   All echelons involved in the operation participate in the full-dress rehearsal.

12-19.   Moving a large part of the force may create an OPSEC risk by attracting unwanted enemy attention. Commanders develop a plan to protect the rehearsal from enemy information collection. Sometimes they develop an alternate plan, including graphics and radio frequencies, that rehearses selected actions without compromising the actual OPORD. Commanders take care not to confuse subordinates when doing this.

12-20.   Terrain management for a full-dress rehearsal is challenging. Units identify, secure, clear, and maintain the rehearsal area throughout the rehearsal.

KEY LEADER REHEARSAL

12-21.   Circumstances may prohibit a rehearsal with all members of the unit. A key leader rehearsal involves only key leaders of the organization and its subordinate units. It normally takes fewer resources than a full-dress rehearsal. Terrain requirements mirror those of a full-dress rehearsal, even though fewer

Soldiers participate. The commander first decides the level of leader involvement. Then the selected leaders rehearse the plan while traversing the actual or similar terrain. Often commanders use this technique to rehearse fire control measures for an engagement area during defensive tasks. Commanders often use a key leader rehearsal to prepare key leaders for a full-dress rehearsal. The key leader rehearsal may require developing a rehearsal plan that mirrors the actual plan but fits the terrain of the rehearsal.

12-22. Often, small-scale replicas of terrain or buildings substitute for the actual area of operations. Leaders not only explain their plans, but also walk through their actions or move replicas across the rehearsal area or sand table. This is called a rock drill. It reinforces the backbrief given by subordinates, since everyone can see the concept of operations and sequence of tasks.

12-23. A key leader rehearsal normally requires less time than a full-dress rehearsal. Commanders consider how much time their subordinates need to plan and prepare when deciding whether to conduct a reduced-force rehearsal.

12-24. A small unit from the echelons involved can perform a full-dress rehearsal as part of a larger organization's key leader rehearsal.

12-25. A key leader rehearsal is less likely to present OPSEC risks than a full-dress rehearsal because it has fewer participants. However, it requires the same number of radio transmissions as for a full-dress rehearsal.

12-26. Terrain management for the key leader rehearsal can be as difficult as for the full-dress rehearsal. Units identify, secure, clear, and maintain the rehearsal area throughout the rehearsal.

TERRAIN-MODEL REHEARSAL

12-27. The terrain-model rehearsal is the most popular rehearsal method. It takes less time and fewer resources than a full-dress or reduced-force rehearsal. An accurately constructed terrain model helps subordinate leaders visualize the commander's intent and concept of operations. When possible, commanders place the terrain model where it overlooks the actual terrain of the area of operations. The model's orientation coincides with that of the terrain. The size of the terrain model can vary from small (using markers to represent units) to large (on which the participants can walk). A large model helps reinforce the participants' perception of unit positions on the terrain.

12-28. Often, constructing the terrain model consumes the most time during this technique. Units require a clear SOP that states how to build the model so it is accurate, large, and detailed enough to conduct the rehearsal. A good SOP also establishes staff responsibility for building the terrain model and a timeline for its completion.

12-29. Because a terrain model is geared to the echelon conducting the rehearsal, multi-echelon rehearsals using this technique are difficult.

12-30. This rehearsal can present OPSEC risks if the area around the rehearsal site is not secured. Assembled commanders and their vehicles can draw enemy attention. Units must sanitize the terrain model after completing the rehearsal.

12-31. Terrain management is less difficult than with the previous rehearsal types. A good site is easy for participants to find, yet it is concealed from the enemy. An optimal location overlooks the terrain where the unit will execute the operation.

DIGITAL TERRAIN-MODEL REHEARSAL

12-32. Digital terrain models are virtual representations of the area of operations. Units drape high-resolution imagery over elevation data thereby creating a fly-through or walk-through. Holographic imagery produces the view in three dimensions. Often, the model hot links graphics, detailed information, unmanned aircraft systems, and ground imagery to key points providing more insight into the plan. The unit geospatial engineers or imagery analysts can assist in digital model creation. Detailed city models already exist for many world cities.

12-33.   The time it takes to create the digital three-dimensional model depends on the amount of available data on the terrain being modeled.

12-34.   Of all the echelons involved, this type of rehearsal best suits small units, although with a good local area network, a wider audience can view the graphics. All echelons may be provided copies of the digital model to take back to their headquarters for a more detailed examination.

12-35.   If not placed on a computer network, there is limited OPSEC risk because it does not use a large physical site that requires securing and leaders can conduct the rehearsal under cover. However, if placed on a computer network, digital terrain models can be subject to enemy exploitation due to inherent vulnerabilities of networks.

12-36.   This space requires the least terrain of all rehearsals. Using tents or enclosed areas conceals the rehearsal from the enemy.

SKETCH-MAP REHEARSAL

12-37.   Commanders can use the sketch-map technique almost anywhere, day or night. The procedures are the same as for a terrain-model rehearsal except the commander uses a sketch map in place of a terrain model. Large sketches ensure all participants can see as each participant walks through execution of the operation. Participants move markers on the sketch to represent unit locations and maneuvers.

12-38.   Sketch-map rehearsals take less time than terrain-model rehearsals and more time than map rehearsals.

12-39.   Units tailor a sketch map to the echelon conducting the rehearsal. Multi-echelon rehearsals using this technique are difficult.

12-40.   This rehearsal can present OPSEC risks, if the area around the rehearsal site is not secured. Assembled commanders and their vehicles can draw enemy attention. Units must sanitize, secure, or destroy the sketch map after use.

12-41.   This technique requires less terrain than a terrain-model rehearsal. A good site ensures participants can easily find it yet stay concealed from the enemy. An optimal location overlooks the terrain where the unit will execute the operation.

MAP REHEARSAL

12-42.   A map rehearsal is similar to a sketch-map rehearsal except the commander uses a map and operation overlay of the same scale used to plan the operation.

12-43.   The map rehearsal itself consumes the most time. A map rehearsal is normally the easiest technique to set up since it requires only maps and graphics for current operations.

12-44.   Units tailor a map rehearsal's operation overlay to the echelon conducting the rehearsal. Multi-echelon rehearsals using this technique are difficult.

12-45.   This rehearsal can present OPSEC risks, if the area around the rehearsal site is not secured. Assembled commanders and their vehicles can draw enemy attention.

12-46.   This technique requires the least terrain of all rehearsals. A good site ensures participants can easily find it yet stay concealed from the enemy. An optimal location overlooks the terrain where the unit will execute the operation.

NETWORK REHEARSAL

12-47.   Units conduct network rehearsals over wide-area networks or local area networks. Commanders and staffs practice these rehearsals by talking through critical portions of the operation over communications networks in a sequence the commander establishes. The organization rehearses only the critical parts of the operation. These rehearsals require all information systems needed to execute that portion of the operation. All participants require working information systems, the OPORD, and graphics. Command posts can rehearse battle tracking during network rehearsals.

12-48. This technique can be time efficient, if units provide clear SOPs. However, if the organization has unclear SOPs, has units not operating on the network, or has units without working communications, this technique can be time-consuming.

12-49. This technique lends itself to multi-echelon rehearsals. Participation is limited only by the commander's intent and the capabilities of the command's information systems.

12-50. If a unit executes a network rehearsal from current unit locations, the OPSEC risk may increase. The enemy may monitor the increased volume of transmissions and potentially compromise information. To avoid such compromise, organizations use different frequencies from those planned for the operation. Using wire systems is an option, but this does not exercise the network systems, which is the strong point of this technique.

12-51. If a network rehearsal is executed from unit locations, terrain considerations are minimal. If a separate rehearsal area is required, considerations are similar to those of a reduced-force rehearsal.

REHEARSAL RESPONSIBILITIES

12-52. This discussion addresses responsibilities for conducting rehearsals based on the combined arms rehearsal. Responsibilities are similar for other types of rehearsals.

REHEARSAL PLANNING

12-53. Commanders and chiefs of staff (executive officers at lower echelons) plan rehearsals.

Commander

12-54. Commanders provide certain information as part of the commander's guidance during the initial mission analysis. They may revise the following information when they select a course of action:

- Rehearsal type.
- Rehearsal technique.
- Location.
- Attendees.
- Enemy course of action to be portrayed.

Chief of Staff (Executive Officer)

12-55. The chief of staff (executive officer) (COS [XO]) ensures all rehearsals are included in the organization's time-management SOP. The COS (XO) responsibilities include—

- Publishing the rehearsal time and location in the OPORD or warning order (WARNORD).
- Conducting staff rehearsals.
- Determining rehearsal products, based on type, technique, and mission variables.
- Coordinating liaison officer attendance from adjacent units.

REHEARSAL PREPARATION

12-56. Everyone involved in executing or supporting the rehearsal has responsibilities during preparation.

Commander

12-57. Commanders prepare to rehearse operations with events phased in proper order, from start to finish. Under time-constrained conditions, this often proves difficult. Commanders—

- Identify and prioritize key events to rehearse.
- Allocate time for each event.
- Perform personal preparation, including reviews of—

- Task organization completeness.
- Personnel and materiel readiness.
- Organizational level of preparation.

Chief of Staff (Executive Officer)

12-58.   The COS (XO) through war-gaming and coordination with the commander—
- Prepares to serve as the rehearsal director.
- Coordinates time for key events requiring rehearsal.
- Establishes rehearsal time limits per the commander's guidance and mission variables.
- Verifies rehearsal site preparation. A separate rehearsal site may be required for some events, such as a possible obstacle site. A good rehearsal site includes—
  - Appropriate markings and associated training aids.
  - Parking areas.
  - Local security.
- Determines the method for controlling the rehearsal and ensuring its logical flow, such as a script.

Subordinate Leaders

12-59.   Subordinate leaders complete their planning. This planning includes—
- Completing unit OPORDs.
- Identifying issues derived from the higher headquarters' OPORD.
- Providing a copy of their unit OPORD with graphics to the higher headquarters.
- Performing personal preparation similar to that of the commander.
- Ensuring they and their subordinates bring all necessary equipment.

Conducting Headquarters Staff

12-60.   Conducting headquarters staff members—
- Develop an OPORD with necessary overlays.
- Deconflict all subordinate unit graphics. Composite overlays are the first step for leaders to visualize the organization's overall plan.
- Publish composite overlays at the rehearsal, including, at a minimum—
  - Movement and maneuver.
  - Intelligence.
  - Fires.
  - Sustainment.
  - Signal operations.
  - Protection.

REHEARSAL EXECUTION

12-61.   During the rehearsal execution, the commander, COS (XO), assistants, subordinate leaders, recorder, and staff from the conducting headquarters have specific responsibilities.

Commander

12-62.   Commanders command the rehearsal just as they will command the operation. They maintain the focus and level of intensity, allowing no potential for subordinate confusion. Although the staff refines the OPORD, it belongs to the commander. The commander uses the order to conduct operations. An effective rehearsal is not a commander's brief to subordinates. It validates synchronization—the what, when, and where—of tasks that subordinate units will perform to execute the operation and achieve the commander's intent.

Chief of Staff (Executive Officer)

12-63. Normally, the COS (XO) serves as the rehearsal director. This officer ensures each unit will accomplish its tasks at the right time and cues the commander to upcoming decisions. The chief of staff's (executive officer's) script is the execution matrix and the decision support template. The COS (XO) as the rehearsal director—

- Starts the rehearsal on time.
- Has a formal roll call.
- Ensures everyone brings the necessary equipment, including organizational graphics and previously issued orders.
- Validates the task organization. Linkups must be complete or on schedule, and required materiel and personnel must be on hand. The importance of this simple check cannot be overemphasized.
- Ensures synchronization of the operational framework being used—deep-close-security, decisive-shaping-sustaining, or main and supporting efforts. (See ADRP 3-0 for more information on the three operational frameworks.)
- Rehearses the synchronization of combat power from flank and higher organizations. These organizations often exceed the communications range of the commander and assistant chief of staff, operations (G-3 [S-3]) when they are away from the command post.
- Synchronizes the timing and contribution of each warfighting function.
- For each decisive point, defines conditions required to—
    - Commit the reserve or striking forces.
    - Move a unit.
    - Close or emplace an obstacle.
    - Fire at planned targets.
    - Move a medical unit, change a supply route, and alert specific observation posts.
- Disciplines leader movements, enforces brevity, and ensures completeness.
- Keeps within time constraints.
- Ensures that the most important events receive the most attention.
- Ensures that absentees and flank units receive changes to the OPORD and transmits changes to them as soon as practical.
- Communicates the key civil considerations of the operation.

Assistant Chief of Staff, G-3 (S-3)

12-64. The G-3 (S-3) assists the commander with the rehearsal. The G-3 (S-3)—

- Portrays the friendly scheme of maneuver.
- Ensures subordinate unit actions comply with the commander's intent.
- Normally provides the recorder.

Assistant Chief of Staff, G-2 (S-2)

12-65. The assistant chief of staff, intelligence (G-2 [S-2]) portrays the enemy forces and other variables of the operational environment during rehearsals. The G-2 (S-2) bases actions on the enemy course of action that the commander selected during the MDMP. The G-2 (S-2)—

- Provides participants with current intelligence.
- Portrays the best possible assessment of the enemy course of action.
- Communicates the enemy's presumed concept of operations, desired effects, and end state.

- Explains other factors of the operational environment that may hinder or complicate friendly actions.
- Communicates the key civil considerations of the operation.

Subordinate Leaders

12-66. Subordinate unit leaders, using an established format, effectively articulate their units' actions and responsibilities as well as record changes on their copies of the graphics or OPORD.

Recorder

12-67. The recorder is normally a representative from the G-3 (S-3). During the rehearsal, the recorder captures all coordination made during execution and notes unresolved problems. At the end of the rehearsal, the recorder—

- Presents any unresolved problems to the commander for resolution.
- Restates any changes, coordination, or clarifications directed by the commander.
- Estimates when a written fragmentary order (FRAGORD) codifying the changes will follow.

Conducting Headquarters Staff

12-68. The staff updates the OPORD, decision support template, and execution matrix based on the decisions of the commander.

REHEARSAL ASSESSMENT

12-69. The commander establishes the standard for a successful rehearsal. A properly executed rehearsal validates each leader's role and how each unit contributes to the overall operation—what each unit does, when each unit does it relative to times and events, and where each unit does it to achieve desired effects. An effective rehearsal ensures commanders have a common vision of the enemy, their own forces, the terrain, and the relationships among them. It identifies specific actions requiring immediate staff resolution and informs the higher commander of critical issues or locations that the commander, COS (XO), or G-3 (S-3) must personally oversee.

12-70. The commander (or rehearsal director in the commander's absence) assesses and critiques all parts of the rehearsal. Critiques center on how well the operation achieves the commander's intent and on the coordination necessary to accomplish that end. Usually, commanders leave the internal execution of tasks within the rehearsal to the subordinate unit commander's judgment and discretion.

REHEARSAL DETAILS

12-71. All participants have responsibilities before, during, and after a rehearsal. Before a rehearsal, the rehearsal director states the commander's expectations and orients the other participants on details of the rehearsal, as necessary. During a rehearsal, all participants rehearse their roles in the operation. They make sure they understand how their actions support the overall operation and note any additional coordination required. After a rehearsal, participants ensure they understand any changes to the OPORD and coordination requirements, and they receive all updated staff products.

12-72. Commanders do not normally address small problems that arise during rehearsals. Instead, the G-3 (S-3) recorder keeps a record of these problems. This ensures the commander does not interrupt the rehearsal's flow. If the problem remains at the end of the rehearsal, the commander resolves it then. If the problem jeopardizes mission accomplishment, the staff accomplishes the coordination necessary to resolve it before the participants disperse. Identifying and solving such problems is a major reason for conducting rehearsals. If commanders do not make corrections while participants are assembled, they may lose the opportunity to do so. Coordinating among dispersed participants and disseminating changes to them often proves more difficult than accomplishing these actions in person.

BEFORE THE REHEARSAL

12-73. Before the rehearsal, the rehearsal director calls the roll and briefs participants on information needed for execution. The briefing begins with an introduction, overview, and orientation. It includes a discussion of the rehearsal script and ground rules. The detail of this discussion is based on participants' familiarity with the rehearsal SOP.

12-74. Before the rehearsal, the staff develops an OPORD with at least the basic five paragraphs and necessary overlays. The staff may not publish annexes; however, responsible staff officers should know their content.

Introduction and Overview

12-75. Before the rehearsal, the rehearsal director introduces all participants as needed. Then, the rehearsal director (normally the COS [XO]) gives an overview of the briefing topics, rehearsal subjects and sequence, and timeline, specifying the no-later-than ending time. The rehearsal director explains after action reviews, describes how and when they occur, and discusses how to incorporate changes into the OPORD. The director explains any constraints, such as pyrotechnics use, light discipline, weapons firing, or radio silence. For safety, the rehearsal director ensures all participants understand safety precautions and enforces their use. Last, the director emphasizes results and states the commander's standard for a successful rehearsal. Subordinate leaders state any results of planning or preparation (including rehearsals) they have already conducted. If a subordinate recommends a change to the OPORD, the rehearsal director acts on the recommendation before the rehearsal begins, if possible. If not, the commander resolves the recommendation with a decision before the rehearsal ends.

Orientation

12-76. The rehearsal director orients the participants to the terrain or rehearsal medium. The rehearsal director identifies orientation using magnetic north on the rehearsal medium and symbols representing actual terrain features. After explaining any graphic control measures, obstacles, and targets, the rehearsal director issues supplemental materials, if needed.

Rehearsal Script

12-77. An effective means for the rehearsal director to control rehearsals is the use of a script. It keeps the rehearsal on track. The script provides a checklist so the organization addresses all warfighting functions and outstanding issues. It has two major parts: the agenda and response sequence.

Agenda

12-78. An effective rehearsal follows a prescribed agenda that everyone knows and understands. This agenda includes, but is not limited to—

- Roll call.
- Participant orientation to the terrain.
- Location of local civilians.
- Enemy situation brief.
- Friendly situation brief.
- Description of expected enemy actions.
- Discussion of friendly unit actions.
- A review of notes made by the recorder.

12-79. The execution matrix, decision support template, and OPORD outline the rehearsal agenda. These tools, especially the execution matrix, both drive and focus the rehearsal. The commander and staff use them to control the operation's execution. Any templates, matrixes, or tools developed within each of the warfighting functions should tie directly to the supported unit's execution matrix and decision support template. Examples include an intelligence synchronization matrix or fires execution matrix.

12-80. An effective rehearsal realistically and quickly portrays the enemy force and other variables of the operational environment without distracting from the rehearsal. One technique for doing this has the G-2 (S-2) preparing an actions checklist. It lists a sequence of events much like the one for friendly units but from the enemy or civilian perspective.

Response Sequence

12-81. Participants respond in a logical sequence: either by warfighting function or by unit as the organization deploys, from front to rear. The commander determines the sequence before the rehearsal. The staff posts the sequence at the rehearsal site, and the rehearsal director may restate it.

12-82. Effective rehearsals allow participants to visualize and synchronize the concept of operations. As the rehearsal proceeds, participants talk through the concept of operations. They focus on key events and the synchronization required to achieve the desired effects. The commander leads the rehearsal and gives orders during the operation. Subordinate commanders enter and leave the discussion at the time they expect to begin and end their tasks or activities during the operation. This practice helps the commander assess the adequacy of synchronization. Commanders do not "re-war-game" unless absolutely necessary to ensure subordinate unit commanders understand the plan.

12-83. The rehearsal director emphasizes integrating fires, events that trigger different branch actions, and actions on contact. The chief of fires (fire support officer) or firing unit commander states when to initiate fires, who to fire them, from where the firing comes, the ammunition available, and the desired target effect. Subordinate commanders state when they initiate fires per their fire support plans. The rehearsal director speaks for any absent staff section and ensures the rehearsal addresses all actions on the synchronization matrix and decision support template at the proper time or event.

12-84. The rehearsal director ensures that the rehearsal includes key sustainment and protection actions at the appropriate times. Failure to do so reduces the value of the rehearsal as a coordination tool. The staff officer with coordinating staff responsibility inserts these items into the rehearsal. Special staff officers should brief by exception when a friendly or enemy event occurs within their area of expertise. Summarizing these actions at the end of the rehearsal can reinforce coordination requirements identified during the rehearsal. The staff updates the decision support template and gives a copy to each participant. Under time-constrained conditions, the conducting headquarters staff may provide copies before the rehearsal and rely on participants to update them with pen-and-ink changes. (See table 12-1.)

Table 12-1. Example sustainment and protection actions for rehearsals

• Casualty evacuation routes.	• Support area displacement times and locations.
• Ambulance exchange point locations.	• Detainee collection points.
• Refuel-on-the-move points.	• Aviation support.
• Class IV and Class V resupply points.	• Military police actions.
• Logistics release points.	

Ground Rules

12-85. After discussing the rehearsal script, the rehearsal director—
- States the standard (what the commander will accept) for a successful rehearsal.
- Ensures everyone understands the parts of the OPORD to rehearse. If the unit will not rehearse the entire operation, the rehearsal director states the events to be rehearsed.
- Quickly reviews the rehearsal SOP if all participants are not familiar with it. An effective rehearsal SOP states—
  - Who controls the rehearsal.
  - Who approves the rehearsal venue and its construction.

   ▪   When special staff officers brief the commander.

   ▪   The relationship between how the execution matrix portrays events and how units rehearse events.

   ●   Establishes the timeline that designates the rehearsal starting time in relation to H-hour. For example, begin the rehearsal by depicting the anticipated situation one hour before H-hour. One event executed before rehearsing the first event is deployment of forces.

   ●   Establishes the time interval to begin and track the rehearsal. For example, the rehearsal director may specify that a ten-minute interval equates to one hour of actual time.

   ●   Updates friendly and enemy activities as necessary. For example, the rehearsal director describes any ongoing reconnaissance.

The rehearsal director concludes the orientation with a call for questions.

DURING THE REHEARSAL

12-86.   Once the rehearsal director finishes discussing the ground rules and answering questions, the G-3 (S-3) reads the mission statement, the commander reads the commander's intent, and the G-3 (S-3) establishes the current friendly situation. The rehearsal then begins, following the rehearsal script.

12-87.   Paragraphs 12-88 through 12-101 outline a generic set of rehearsal steps developed for combined arms rehearsals. However, with a few modifications, these steps support any rehearsal technique. The products depend on the rehearsal type.

Step 1–Enemy Forces Deployed

12-88.   The G-2 (S-2) briefs the current enemy situation and operational environment and places markers on the map or terrain board (as applicable) indicating where enemy forces and other operationally significant groups or activities would be before the first rehearsal event. The G-2 (S-2) then briefs the most likely enemy course of action and operational context. The G-2 (S-2) also briefs the status of information collection operations (for example, citing any patrols still out or any observation post positions).

Step 2–Friendly Forces Deployed

12-89.   The G-3 (S-3) briefs friendly maneuver unit dispositions, including security forces, as they are arrayed at the start of the operation. Subordinate commanders and other staff officers brief their unit positions at the starting time and any particular points of emphasis. For example, the chemical, biological, radiological, and nuclear (CBRN) officer states the mission-oriented protective posture level, and the chief of fires (fire support officer) or fires unit commander states the range of friendly and enemy artillery. Other participants place markers for friendly forces, including adjacent units, at the positions they will occupy at the start of the operation. As participants place markers, they state their task and purpose, task organization, and strength.

12-90.   Sustainment and protection units brief positions, plans, and actions at the starting time and at points of emphasis the rehearsal director designates. Subordinate units may include forward arming and refueling points, refuel-on-the-move points, communications checkpoints, security points, or operations security procedures that differ for any period during the operation. The rehearsal director restates the commander's intent, if necessary.

Step 3–Initiate Action

12-91.   The rehearsal director states the first event on the execution matrix. Normally this involves the G-2 (S-2) moving enemy markers according to the most likely course of action. The depiction must tie enemy actions to specific terrain or to friendly unit actions. The G-2 (S-2) portrays enemy actions based on the situational template developed for staff war-gaming.

12-92.   As the rehearsal proceeds, the G-2 (S-2) portrays the enemy and other operational factors and walks through the most likely enemy course of action (per the situational template). The G-2 (S-2) stresses reconnaissance routes, objectives, security force composition and locations, initial contact, initial fires (artillery, air, and attack helicopters), probable main force objectives or engagement areas, and likely commitment of reserve forces.

Step 4–Decision Point

12-93.   When the rehearsal director determines that a particular enemy movement or reaction is complete, the commander assesses the situation to determine if a decision point has been reached. Decision points are taken directly from the decision support template.

12-94.   If the commander determines the unit is not at a decision point and not at the end state, the commander directs the rehearsal director to continue to the next event on the execution matrix. Participants use the response sequence (see paragraphs 12-81 through 12-84) and continue to act out and describe their units' actions.

12-95.   When the rehearsal reaches conditions that establish a decision point, the commander decides whether to continue with the current course of action or select a branching course of action. If electing the current course of action, the commander directs the rehearsal director to move to the next event in the execution matrix. If selecting a branch, the commander states the reason for selecting that branch, states the first event of that branch, and continues the rehearsal until the organization has rehearsed all events of that branch. As the unit reaches decisive points, the rehearsal director states the conditions required for success.

12-96.   When it becomes obvious that the operation requires additional coordination to ensure success, participants immediately begin coordinating. This is one of the key reasons for rehearsals. The rehearsal director ensures that the recorder captures the coordination and any changes and all participants understand the coordination.

Step 5–End State Reached

12-97.   Achieving the desired end state completes that phase of the rehearsal. In an attack, this will usually be when the unit is on the objective and has finished consolidation and casualty evacuation. In the defense, this will usually be after the decisive action (such as committing the reserve or striking force), the final destruction or withdrawal of the enemy, and casualty evacuation is complete. In stability tasks, this usually occurs when a unit achieves the targeted progress within a designated line of effort.

Step 6–Reset

12-98.   At this point, the commander states the next branch to rehearse. The rehearsal director resets the situation to the decision point where that branch begins and states the criteria for a decision to execute that branch. Participants assume those criteria have been met and then refight the operation along that branch until they attain the desired end state. They complete any coordination needed to ensure all participants understand and can meet any requirements. The recorder records any changes to the branch.

12-99.   The commander then states the next branch to rehearse. The rehearsal director again resets the situation to the decision point where that branch begins, and participants repeat the process. This continues until the rehearsal has addressed all decision points and branches that the commander wants to rehearse.

12-100. If the standard is not met and time permits, the commander directs participants to repeat the rehearsal. The rehearsal continues until participants are prepared or until the time available expires. (Commanders may allocate more time for a rehearsal but must assess the effects on subordinate commanders' preparation time.) Successive rehearsals, if conducted, should be more complex and realistic.

12-101. At the end of the rehearsal, the recorder restates any changes, coordination, or clarifications that the commander directed and estimates how long it will take to codify changes in a written FRAGORD.

AFTER THE REHEARSAL

12-102. After the rehearsal, the commander leads an after action review. The commander reviews lessons learned and makes the minimum required modifications to the existing plan. (Normally, a FRAGORD effects these changes.) Changes should be refinements to the OPORD; they should not be radical or significant. Changes not critical to the operation's execution may confuse subordinates and hinder the synchronization of the plan. The commander issues any last minute instructions or reminders and reiterates the commander's intent.

12-103. Based on the commander's instructions, the staff makes any necessary changes to the OPORD, decision support template, and execution matrix based on the rehearsal results. Subordinate commanders incorporate these changes into their units' OPORDs. The COS (XO) ensures the changes are briefed to all leaders or liaison officers who did not participate in the rehearsal.

12-104. A rehearsal provides the final opportunity for subordinates to identify and fix unresolved problems. The staff ensures that all participants understand any changes to the OPORD and that the recorder captures all coordination done at the rehearsal. All changes to the published OPORD are, in effect, verbal FRAGORDs. As soon as possible, the staff publishes these verbal FRAGORDs as a written FRAGORD that changes the operation order.

Chapter 13

Liaison

This chapter discusses responsibilities of liaison officers and teams. It addresses requirements distinct to Army operations and unified action. It includes liaison checklists and an example outline for a liaison officer handbook.

ROLE OF LIAISON

13-1. *Liaison* is that contact or intercommunication maintained between elements of military forces or other agencies to ensure mutual understanding and unity of purpose and action (JP 3-08). Most commonly used for establishing and maintaining close communications, liaison continuously enables direct, physical communications between commands and with unified action partners. Commanders use liaison during operations and normal daily activities to help facilitate a shared understanding and purpose among organizations, preserve freedom of action, and maintain flexibility. Liaison provides commanders with relevant information and answers to operational questions, thus enhancing the commander's situational understanding.

13-2. Liaison activities augment the commander's ability to synchronize and focus combat power. They include establishing and maintaining physical contact and communications between elements of military forces and nonmilitary agencies during unified action. Liaison activities ensure—

- Cooperation and understanding among commanders and staffs of different headquarters.
- Coordination on tactical matters to achieve unity of effort.
- Synchronization of lethal and nonlethal effects.
- Understanding of implied or inferred coordination measures to achieve synchronized results.

LIAISON OFFICER

13-3. A liaison officer (LNO) represents a commander or staff officer. LNOs transmit information directly, bypassing headquarters and staff layers. A trained, competent, trusted, and informed LNO (either a commissioned or a noncommissioned officer [NCO]) is the key to effective liaison. LNOs must have the commander's full confidence and experience for the mission. At higher echelons, the complexity of operations often requires an increase in the rank required for LNOs. (See table 13-1.)

Table 13-1. Senior liaison officer rank by echelon

Senior liaison officer rank by echelon	Recommended rank
Multinational or joint force commander[1]	Colonel
Corps	Lieutenant Colonel
Division	Major
Brigade, regiment, or group	Captain
Battalion	Lieutenant
[1]These include joint force commanders and functional component commanders and may also include major interagency and international organizations.	

13-4. The LNO's parent unit or unit of assignment is the sending unit. The unit or activity that the LNO is sent to is the receiving unit, which may be a host nation. An LNO normally remains at the receiving unit until recalled. LNOs represent the commander and they—

- Understand how the commander thinks and interpret the commander's messages.
- Convey the commander's intent, guidance, mission, and concept of operations.
- Represent the commander's position.

13-5. As a representative, the LNO has access to the commander consistent with the duties involved. However, for routine matters, LNOs work for and receive direction from the chief of staff or executive officer (COS [XO]). Using one officer to perform a liaison mission conserves manpower while guaranteeing a consistent, accurate flow of information. However, continuous operations may require a liaison team or liaison detachment.

13-6. The professional capabilities and personal characteristics of an effective LNO encourage confidence and cooperation with the commander and staff of the receiving unit. In addition to the discussion in paragraph 13-4, effective LNOs—

- Know the sending unit's mission; current and future operations; logistics status; organization; disposition; capabilities; and tactics, techniques, and procedures.
- Appreciate and understand the receiving unit's tactics, techniques, and procedures; organization; capabilities; mission; doctrine; staff procedures; and customs.
- Are familiar with—
  - Requirements for and purpose of liaison.
  - The liaison system and its reports, documents, and records.
  - Liaison team training.
- Observe the established channels of command and staff functions.
- Are tactful.
- Possess familiarity with local culture and language, and have advanced regional expertise if possible.

LIAISON ELEMENTS

13-7. Commanders organize liaison elements based on the mission variables (mission, enemy, terrain and weather, troops and support available, time available, and civil considerations [METT-TC]) and echelon of command. (See appendix A for more details.) Two command liaison teams are authorized in division, corps, and theater army headquarters. Common ways to organize liaison elements include, but are not limited to—

- A single LNO.
- A liaison team consisting of one or two LNOs, or an LNO and a liaison NCO in charge, clerical personnel, and communications personnel along with their equipment.
- Couriers (messengers) responsible for the secure physical transmission and delivery of documents and other materials.
- A digital liaison detachment comprised of several teams with expertise and equipment in specialized areas, such as intelligence, operations, fire support, air defense, and sustainment.

DIGITAL LIAISON DETACHMENTS

13-8. Digital liaison detachments provide Army commanders units to conduct liaison with major subordinate or parallel headquarters. Digital liaison detachments consist of staff officers with a broad range of expertise who are capable of analyzing the situation, facilitating coordination between multinational forces, and assisting in cross-boundary information flow and operational support. These 30-Soldier teams are essential not only for routine liaison, but also for advising and assisting multinational partners in conducting planning and operations at intermediate tactical levels. These detachments can operate as a single entity for liaison with a major multinational headquarters, or provide two smaller teams for digital connectivity and liaison with smaller multinational headquarters. Commanders can also tailor digital liaison detachments to match a given mission. The basis of digital liaison detachments allocation is five per committed theater Army, one per corps and division serving as a joint task force headquarters, or as approved by the Department of the Army. The support requirement for a coalition during counterinsurgency or foreign internal defense is one digital liaison detachment for each multinational headquarters (division or above) and one for the host-nation Ministry of Defense.

LIAISON PRACTICES

13-9. When possible, liaison is reciprocal among higher, lower, supporting, supported, and adjacent organizations. Each organization sends a liaison element to the other. It must be reciprocal when U.S. forces are placed under control of a headquarters of a different nationality and vice versa, or when brigade-sized and larger formations of different nationalities are adjacent. When not reciprocal, the following practices apply to liaison where applicable:

- Higher-echelon units establish liaison with lower echelons.
- In contiguous operations units on the left establish liaison with units on their right.
- In contiguous operations units of the same echelon establish liaison with those to their front.
- In noncontiguous operations units establish liaison with units within closest proximity.
- Supporting units establish liaison with units they support.
- Units not in contact with the enemy establish liaison with units in contact with the enemy.
- During a passage of lines, the passing unit establishes liaison with the stationary unit.
- During a relief in place, the relieving unit establishes liaison with the unit being relieved.

If liaison is broken, both units act to reestablish it. However, the primary responsibility rests with the unit originally responsible for establishing liaison.

LIAISON RESPONSIBILITIES

13-10.   Both sending and receiving units have liaison responsibilities before, during, and after operations.

SENDING UNIT

13-11.   The sending unit's most important tasks include selecting and training the best qualified Soldiers for liaison duties. Liaison personnel should have the characteristics and qualifications discussed in paragraphs 13-3 through 13-6. (See figure 13-1 for an example outline for an LNO handbook. See figure 13-2 on page 13-4 for sample questions that LNOs should be able to answer. See figure 13-3 on page 13-5 for a sample LNO packing list.)

Sample Liaison Officer Handbook Outline

Table of contents.

Sending unit's tasking order.

Purpose statement.

Introduction statement.

Definitions.

Scope statement.

Responsibilities and guidelines for conduct.

Actions to take before departing from the sending unit.

Actions to take on arriving at the receiving unit.

Actions to take during liaison operations at the receiving unit.

Actions to take before departing from the receiving unit.

Actions to take upon returning to the sending unit.

Figure 13-1. Example liaison officer handbook outline

Sample questions. *Liaison officers should be able to answer the following questions:*

Does the sending unit have a copy of the receiving unit's latest operation plan, operation order, and fragmentary order?

Does the receiving unit's plan support the plan of the higher headquarters? This includes sustainment as well as the tactical concept. Are main supply routes and required supply rates known?

Can the controlled supply rate support the receiving unit's plan?

What are the receiving unit's commander's critical information requirements?

At what time, phase, or event are they expected to change? Are there any items the commander's critical information requirements do not contain with which the sending unit can help?

Which sending commander decisions are critical to executing the receiving unit operation?

What are the "no-later-than" times for those decisions?

What assets does the unit need to acquire to accomplish its mission? How would the unit use them?

How do they support attaining the more senior commander's intent? From where can the unit obtain them? Higher headquarters? Other Services? Multinational partners?

How do units use aviation assets?

How can the liaison officers communicate with the sending unit? Are telephones, radios, facsimile machines, computers, and other information systems available? Where are they located? Which communications are secure?

What terrain did the unit designate as key? Decisive?

What weather conditions would have a major impact on the operation?

What effect would a chemical, biological, radiological, and nuclear environment have on the operation?

What effect would large numbers of refugees or enemy prisoners of war have on the receiving unit's operations?

What is the worst thing that could happen during execution of the current operation?

How would a unit handle a passage of lines by other units through the force?

What conditions would cause the unit to request operational control of a multinational force?

If the unit is placed under operational control of a larger multinational force, or given operational control of a smaller such force, what special problems would it present?

If going to a multinational force headquarters, how do the tactical principles and command concepts of that force differ from those of U.S. forces?

What host-nation support is available to the sending unit?

What are the required reports from higher and sending units' standard operating procedures?

Figure 13-2. Examples of liaison officer questions

Example recommended packing list:

Credentials (including courier card, permissive jump orders, if qualified). Blank forms as required.

References.

Excerpts of higher and sending headquarters' operation orders and plans.

Sending unit standard operating procedures.

Sending unit's command diagrams and recapitulation of major systems.

The unit modified table of equipment, unit status report (if its classification allows), and mission briefings. The assistant chief of staff, operations (G-3 [S-3]) and the force modernization officer are excellent sources of these.

Computers and other information systems required for information and data exchange (for example, command post of the future, SECRET Internet Protocol Router Network, and Nonsecure Internet Protocol Router Network devices.)

Automated network control device.

Communications equipment.

Sending unit telephone book.

List of commanders and staff officers.

Telephone calling (credit) card.

Cell phone.

Movement table.

Administrative equipment (for example, pens, paper, scissors, tape, and hole punch).

Map and chart equipment (for example, pens, pins, protractor, straight edge, scale, distance counter, acetate, and unit markers).

Tent and accessories (camouflage net, cots, and stove, as appropriate).

Foreign phrase book and dictionary.

Local currency as required.

Rations and water.

Weapons and ammunition.

Night-vision device.

Figure 13-3. Example recommended packing list

13-12. The sending unit describes the liaison team to the receiving unit providing number and types of vehicles and personnel, equipment, call signs, and frequencies. The LNO or liaison team also requires—

- Point-to-point transportation, as required.
- Identification and appropriate credentials for the receiving unit.
- Appropriate security clearance, courier orders, and information systems accredited for use on the receiving unit's network.
- The standard operating procedures (SOPs) outlining the missions, functions, procedures, and duties of the sending unit's liaison section.
- If the receiving unit is multinational, it may provide communications equipment and personnel.

13-13. The movement from the sending unit to the receiving unit requires careful planning and coordination. (See figure 13-4 on page 13-6 for a list of tasks for liaison personnel to perform before departing the sending unit.)

Example liaison checklist before departure from sending unit:

Understand what the sending commander wants the receiving commander to know.

Receive a briefing from operations, intelligence, and other staff elements on current and future operations.

Receive and understand the tasks from the sending unit staff.

Obtain the correct maps, traces, and overlays.

Arrange for transport, communications and cryptographic equipment, codes, signal instructions, and the challenge and password—including their protection and security. Arrange for replacement of these items, as necessary.

Complete route-reconnaissance and time-management plans so the liaison team arrives at the designated location on time.

Ensure that liaison team and interpreters have security clearances and access appropriate for the mission.

Verify that the receiving unit received the liaison team's security clearances and will grant access to the level of information the mission requires.

Verify courier orders.

Know how to destroy classified information in case of an emergency during transit or at the receiving unit.

Inform the sending unit of the liaison officer's departure time, route, arrival time, and, when known, the estimated time and route of return.

Pick up all correspondence designated for the receiving unit.

Conduct a radio check.

Know the impending moves of the sending and receiving units.

Bring accredited information systems needed to support liaison operations.

Pack adequate rations and water for use in transit.

Arrange for the liaison party's departure.

Figure 13-4. Liaison checklist—before departing the sending unit

RECEIVING UNIT

13-14. The receiving unit—
- Provides the sending unit with the LNO's reporting time, place, point of contact, recognition signal, and password.
- Provides details of any tactical movement and logistics information relevant to the LNO's mission, especially while the LNO is in transit.
- Ensures that the LNO has access to the commander, the COS (XO), and other officers, as required.
- Gives the LNO an initial briefing of the unit battle rhythm and allows the LNO access necessary to remain informed of current operations.
- Protects the LNO while at the receiving unit.
- Publishes an SOP outlining the missions, functions, procedures to request information, information release restrictions and clearance procedures, and duties of the LNO or team at the receiving unit.
- If possible, provides access to communications equipment (and operating instructions, as needed) when the LNO needs to communicate using the receiving unit's equipment.

- Provides adequate workspace for the LNO.
- Provides administrative and logistic support, or agreed to host-nation support.

DURING THE TOUR

13-15.   During the tour, LNOs have specific duties. LNOs inform the receiving unit's commander or staff of the sending unit's needs or requirements. Due to the numbers of LNOs in the headquarters, sending units guard against inundating the receiving unit with formal requests for information. By virtue of their location in the headquarters and knowledge of the situation, LNOs can rapidly answer questions from the sending unit and keep the receiving unit from wasting planning time answering requests for information. (See figure 13-5 on page 13-8 for a summary of LNO duties.) During the liaison tour, LNOs—

- Arrive at the designated location on time.
- Promote cooperation between the sending and receiving units.
- Accomplish their mission without becoming overly involved in the receiving unit's staff procedures or actions; however, they may assist higher echelon staffs in war-gaming.
- Follow the receiving unit's communications procedures.
- Actively obtain information without interfering with the receiving unit's operations.
- Facilitate understanding of the sending unit's commander's intent.
- Help the sending unit's commander assess current and future operations.
- Remain informed of the sending unit's current situation and provide that information to the receiving unit's commander and staff.
- Quickly inform the sending unit of the receiving unit's upcoming missions, tasks, and orders.
- Ensure the sending unit has a copy of the receiving unit's SOP.
- Inform the receiving unit's commander or COS (XO) of the content of reports transmitted to the sending unit.
- Keep a record of their reports, listing everyone met (including each person's name, rank, duty position, and telephone number) as well as key staff members and their telephone numbers.
- Attempt to resolve issues within the receiving unit before involving the sending unit.
- Notify the sending unit promptly if unable to accomplish the liaison mission.
- Report their departure to the receiving unit's commander at the end of their mission.

Example liaison duties during the tour:

Arrive at least two hours before any scheduled briefings.

Check in with security and complete any required documentation.

Report to and present credentials to the chief of staff (executive officer) or supervisor, as appropriate.

Arrange for an office call with the commander.

Meet coordinating and special staff officers.

Notify the sending unit of arrival (use the liaison establishment report).

Visit staff elements, brief them on the sending unit's situation, and collect information from them.

Deliver all correspondence designated for the receiving unit.

Annotate on all overlays the security classification, title, map scale, grid intersection points, and effective date-time group, when received, and from whom received.

Pick up all correspondence for the sending unit when departing the receiving unit.

Inform the receiving unit of the liaison officer's departure time, return route, and expected arrival time at the sending unit.

Submit a liaison disestablishment report to the sending unit when departing.

Figure 13-5. Liaison duties—during the liaison tour

13-16.   Once a deploying liaison team or detachment arrives and sets up communications at the receiving unit, it submits a liaison establishment report to the sending unit. This report informs the sending unit's command that the detachment is ready to conduct liaison, and it establishes exactly what systems are available. A re-deploying team or detachment submits a liaison disestablishment report to the sending unit as its last action prior to disconnecting its digital devices. This report informs the command that the element is leaving the network and is no longer capable of conducting liaison at any level beyond unsecure voice. (See unit SOPs for the liaison establishment report and the liaison disestablishment report formats.)

AFTER THE TOUR

13-17.   After returning to the sending unit, LNOs promptly transmit the receiving unit's requests to the sending unit's commander or staff, as appropriate. (See figure 13-6 for a list of tasks to perform after completing a liaison tour.)

Example liaison duties after the tour:

Deliver all correspondence.

Brief the chief of staff (executive officer) and appropriate staff elements.

Prepare the necessary reports.

Clearly state what they did and did not learn from the mission.

Figure 13-6. Liaison duties—after the liaison tour

13-18.   Accuracy is important. Effective LNOs provide clear, concise, complete information. If the accuracy of information is in doubt, they quote the source and include the source in the report. LNOs limit their remarks to mission-related observations.

LIAISON CONSIDERATIONS

13-19.   Joint, interagency, and multinational operations require greater liaison efforts than most other operations.

JOINT OPERATIONS

13-20. Current joint information systems do not meet all operational requirements. Few U.S. military information systems are interoperable. Army liaison teams and detachments require information systems that can rapidly exchange information between commands to ensure Army force operations are synchronized with operations of the joint force and its Service components.

INTERAGENCY OPERATIONS

13-21. Army forces may participate in interagency operations across the range of military operations, especially when conducting stability or defense support of civil authorities tasks. Frequently, Army forces conduct operations in cooperation with or in support of civilian government agencies. Relations in these operations are rarely based on standard military command and support relationships; rather, national laws or specific agreements for each situation govern the specific relationships in interagency operations. Defense support of civil authorities provides an excellent example. Federal military forces that respond to a domestic disaster will support the Federal Emergency Management Agency, while National Guard forces working in state active duty status (Title 32 United States Code) or conducting National Guard defense support of civil authorities will support that state's emergency management agency. National Guard forces federalized under Title 10 United States Code will support the Federal Emergency Management Agency. The goal is always unity of effort between military forces and civilian agencies, although unity of command may not be possible. Effective liaison and continuous coordination become keys to mission accomplishment. (See FM 3-28.)

13-22. Some missions require coordination with nongovernmental organizations. While no overarching interagency doctrine delineates or dictates the relationships and procedures governing all agencies, departments, and organizations in interagency operations, the National Response Framework provides some guidance. Effective liaison elements work toward establishing mutual trust and confidence, continuously coordinating actions to achieve cooperation and unity of effort. (See JP 3-08.) In these situations, LNOs and their teams require a broader understanding of the interagency environment, responsibilities, motivations, and limitations of nongovernmental organizations, and the relationships these organizations have with the U.S. military.

MULTINATIONAL OPERATIONS

13-23. Army units often operate as part of a multinational force. Interoperability is an essential requirement for multinational operations. The North Atlantic Treaty Organization (NATO) defines interoperability as the ability to operate in synergy in the execution of assigned tasks. Interoperability is also the condition achieved among communications-electronics systems or items of satisfactory communication between them and their users. The degree of interoperability should be defined when referring to specific cases. Examples of interoperability include the deployment of a computer network (such as the Combined Enterprise Network Theater Information Exchange System) to facilitate inter-staff communication. Nations whose forces are interoperable can operate together effectively in numerous ways. Less interoperable forces have correspondingly fewer ways to work together. Although frequently identified with technology, important areas of interoperability include doctrine, procedures, communications, and training. Factors that enhance interoperability include planning for interoperability, conducting multinational training exercises, staff visits to assess multinational capabilities, a command atmosphere that rewards sharing information, and command emphasis on a constant effort to eliminate the sources of confusion and misunderstanding.

13-24. The multinational forces with which the U.S. operates may not have information systems that are compatible with U.S. or other systems. Some nations may lack computerized information systems. Reciprocal liaison is especially important under these conditions. Mutual trust and confidence is the key to making these multinational operations successful. Liaison during multinational operations includes explicit coordination of doctrine as well as tactics, techniques, and procedures. Effective liaison requires patience and tact during personal interactions. The liaison officer needs a thorough understanding of the strategic, operational, and tactical aims of the international effort. Foreign disclosure limitations often require special communications and liaison arrangements to address cultural differences and sensitivities as well as to ensure explicit understanding throughout the multinational force. Two structural enhancements that improve the coordination of multinational forces are liaison networks and coordination centers.

13-25. A multinational coordination center or coalition coordination center is a means of increasing multinational coordination. U.S. commanders routinely create such a center in the early stages of any coalition effort, especially one that is operating under a parallel command structure. It is a proven means of integrating the participating nations' military forces into the coalition planning and operations processes, enhancing coordination and cooperation, and supporting an open and full interaction within the coalition structure. Normally, the multinational coordination center focuses upon coordination of coalition force operations, which will most likely involve classified information. (See JP 3-16 for more information on multinational operations.)

Chapter 14

Decisionmaking in Execution

This chapter describes the major activities of execution to include assessing, decisionmaking, and directing action. The chapter concludes with a discussion of the rapid decisionmaking and synchronization process.

ACTIVITIES OF EXECUTION

14-1. Planning and preparation accomplish nothing if the command does not execute effectively. *Execution is putting a plan into action by applying combat power to accomplish the mission* (ADRP 5-0). In execution, commanders, supported by their staffs, focus their efforts on translating decisions into actions. Inherent in execution is deciding whether to execute planned actions, such as changing phases or executing a branch plan. Execution also includes adjusting the plan based on changes in the situation and an assessment of the operation's progress. (See ADRP 5-0 for fundamentals of execution).

14-2. Throughout execution, commanders, supported by their staffs, assess the operation's progress, make decisions, and direct the application of combat power to seize, retain, and exploit the initiative. Major activities of execution include—

- Assessment: Monitoring current operations and evaluating progress.
- Decisionmaking: Making decisions to exploit opportunities or counter treats.
- Directing action: Apply combat power at decisive points and times.

ASSESSMENT DURING EXECUTION

14-3. During execution, continuous assessment is essential. Assessment involves a deliberate comparison of forecasted outcomes to actual events, using criteria to judge operational progress towards success. The commander and staff assess the probable outcome of the operation to determine whether changes are necessary to accomplish the mission, take advantage of opportunities, or react to unexpected threats. Commanders also assess the probable outcome of current operations in terms of their impact on potential future operations in order to develop concepts for these operations early. (See ADRP 5-0 for fundamentals of assessment. See chapter 15 for details for building an assessment plan).

14-4. Assessment includes both monitoring the situation and evaluating progress. During monitoring, commanders and staffs collect and use relevant information to develop a clear understanding of the command's current situation. Commanders and staffs also evaluate the operation's progress in terms of measures of performance (MOPs) and measures of effectiveness (MOEs). This evaluation helps commanders assess progress and identify variances—the difference between the actual situation and what the plan forecasted the situation would be at that time or event. Identifying variances and their significance leads to determining if a decision is required during execution.

DECISIONMAKING DURING EXECUTION

14-5. When operations are progressing satisfactorily, variances are minor and within acceptable levels. Commanders who make this evaluation—explicitly or implicitly—allow operations to continue according to plan. This situation leads to execution decisions included in the plan. Execution decisions implement a planned action under circumstances anticipated in the order. An execution decision is normally tied to a decision point.

14-6. An assessment may determine that the operation as a whole, or one or more of its major actions, is not progressing according to expectations. Variances of this magnitude present one of two situations:

- Significant, unforeseen opportunities to achieve the commander's intent.
- Significant threats to the operation's success.

In either case, the commander makes an adjustment decision. An adjustment decision is the selection of a course of action that modifies the order to respond to unanticipated opportunities or threats. An adjustment decision may include a decision to reframe the problem and develop an entirely new plan.

14-7. Executing, adjusting, or abandoning the original operation is part of decisionmaking in execution. By fighting the enemy and not the plan, successful commanders balance the tendency to abandon a well-conceived plan too soon against persisting in a failing effort too long. Effective decisionmaking during execution—

- Relates all actions to the commander's intent and concept of operations.
- Is comprehensive, maintaining integration of combined arms rather than dealing with separate functions.
- Relies heavily on intuitive decisionmaking by commanders and staffs to make rapid adjustments.

DIRECTING ACTION

14-8. To implement execution or adjustment decisions, commanders direct actions that apply combat power. Based on the commander's decision and guidance, the staff resynchronizes the operation to mass the maximum effects of combat power to seize, retain, and exploit the initiative. This involves synchronizing the operations in time, space, and purpose and issuing directives to subordinates. (See table 14-1 for a summary of a range of possible actions with respect to decisions made during execution.)

Table 14-1. Decision types and related actions

Decision types		Actions
Execution decisions	**Minor variances from the plan** Operation proceeding according to plan. Variances are within acceptable limits.	**Execute planned actions** • Commander or designee decides which planned actions best meet the situation and directs their execution. • Staff issues fragmentary order. • Staff completes follow-up actions.
	Anticipated situation Operation encountering variances within the limits for one or more branches or sequels anticipated in the plan.	**Execute a branch or sequel** • Commander or staff review branch or sequel plan. • Commander receives assessments and recommendations for modifications to the plan, determines the time available to refine it, and either issues guidance for further actions or directs execution of a branch or sequel. • Staff issues fragmentary order. • Staff completes follow-up actions.
Adjustment decisions	**Unanticipated situation— friendly success** Significant, unanticipated positive variances result in opportunities to achieve the end state in ways that differ significantly from the plan.	**Make an adjustment decision** • Commander recognizes the opportunity or threat and determines time available for decisionmaking. • Based on available planning time, commanders determine if they want to reframe the problem and develop a new plan. In these instances, the decision initiates planning. Otherwise, the commander directs the staff to refine a single course of action or directs actions by subordinates to exploit the opportunity or counter the threat and exercise initiative within the higher commander's intent. • Commander normally does not attempt to restore the plan. • Commander issues a verbal warning or fragmentary order to subordinate commanders. • Staff resynchronizes operation, modifies measures of effectiveness, and begins assessing the operation for progress using new measures of effectiveness.
	Unanticipated situation— enemy threat Significant, unanticipated negative variances impede mission accomplishment.	

RAPID DECISIONMAKING AND SYNCHRONIZATION PROCESS

14-9. The rapid decisionmaking and synchronization process is a technique that commanders and staffs commonly use during execution. While identified here with a specific name and method, the approach is not new; its use in the Army is well established. Commanders and staffs develop this capability through training and practice. When using this technique, the following considerations apply:

- *Rapid* is often more important than *process*.
- Much of it may be mental rather than written.
- It should become a battle drill for the current operations integration cells, future operations cells, or both.

14-10. While the military decisionmaking process (MDMP) seeks the optimal solution (see chapter 9), the rapid decisionmaking and synchronization process seeks a timely and effective solution within the commander's intent, mission, and concept of operations. Using the rapid decisionmaking and synchronization process lets leaders avoid the time-consuming requirements of developing decision criteria and comparing courses of action (COAs). Operational and mission variables continually change during execution. This often invalidates or weakens COAs and decision criteria before leaders can make a decision. Under the rapid decisionmaking and synchronization process, leaders combine their experience and intuition to quickly reach situational understanding. Based on this, they develop and refine workable COAs.

14-11. The rapid decisionmaking and synchronization process facilitates continuously integrating and synchronizing the warfighting functions to address ever-changing situations. It meets the following criteria for making effective decisions during execution:

- It is comprehensive, integrating all warfighting functions. It is not limited to any one warfighting function.
- It ensures all actions support the decisive operation by relating them to the commander's intent and concept of operations.
- It allows rapid changes to the order or mission.
- It is continuous, allowing commanders to react immediately to opportunities and threats.

14-12. The rapid decisionmaking and synchronization process is based on an existing order and the commander's priorities as expressed in the order. The most important of these control measures are the commander's intent, concept of operations, and commander's critical information requirements (CCIRs). The rapid decisionmaking and synchronization process includes five steps. The first two may be performed in any order, including concurrently. The last three are performed interactively until commanders identify an acceptable course of action. (See figure 14-1.)

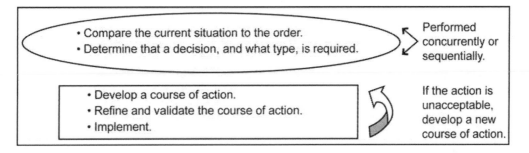

Figure 14-1. Rapid decisionmaking and synchronization process

COMPARE THE CURRENT SITUATION TO THE ORDER

14-13. Commanders and staffs identify likely variances during planning and identify options that will be present and actions that will be available when each variance occurs. During execution, commanders and staffs monitor the situation to identify changes in conditions. Then they ask if these changes affect the overall conduct of operations or their part in them and if the changes are significant. Finally, they identify if the changed conditions represent variances from the order—especially opportunities and risks. Staff members use running estimates to look for indicators of variances that affect their areas of expertise. (See table 14-2 for examples of indicators.)

14-14.   Staff members are particularly alert for answers to CCIRs that support anticipated decisions. They also watch for exceptional information—information that would have answered one of the CCIRs if the requirement for it had been foreseen and stated as one of the CCIRs. Exceptional information usually reveals a need for an adjustment decision.

14-15.   When performing the rapid decisionmaking and synchronization process, the current operations integration cell first compares the current situation to the one envisioned in the order. It may obtain assistance from the assessment section or the red team section in this analysis. If the situation requires greater analysis, the chief of staff or executive officer (COS [XO]) may task the future operations cell (where authorized) or the plans cell to perform this analysis. At echelons with no future operations cell, the plans cell or the current operations integration cell performs this function.

Table 14-2. Examples of change indicators

Types	Indicators	
General	• Answer to a commander's critical information requirement. • Identification of an information requirement. • Change in mission. • Change in organization of unit. • Change in leadership of unit. • Signing or implementation of peace treaty or other key political arrangement.	• Change in capabilities of subordinate unit. • Change in role of host-nation military force. • Climate changes or natural disasters impacting on the population, agriculture, industry. • Upcoming local election. • Changes in key civilian leadership.
Intelligence	• Identification of enemy main effort. • Identification of enemy reserves or counterattack. • Indications of unexpected enemy action or preparation. • Increase in enemy solicitation of civilians for intelligence operations. • Identification of an information requirement. • Insertion of manned surveillance teams. • Disruption of primary and secondary education system. • Unexplained disappearance of key members of intelligence community.	• Enemy electronic attack use. • Indicators of illicit economic activity. • Identification of threats from within the population. • Increased unemployment within the population. • Interference with freedom of religious worship. • Identification of high-value targets. • Unmanned aircraft system launch. • Answer to a priority intelligence requirement. • Enemy rotary-wing or unmanned aircraft system use.
Movement and Maneuver	• Success or failure in breaching or gap crossing operations. • Capture of significant numbers of enemy prisoners of war, enemy command posts, supply points, or artillery units. • Establishment of road blocks along major traffic routes. • Unexplained displacement of neighborhoods within a given sector.	• Success or failure of a subordinate unit task. • Modification of an airspace control measure. • Numbers of dislocated civilians sufficient to affect friendly operations. • Damages to civilian infrastructure affecting friendly mobility. • Loss of one or more critical transportation systems.
Fires	• Receipt of an air tasking order. • Battle damage assessment results. • Unplanned repositioning of firing units. • Identification of high-payoff targets. • Identification of an information requirement.	• Execution of planned fires. • Modification of a fire support coordination measure. • Effective enemy counterfire. • Negative effects of fires on civilians. • Destruction of any place of worship by friendly fire.

Table 14-2. Examples of change indicators (continued)

Types	Indicators	
Protection	• Chemical, biological, radiological, nuclear report or other indicators of enemy chemical, biological, radiological, nuclear use. • Report or other indicators of enemy improvised explosive device use. • Indicators of coordinated enemy actions against civilians or friendly forces. • Increased criminal activity in a given sector. • Increase in organized protests or riots.	• Identification of threats to communications or computer systems. • Reports of enemy targeting critical host-nation infrastructure. • Identification of threat to base or sustainment facilities. • Escalation of force incidents. • Loss of border security.
Sustainment	• Significant loss of capability in any class of supply. • Opening or closing of civilian businesses within a given area. • Identification of significant incidences of disease and nonbattle injury casualties. • Closing of major financial institutions. • Mass casualties. • Receipt of significant resupply. • Disruption of one or more essential civil services (such as water or electricity). • Contact on a supply route. • Answer to a friendly force information requirement. • Mass detainees.	• Degradations to essential civilian infrastructure by threat actions. • Civilian mass casualty event beyond capability of host-nation resources. • Identification of significant shortage in any class of supply. • Outbreak of epidemic or famine within the civilian population. • Medical evacuation launch. • Dislocated civilian event beyond capability of host-nation resources. • Disruption of key logistics lines of communication. • Changes in availability of host-nation support.
Mission Command	• Impending changes in key military leadership. • Interference with freedom of the press or news media. • Receipt of a fragmentary order or warning order from higher headquarters.	• Effective adversary information efforts on civilians. • Loss of civilian communications nodes. • Loss of contact with a command post or commander. • Jamming or interference.

DETERMINE THE TYPE OF DECISION REQUIRED

14-16. When a variance is identified, the commander directs action while the chief of operations leads chiefs of the current operations integration cell and selected functional cells in quickly comparing the current situation to the expected situation. This assessment accomplishes the following:

- Describes the variance.
- Determines if the variance provides a significant opportunity or threat and examines the potential of either.
- Determines if a decision is needed by identifying if the variance:
  - Indicates an opportunity that can be exploited to accomplish the mission faster or with fewer resources.
  - Directly threatens the decisive operation's success.
  - Threatens a shaping operation such that it may threaten the decisive operation directly or in the near future.
  - Can be addressed within the commander's intent and concept of operations. (If so, determine what execution decision is needed.)
  - Requires changing the concept of operations substantially. (If so, determine what adjustment decision or new approach will best suit the circumstances.)

14-17. For minor variances, the chief of operations works with other cell chiefs to determine whether changes to control measures are needed. If so, they determine how those changes affect other warfighting functions. They direct changes within their authority (execution decisions) and notify the COS (XO) and the affected command post cells and staff elements.

14-18. Commanders intervene directly in cases that affect the overall direction of the unit. They describe the situation, direct their subordinates to provide any additional information they need, and order either implementation of planned responses or development of an order to redirect the force.

DEVELOP A COURSE OF ACTION

14-19. If the variance requires an adjustment decision, the designated integrating cell and affected command post cell chiefs recommend implementation of a COA or obtain the commander's guidance for developing one. They use the following conditions to screen possible COAs:

- Mission.
- Commander's intent.
- Current dispositions and freedom of action.
- CCIRs.
- Limiting factors, such as supply constraints, boundaries, and combat strength.

14-20. The new options must conform to the commander's intent. Possible COAs may alter the concept of operations and CCIRs, if they remain within the commander's intent. However, the commander approves changes to the CCIRs. Functional cell chiefs and other staff leaders identify areas that may be affected within their areas of expertise by proposed changes to the order or mission. Course of action considerations include, but are not limited to, those shown in table 14-3.

14-21. The commander is as likely as anyone else to detect the need for change and to sketch out the options. Whether the commander, COS (XO), or chief of operations does this, the future operations cell is often directed to further develop the concept and draft the order. The chief of operations and the current operations integration cell normally lead this effort, especially if the response is needed promptly or the situation is not complex. The commander or COS (XO) is usually the decisionmaking authority, depending on the commander's delegation of authority.

14-22. Commanders may delegate authority for execution decisions to their deputies, COSs (XOs), or their operations officers. They retain personal responsibility for all decisions and normally retain the authority for approving adjustment decisions.

14-23. When reallocating resources or priorities, commanders assign only minimum essential assets to shaping operations. They use all other assets to weight the decisive operation. This applies when allocating resources for the overall operation or within a warfighting function.

14-24. Commanders normally direct the future operations cell or the current operations integration cell to prepare a fragmentary order (FRAGORD) setting conditions for executing a new COA. When lacking time to perform the MDMP, or quickness of action is desirable, commanders make an immediate adjustment decision—using intuitive decisionmaking—in the form of a focused COA. Developing the focused COA often follows mental war-gaming by commanders until they reach an acceptable COA. If time is available, commanders may direct the plans cell to develop a new COA using the MDMP, and the considerations for planning become operative. (See table 14-3.)

Table 14-3. Course of action considerations

Types	Actions	
Intelligence	• Modifying priority intelligence requirements and other intelligence requirements. • Updating named areas of interest and target areas of interest. • Updating the intelligence estimate.	• Updating the enemy situation template and enemy course of action statements. • Modifying the information collection plan. • Confirming or denying threat course of action.
Movement and Maneuver	• Assigning new objectives. • Assigning new tasks to subordinate units. • Adjusting terrain management. • Employing obscurants.	• Modifying airspace control measures. • Making unit boundary changes. • Emplacing obstacles. • Clearing obstacles. • Establishing and enforcing movement priority.
Fires	• Delivering fires against targets or target sets. • Modifying the high-payoff target list and the attack guidance matrix.	• Modifying radar zones. • Modifying the priority of fires. • Modifying fire support coordination measures.
Protection	• Moving air defense weapons systems. • Establishing decontamination sites. • Conducting chemical, biological, radiological, and nuclear reconnaissance. • Establish movement corridors on critical lines of communications.	• Changing air defense weapons control status. • Enhancing survivability through engineer support. • Revising and updating personnel recovery coordination. • Reassigning or repositioning response forces.
Sustainment	• Prioritizing medical evacuation assets. • Repositioning logistics assets. • Positioning and prioritizing detainee and resettlement assets.	• Repositioning and prioritizing general engineering assets. • Modifying priorities. • Modifying distribution.
Mission Command	• Moving communications nodes. • Moving command posts. • Modifying information priorities for employing information as combat power. • Adjusting themes and messages to support the new decision. • Adjusting measures for minimizing civilian interference with operations. • Revising recommended protected targets. • Recommending modifications of stability tasks, including employment of civil affairs operations and other units, to perform civil affairs operations tasks.	

REFINE AND VALIDATE THE COURSE OF ACTION

14-25. Once commanders describe the new COA, the current operations integration cell conducts an analysis to validate its feasibility, suitability, and acceptability. If acceptable, the COA is refined to resynchronize the warfighting functions enough to generate and apply the needed combat power. Staffs with a future operations cell may assign that cell responsibility for developing the details of the new COA and drafting a fragmentary order to implement it. The commander or COS (XO) may direct an "on-call" operations synchronization meeting to perform this task and ensure rapid resynchronization.

14-26. Validation and refinement are done quickly. Normally, the commander and staff officers conduct a mental war game of the new COA. They consider potential enemy reactions, the unit's counteractions, and secondary effects that might affect the force's synchronization. Each staff member considers the following items:

- Is the new COA feasible in terms of my area of expertise?
- How will this action affect my area of expertise?
- Does it require changing my information requirements?
    - Should any of the information requirements be nominated as a CCIR?
    - What actions within my area of expertise does this change require?
    - Will this COA require changing objectives or targets nominated by staff members?
- What other command post cells and elements does this action affect?
- What are potential enemy reactions?
- What are the possible friendly counteractions?
    - Does this counteraction affect my area of expertise?
    - Will it require changing my information requirements?
    - Are any of my information requirements potential CCIRs?
    - What actions within my area of expertise does this counteraction require?
    - Will it require changing objectives or targets nominated by staff members?
    - What other command post cells and elements does this counteraction affect?

14-27.   The validation and refinement will show if the COA will solve the problem adequately. If it does not, the COS or chief of operations modifies it through additional analysis or develops a new COA. The COS (XO) informs the commander of any changes made to the COA.

IMPLEMENT

14-28.   When a COA is acceptable, the COS (XO) recommends implementation to the commander or implements it directly, if the commander has delegated that authority. Implementation normally requires a FRAGORD; in exceptional circumstances, it may require a new operation order (OPORD). That order changes the concept of operations (in adjustment decisions), resynchronizes the warfighting functions, and disseminates changes to control measures. The staff uses warning orders (WARNORDs) to alert subordinates to a pending change. The staff also establishes sufficient time for the unit to implement the change without losing integration or being exposed to unnecessary tactical risk.

14-29.   Commanders often issue orders to subordinates verbally in situations requiring quick reactions. At battalion and higher echelons, written FRAGORDs confirm verbal orders to ensure synchronization, integration, and notification of all parts of the force. If time permits, leaders verify that subordinates understand critical tasks. Verification methods include the confirmation brief and backbrief. These are conducted both between commanders and within staff elements to ensure mutual understanding.

14-30.   After the analysis is complete, the current operations integration cell and command post cell chiefs update decision support templates and synchronization matrixes. When time is available, the operations officer or chief of operations continues this analysis to the operation's end to complete combat power integration. Staff members begin the synchronization needed to implement the decision. This synchronization involves collaboration with other command post cells and subordinate staffs. Staff members determine how actions in their areas of expertise affect others. They coordinate those actions to eliminate undesired effects that might cause friction. The cells provide results of this synchronization to the current operations integration cell and the common operational picture.

Chapter 15

Assessment Plans

This chapter provides information on assessment and its role in the operations process. Next, it describes the assessment process and defines key assessment terms. This chapter concludes by describing a methodology for developing formal assessment plans. (See ADRP 5-0 for more information on the fundamentals of assessment.)

ASSESSMENT AND THE OPERATIONS PROCESS

15-1. *Assessment* is the determination of the progress toward accomplishing a task, creating a condition, or achieving an objective (JP 3-0). Assessment precedes and guides the other activities of the operations process. Assessment involves deliberately comparing forecasted outcomes with actual events to determine the overall effectiveness of force employment. More specifically, assessment helps the commander determine progress toward attaining the desired end state, achieving objectives, and performing tasks. It also involves continuously monitoring and evaluating the operational environment to determine what changes might affect the conduct of operations.

15-2. Throughout the operations process, commanders integrate their own assessments with those of the staff, subordinate commanders, and other unified action partners. Primary tools for assessing progress of an operation include the operation order (OPORD), the common operational picture, personal observations, running estimates, and the assessment plan.

15-3. Assessment occurs at all echelons. The situation and echelon dictate the focus and methods leaders use to assess. Normally, commanders assess those specific operations or tasks that they were directed to accomplish. This properly focuses collection and assessment at each echelon, reduces redundancy, and enhances the efficiency of the overall assessment process.

15-4. For units with a staff, assessment becomes more formal at each higher echelon. Assessment resources (including staff officer expertise and time available) proportionally increase from battalion to brigade, division, corps, and theater army. The analytic resources and level of expertise of staffs available at higher echelon headquarters include a dedicated core group of analysts. This group specializes in operations research and systems analysis, formal assessment plans, and various assessment products. Division, corps, and theater army headquarters, for example, have fully resourced plans, future operations, and current operations integration cells. They have larger intelligence staffs and more staff officers trained in operations research and systems analysis. Assessment at brigade echelon and lower is usually less formal, often relying on direct observations and the judgment of commanders and their staffs.

15-5. Often, time available for detailed analysis and assessment is shorter at lower echelons. Additionally, lower echelon staffs are progressively smaller and have less analytic capability at each lower echelon. As such, assessment at these echelons focuses on the near term and relies more on direct observation and judgments than on detailed assessment plans and methods.

15-6. For small units (those without a staff), assessment is mostly informal. Small-unit leaders focus on assessing their unit's readiness—personnel, equipment, supplies, and morale—and their unit's ability to perform assigned tasks. Leaders also determine whether the unit has completed assigned tasks. If those tasks have not produced the desired results, leaders explore why they have not and consider what improvements could be made for unit operations. As they assess and learn, small units change their tactics, techniques, and procedures based on their experiences. In this way, even the lowest echelons in the Army follow the assessment process.

THE ASSESSMENT PROCESS

15-7. Assessment is continuous; it precedes and guides every operations process activity and concludes each operation or phase of an operation. Broadly, assessment consists of, but is not limited to, the following activities—

- Monitoring the current situation to collect relevant information.
- Evaluating progress toward attaining end state conditions, achieving objectives, and performing tasks.
- Recommending or directing action for improvement.

MONITORING

15-8. *Monitoring* is continuous observation of those conditions relevant to the current operation (ADRP 5-0). Monitoring within the assessment process allows staffs to collect relevant information, specifically that information about the current situation that can be compared to the forecasted situation described in the commander's intent and concept of operations. Progress cannot be judged, nor effective decisions made, without an accurate understanding of the current situation.

15-9. Staff elements record relevant information in running estimates. Staff elements maintain a continuous assessment of current operations to determine if they are proceeding according to the commander's intent, mission, and concept of operations. In their running estimates, staff elements use this new information and these updated facts and assumptions as the basis for evaluation.

EVALUATING

15-10. The staff analyzes relevant information collected through monitoring to evaluate the operation's progress toward attaining end state conditions, achieving objectives, and performing tasks. *Evaluating* is using criteria to judge progress toward desired conditions and determining why the current degree of progress exists (ADRP 5-0). Evaluation is at the heart of the assessment process where most of the analysis occurs. Evaluation helps commanders determine what is working and what is not working, and it helps them gain insights into how to better accomplish the mission.

15-11. Criteria in the forms of measures of effectiveness (MOEs) and measures of performance (MOPs) aid in evaluating progress. MOEs help determine if a task is achieving its intended results. MOPs help determine if a task is completed properly. MOEs and MOPs are simply criteria—they do not represent the assessment itself. MOEs and MOPs require relevant information in the form of indicators for evaluation.

15-12. A *measure of effectiveness* is a criterion used to assess changes in system behavior, capability, or operational environment that is tied to measuring the attainment of an end state, achievement of an objective, or creation of an effect (JP 3-0). MOEs help measure changes in conditions, both positive and negative. MOEs are commonly found and tracked in formal assessment plans. MOEs help to answer the question "Are we doing the right things?"

15-13. A *measure of performance* is a criterion used to assess friendly actions that is tied to measuring task accomplishment (JP 3-0). MOPs help answer questions such as "Was the action taken?" or "Were the tasks completed to standard?" A MOP confirms or denies that a task has been properly performed. MOPs are commonly found and tracked at all echelons in execution matrixes. MOPs are also commonly used to evaluate training. MOPs help to answer the question "Are we doing things right?" There is no direct hierarchical relationship among MOPs to MOEs. Measures of performance do not feed MOEs, or combine in any way to produce MOEs—MOPs simply measure the performance of a task.

15-14. In the context of assessment, an *indicator* is an item of information that provides insight into a measure of effectiveness or measure of performance (ADRP 5-0). Indicators take the form of reports from subordinates, surveys and polls, and information requirements. Indicators help to answer the question "What is the current status of this MOE or MOP?" A single indicator can inform multiple MOPs and MOEs. (See table 15-1 for additional information concerning MOEs, MOPs, and indicators.)

Table 15-1. Assessment measures and indicators

Measure of effectiveness (MOE)	Measure of performance (MOP)	Indicator
Used to measure attainment of an end state condition, achievement of an objective, or creation of an effect.	Used to measure task accomplishment.	Used to provide insight into a MOE or MOP.
Answers the question: Are we doing the right things?	Answers the question: Are we doing things right?	Answers the question: What is the status of this MOE or MOP?
Measures *why* (purpose) in the mission statement.	Measures *what* (task completion) in the mission statement.	Information used to make measuring *what* or *why* possible.
No direct hierarchical relationship to MOPs.	No direct hierarchical relationship to MOEs.	Subordinate to MOEs and MOPs.
Often formally tracked in formal assessment plans.	Often formally tracked in execution matrixes.	Often formally tracked in formal assessment plans.
Typically challenging to choose the appropriate ones.	Typically simple to choose the appropriate ones.	Typically as challenging to select appropriately as the supported MOE or MOP.

RECOMMENDING OR DIRECTING ACTION

15-15. Monitoring and evaluating are critical activities; however, assessment is incomplete without recommending or directing action. Assessment may diagnose problems, but unless it results in recommended adjustments, its use to the commander is limited.

15-16. When developing recommendations, staffs draw from many sources and consider their recommendations within the larger context of the operation. While several ways to improve a particular aspect of the operation might exist, some recommendations could impact other aspects of the operation. As with all recommendations, staffs should address any future implications.

ASSESSMENT PLAN DEVELOPMENT

15-17. Critical to the assessment process is developing an assessment plan. Units use assessment working groups to develop assessment plans when appropriate. A critical element of the commander's planning guidance is determining which assessment plans to develop. An assessment plan focused on attainment of end state conditions often works well. It is also possible, and may be desirable, to develop an entire formal assessment plan for an intermediate objective, a named operation subordinate to the base operation plan, or a named operation focused solely on a single line of operations or geographic area. The time, resources, and added complexity involved in generating an assessment plan strictly limit the number of such efforts.

15-18. Commanders and staffs integrate and develop an assessment plan within the military decisionmaking process (MDMP). As the commander and staff begin mission analysis, they also need to determine how to measure progress towards the operation's end state.

15-19. Effective assessment incorporates both quantitative (observation-based) and qualitative (judgment-based) indicators. Human judgment is integral to assessment. A key aspect of any assessment is the degree to which it relies upon human judgment and the degree to which it relies upon direct observation and mathematical rigor. Rigor offsets the inevitable bias, while human judgment focuses rigor and processes on intangibles that are often key to success. The appropriate balance depends on the situation—particularly the nature of the operation and available resources for assessment—but rarely lies at the ends of the scale.

ASSESSMENT STEPS

15-20. During planning, the assessment working group develops an assessment plan using six steps:

- Step 1—Gather tools and assessment data.
- Step 2—Understand current and desired conditions.
- Step 3—Develop an assessment framework.
- Step 4—Develop the collection plan.
- Step 5—Assign responsibilities for conducting analysis and generating recommendations.
- Step 6—Identify feedback mechanisms.

Once the assessment working group develops the assessment plan, it applies the assessment process of monitor, evaluate, and recommend or direct continuously throughout preparation and execution.

STEP 1—GATHER TOOLS AND ASSESSMENT DATA

15-21. Planning begins with receipt of mission. The receipt of mission alerts the staffs to begin updating their running estimates and gather the tools necessary for mission analysis and continued planning. Specific tools and information gathered regarding assessment include, but are not limited to—

- The higher headquarters' plan or order, including the assessment annex if available.
- If replacing a unit, any current assessments and assessment products.
- Relevant assessment products (classified or open-source) produced by civilian and military organizations.
- The identification of potential data sources, including academic institutions and civilian subject matter experts.

STEP 2—UNDERSTAND CURRENT AND DESIRED CONDITIONS

15-22. Fundamentally, assessment is about measuring progress toward the desired end state. To do this, commanders and staffs compare current conditions in the area of operations against desired conditions. Army design methodology and the MDMP help commanders and staffs develop an understanding of the current situation. As planning continues, the commander identifies desired conditions that represent the operation's end state.

15-23. Early in planning, commanders issue their initial commander's intent, planning guidance, and commander's critical information requirements (CCIRs). The end state in the initial commander's intent describes the desired conditions the commander wants to achieve. The staff element responsible for the assessment plan identifies each specific desired condition mentioned in the commander's intent. These specific desired conditions focus the overall assessment of the operation. Understanding current conditions and desired conditions forms the basis for building the assessment framework.

STEP 3—DEVELOP AN ASSESSMENT FRAMEWORK

15-24. All plans and orders have a general logic. This logic links tasks to subordinate units to the achievement of objectives, and the achievement of objectives to attainment of the operation's end state. An assessment framework incorporates the logic of the plan and uses measures (MOEs, MOPs, and indicators) as tools to determine progress toward attaining desired end state conditions. (See figure 15-1.)

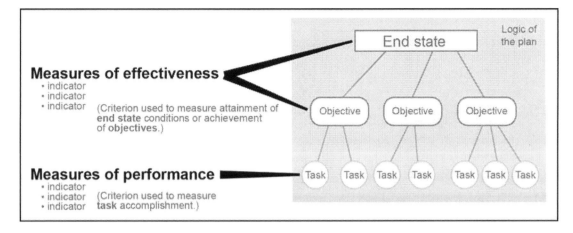

Figure 15-1. Assessment framework

15-25. Developing an assessment framework involves—
*   Selecting and writing measures (MOEs and MOPs).
*   Organizing the measures into an assessment framework.

Selecting and Writing Measures

15-26. Based on their understanding of the plan, members of the staff develop specific MOEs and MOPs (with associated indicators) to evaluate the operations process. Measures of effectiveness are tools used to help measure the attainment of end state conditions, achievement of objectives, or creation of effects. Measures of performance are criteria used to assess friendly actions that are tied to measuring task accomplishment.

Selecting and Writing Measures of Effectiveness

15-27. Guidelines for selecting and writing MOEs consist of the following:
*   Select only MOEs that measure the degree to which the desired outcome is achieved.
*   Choose distinct MOEs.
*   Include MOEs from different causal chains.
*   Use the same MOE to measure more than one condition when appropriate.
*   Avoid overburdening subordinates with additional reporting requirements.
*   Structure MOEs so that they have measurable, collectable, and relevant indicators.
*   Write MOEs as statements not questions.
*   Maximize clarity.

15-28. Commanders select only MOEs that measure the degree to which the desired outcome is achieved. There must be an expectation that a given MOE will change as the conditions being measured change.

15-29. Commanders choose MOEs for each condition as distinct from each other as possible. Using similar MOEs can skew the assessment by containing virtually the same MOE twice.

15-30. Commanders include MOEs from differing relevant causal chains for each condition whenever possible. When MOEs have a cause and effect relationship with each other, either directly or indirectly, it decreases their value in measuring a particular condition. Measuring progress towards a desired condition by multiple means adds rigor to the assessment.

15-31. In the example in figure 15-2 on page 15-6 under condition 1, both MOE 1 and MOE 3 have no apparent cause and effect relationship with each other, although both are valid measures of the condition. This adds rigor and validity to the measurement of that condition. MOE 2 does have a cause and effect relationship with MOE 1 and MOE 3, but it is a worthwhile addition because of the direct relevancy and mathematical rigor of that particular source of data. (See figure 15-2.)

End state condition 1: Enemy division X forces prevented from interfering with corps decisive operation.

MOE 1: Enemy division X forces west of phase line blue are defeated.
- **Indicator 1**: Friendly forces occupy objective SLAM (yes or no).
- **Indicator 2**: Number of reports of squad-sized or larger enemy forces in the division area of operations in the past 24 hours.
- **Indicator 3**: Current G-2 assessment of number of enemy division X battalions west of phase line blue.

MOE 2: Enemy division X forces indirect fire systems neutralized.
- **Indicator 1**: Number of indirect fires originating from enemy division X's integrated fires command in the past 24 hours.
- **Indicator 2**: Current G-2 assessment of number of operational 240mm rocket launchers within enemy division X's integrated fires command.

MOE 3: Enemy division X communications systems disrupted.
- **Indicator 1**: Number of electronic transmissions from enemy division X detected in the past 24 hours.
- **Indicator 2**: Number of enemy division X battalion and higher command posts destroyed.

| G-2 | assistant chief of staff, intelligence | mm | millimeter |
| MOE | measure of effectiveness | | |

Figure 15-2. Example end state conditions for a defense

15-32. Commanders use the same MOE to measure more than one condition when appropriate. This sort of duplication in the assessment framework does not introduce significant bias unless carried to the extreme.

15-33. Commanders avoid or minimize additional reporting requirements for subordinate units. In many cases, commanders use information requirements generated by other staff elements as MOEs and indicators in the assessment plan. With careful consideration, commanders and staffs can often find viable alternative MOEs without creating new reporting requirements. Excessive reporting requirements can render an otherwise valid assessment plan onerous and untenable.

15-34. Commanders structure MOEs so that measurable, collectable, and relevant indicators exist for them. A MOE is of no use if the staff cannot actually measure it.

15-35. Commanders write MOEs as statements, not questions. MOEs supply answers to questions rather than the questions themselves.

15-36. Commanders maximize clarity. A MOE describes the sought information precisely, including specifics on time, information, geography, or unit, if needed. Any staff member should be able to read the MOE and understand exactly what information it describes.

Selecting and Writing Measures of Performance

15-37. MOPs are criteria used to assess friendly actions that are tied to measuring task accomplishment. MOPs help to answer questions such as "Was the action taken?" or "Were the tasks completed to standard?" A MOP confirms or denies that a task has been properly performed. MOPs are commonly found and tracked at all levels in execution matrixes.

15-38. In general, operations consist of a series of collective tasks sequenced in time, space, and purpose to accomplish missions. The current operations cells use MOPs in execution matrixes and running estimates to track completed tasks. Evaluating task accomplishment using MOPs is relatively straightforward and often results in a yes or no answer. Examples of MOPs include:
- Route X cleared.
- Generators delivered, are operational, and are secured at villages A, B, and C.

- $15,000 spent for schoolhouse completion.
- Aerial dissemination of 60,000 leaflets over village D.

Selecting and Writing Indicators

15-39. Staffs develop indicators that provide insights into MOEs and MOPs. Staffs can gauge a measurable indicator either quantitatively or qualitatively. Imprecisely defined indicators often pose a problem. For example, staffs cannot measure the indicator "Number of local nationals shopping." The information lacks clear parameters in time or geography. Staffs can measure the revised indicator "Average daily number of local nationals visiting main street market in city X this month." Additionally, staffs should design the indicator to minimize bias. This particularly applies when staffs only have qualitative indicators available for a given MOE. Many qualitative measures are easily biased, and Soldiers must use safeguards to protect objectivity in the assessment process. (See figure 15-3.)

15-40. A collectable indicator has reasonably obtained data associated with the indicator. In some cases, the data may not exist, or the data may be prohibitively difficult to collect. As an example, if condition 2 MOE 2 in figure 15-3 had the indicator "Host-nation medical care availability in city X this month" that indicator is not likely collectable. This number exists, but unless a trusted source tracks and reports it, Soldiers cannot collect it. The revised indicator "Battalion commander's monthly estimate of host-nation medical care availability in city X on a scale of 1 to 5" is collectable. In this case, the staff did not have a quantitative indicator available, so they substituted a qualitative indicator.

End state condition 1: Enemy defeated in the brigade area of operations.

MOE 1: Enemy kidnapping activity in the brigade area of operations disrupted.
- **Indicator 1**: Monthly reported dollars in ransom paid as a result of kidnapping operations.
- **Indicator 2**: Monthly number of reported attempted kidnappings.
- **Indicator 3**: Monthly poll question #23: "Have any kidnappings occurred in your neighborhood in the past 30 days?" Results for provinces ABC only.

MOE 2: Public perception of security in the brigade area of operations improved.
- **Indicator 1**: Monthly poll question #34: "Have you changed your normal activities in the past month because of concerns about your safety and that of your family?" Results for provinces ABC only.
- **Indicator 2**: Monthly kindergarten through high school attendance in provinces ABC as reported by the host-nation ministry of education.
- **Indicator 3**: Monthly number of tips from local nationals reported to the brigade terrorism tips hotline.

MOE 3: Sniper events in the brigade area of operations disrupted.
- **Indicator 1**: Monthly decrease in reported sniper events in the brigade area of operations. (*Note:* It is acceptable to have only one indicator that directly answers a given MOE. Avoid complicating the assessment needlessly when a simple construct suffices.)

Condition 2: Role 1 medical care available to the population in city X.

MOE 1: Public perception of medical care availability improved in city X.
- **Indicator 1**: Monthly poll question #42: "Are you and your family able to visit a doctor or health clinic when you need to?" Results for provinces ABC only.
- **Indicator 2**: Monthly poll question #8: "Do you and your family have important health needs that are not being met?" Results for provinces ABC only.
- **Indicator 3**: Monthly decrease in number of requests for medical care received from local nationals by the brigade.

MOE 2: Battalion commander estimated monthly host-nation medical care availability in battalion area of operations.
- **Indicator 1**: Monthly average of reported battalion commander's estimates (scale of 1 to 5) of host-nation medical care availability in the battalion area of operation.

MOE	measure of effectiveness

Figure 15-3. Example end state conditions for a stability operation

15-41.   An indicator is relevant if it provides insight into a supported MOE or MOP. Commanders must ask pertinent questions, such as—

● Does a change in this indicator actually indicate a change in the MOE?
● What factors unrelated to the MOE could cause this indicator to change?
● How reliable is the correlation between the indicator and the MOE?

15-42.   For example, the indicator "Decrease in monthly weapons caches found and cleared in the division area of operations" is not relevant to the MOE "Decrease in enemy activity in the division area of operations."

15-43.   The indicator could plausibly increase or decrease with a decrease in enemy activity. An increase in friendly patrols could result in greater numbers of caches found and cleared. Staffs may also have difficulty determining when the enemy left the weapons, raising the question of when the enemy activity actually occurred. These factors, unrelated to enemy activity, could artificially inflate the indicator, creating a false impression of increased enemy activity within the assessment framework. In this example, staffs can reliably measure enemy activity levels without considering weapons caches or using the indicator for this MOE.

Organizing the Assessment Framework

15-44.   When organizing the assessment framework, MOEs and MOPs can be applied to the logic of the plan in different ways. The overall approach to using those measures is visually depicted in the assessment framework. That overall approach may be more quantitative or more qualitative, although it rarely resides at the extremes.

15-45.   Regardless of the specific measures chosen and the specific ways they are used in the formal assessment plan, it is imperative that the staff explicitly records the logic it uses to create the assessment framework. Every measure used is chosen for a reason, and that reason must be recorded in a narrative form in the formal assessment plan. Specific measures are combined (or not combined) in specific ways in the formal assessment, and those reasons must be recorded explicitly in the formal assessment plan. Lessons learned have shown that the rationale involved in creating a formal assessment plan can be rapidly lost. Recording the logic in the assessment plan mitigates this risk.

15-46.   One example of organizing an assessment framework is a hierarchical structure which begins with end state conditions, followed by MOEs, and finally indicators. Commanders broadly describe the operation's end state in their commander's intent. The assessment working group then identified specific desired conditions from the commander's intent. Staffs measure each condition by MOEs. The MOEs are in turn informed by indicators.

15-47.   Such a formal assessment framework is simply a tool to assist commanders with estimating progress. Using a formal assessment framework does not imply that commanders mathematically determine the outcomes of military operations. Commanders and staff officers apply judgment to results of mathematical assessment to assess progress holistically. For example, commanders in an enduring operation may receive a monthly formal assessment briefing from their staff. This briefing includes both the products of the formal assessment process as well as the expert opinions of members of the staff, subordinate commanders, and other partners. Commanders combine what they find useful in those two viewpoints with their personal assessment of the operation, consider recommendations, and direct action as needed.

15-48.   A significant amount of human judgment goes into designing such an assessment framework. Choosing MOEs and indicators that accurately measure progress toward each desired condition is an art. Processing elements of the assessment framework requires establishing weights and thresholds for each MOE and indicator. Setting proper weights and thresholds requires operational expertise and judgment. Input from the relevant staff elements and subject matter experts is critical. Staffs record the logic of why the commander chose each MOE and indicator. This facilitates personnel turnover as well as an understanding of the assessment plan among all staff elements.

15-49.   Another approach to organizing an assessment framework focuses on producing narrative assessments for each end state condition, using measures (MOEs, MOPs, and indicators) to argue for or against progress rather than combining the measures in one holistic mathematical model. Under this approach, the assessment working group builds the most convincing arguments they can for and against the achievement of a given end state condition. Significant mathematical rigor may be a part of these arguments, but a holistic mathematical model is rejected. After the for and against arguments are produced, the assessment working group applies judgment to that raw material to create the most plausible narrative assessment for that end state condition. The chair of the assessment working group determines when the collective judgment of the group is in dispute.

STEP 4—DEVELOP THE COLLECTION PLAN

15-50. Each indicator represents an information requirement. In some situations, staffs feed these information requirements into the information collection synchronization process. Then, staffs task information collection assets to collect on these information requirements. In other situations, reports in the unit standard operating procedures (SOPs) may suffice. If not, the unit may develop a new report. Staffs may collect the information requirement from organizations external to the unit. For example, a host nation's central bank may publish a consumer price index for that nation. The assessment plan identifies the source for each indicator as well as the staff member who collects that information. Assessment information requirements compete with other information requirements for resources. When an information requirement is not resourced, staffs cannot collect the associated indicator and must remove it from the plan. Staffs then adjust the assessment framework to ensure that the MOE or MOP is properly worded.

STEP 5—ASSIGN RESPONSIBILITIES FOR CONDUCTING ANALYSIS AND GENERATING RECOMMENDATIONS

15-51. In addition to assigning responsibility for collection, commanders assign staff members to analyze assessment data and develop recommendations. For example, the intelligence officer leads the assessment of enemy forces. The engineer officer leads the effort on assessing infrastructure development. The civil affairs operations officer leads assessment concerning the progress of local and provincial governments. The chief of staff aggressively requires staff principals and subject matter experts to participate in processing the formal assessment and in generating smart, actionable recommendations. The operations research and analysis officer assists the commander and staff with developing both assessment frameworks and the command's assessment process.

STEP 6—IDENTIFY FEEDBACK MECHANISMS

15-52. A formal assessment with meaningful recommendations that is not presented to the appropriate decisionmaker wastes time and energy. The assessment plan identifies the who, what, when, where, and why of that presentation. The commander and staff discuss feedback leading up to and following that presentation as well. Feedback might include which assessment working groups the commander requires and how to act on and follow up on recommendations.

This page intentionally left blank.

Chapter 16

After Action Reviews and Reports

This chapter discusses how to plan for and use the after action review as a learning tool. This chapter concludes with a discussion of and a format for the after action report.

INTRODUCTION TO AFTER ACTION REVIEWS AND REPORTS

16-1. An *after action review* is a guided analysis of an organization's performance, conducted at appropriate times during and at the conclusion of a training event or operation with the objective of improving future performance. It includes a facilitator, event participants, and other observers (ADRP 7-0). Leaders can use action reviews not only for training situations, but also for operations. Leaders can also employ after action reviews during pauses in action, as individual missions are completed, or after phases of the operation as time permits, enabling units to also learn during operations.

16-2. The after action report is the written record of a unit's after action review, or a consolidation of comments, lessons learned, or best practices during the course of an operation or exercise. The after action report documents a unit's actions for historical purposes and highlights key lessons learned and best practices. Commanders systematically collect, use, and share lessons learned and best practices throughout an operation or extended training event. Additionally, lessons learned and best practices are available for consideration and application by the operating and generating force through the Center for Army Lessons Learned. (See paragraphs 16-50 to 16-53 for an after action report discussion and table 16-1 for an after action report format.)

FORMAL AND INFORMAL AFTER ACTION REVIEWS

16-3. Two types of after action reviews exist: formal and informal. Commanders generally conduct formal action reviews after completing a mission. Normally, only informal after action reviews are possible during the conduct of operations.

16-4. Leaders plan formal after action reviews when they complete an operation or otherwise realize they have the need, time, and resources available. Formal after action reviews require more planning and preparation than informal after action reviews. Formal after action reviews require site reconnaissance and selection; coordination for aids (such as terrain models and large-scale maps); and selection, setup, maintenance, and security of the after action review site. During formal after action reviews, the after action review facilitator (unit leader or other facilitator) provides an overview of the operation and focuses the discussion on topics the after action review plan identifies. At the conclusion, the facilitator reviews identified and discussed key points and issues, and summarizes strengths and weaknesses.

16-5. Leaders use informal after action reviews as on-the-spot coaching tools while reviewing Soldier and unit performance during or immediately after execution. Informal after action reviews involve all Soldiers. These after action reviews provide immediate feedback to Soldiers, leaders, and units after execution. Ideas and solutions leaders gathered during informal after action reviews can be applied immediately as the unit continues operations. Successful solutions can be identified and transferred as lessons learned.

16-6. Formal and informal after action reviews generally follow the same format:
- Review what was supposed to happen. The facilitator and participants review what was supposed to happen. This review is based on the commander's intent for the operation, unit operation or fragmentary orders (FRAGORDs), the mission, and the concept of operations.
- Establish what happened. The facilitator and participants determine to the extent possible what actually happened during execution. Unit records and reports form the basis of this

determination. An account describing actual events as closely as possible is vital to an effective discussion. The assistant chief of staff, intelligence (G-2 [S-2]) provides input about the operation from the enemy's perspective.

- Determine what was right or wrong with what happened. Participants establish the strong and weak points of their performance. The facilitator guides discussions so that the conclusions the participants reach are operationally sound, consistent with Army standards, and relevant to the operational environment.

- Determine how the task should be done differently next time. The facilitator helps the chain of command lead the group in determining how participants might perform the task more effectively. The intended result is organizational and individual learning that can be applied to future operations. If successful, this learning can be disseminated as lessons learned.

16-7. Leaders understand that not all tasks will be performed to standard. In their initial planning, they allocate time and other resources for retraining after execution or before the next operation. Retraining allows participants to apply the lessons learned from after action reviews and implement corrective actions. Retraining should be conducted at the earliest opportunity to translate observations and evaluations from after action reviews into performance in operations. Commanders ensure Soldiers understand that training is incomplete until the identified corrections in performance have been achieved. Successful lessons can be identified as lessons learned and disseminated.

16-8. After action reviews are often tiered as multi-echelon leader development tools. Following a session involving all participants, senior commanders may continue after action reviews with selected leaders as extended professional discussions. These discussions usually include a more specific review of leader contributions to the operation's results. Commanders use this opportunity to help subordinate leaders master current skills and prepare them for future responsibilities. After action reviews are opportunities for knowledge transfer through teaching, coaching, and mentoring.

16-9. Commanders conduct a final after action review during recovery after an operation. This after action review may include a facilitator. Unit leaders review and discuss the operation. Weaknesses or shortcomings identified during earlier after action reviews are identified again and discussed. If time permits, the unit conducts training to correct these weaknesses or shortcomings in preparation for future operations.

16-10. Lessons learned can be disseminated in at least three ways. First, participants may make notes to use in retraining themselves and their sections or units. Second, facilitators may gather their own and participants' notes for collation and analysis before dissemination and storage for others to use. Dissemination includes forwarding lessons to other units conducting similar operations as well as to the Center for Army Lessons Learned, doctrinal proponents, and generating force agencies. Third, units should publicize future successful applications of lessons as lessons learned.

BENEFITS OF AFTER ACTION REVIEWS

16-11. After action reviews are the dynamic link between task performance and execution to standard. Through the professional, candid discussion of events, Soldiers can identify what went right and what went wrong during the operation using measures of effectiveness. When appropriate, they can evaluate their performance of tasks using measures of performance. (See chapter 15 for more information on measures of effectiveness and measures of performance).

16-12. The discussion helps Soldiers and leaders identify specific ways to improve unit proficiency. Units achieve the benefits of after action reviews by applying their results. Applications may include organizing observations, insights, and lessons learned; revising how the unit executes tactics, techniques, and procedures; and developing future training.

16-13. After action reviews may reveal problems with unit standard operating procedures. If so, unit leaders revise the procedures and ensure that the unit implements the changes during future operations. Leaders can use the knowledge that after action reviews develop to assess performance, correct deficiencies, and sustain demonstrated task proficiency. These improvements will enhance unit performance in future operations.

CONDUCTING AFTER ACTION REVIEWS

16-14.   Effective after action reviews require planning and preparation. During planning for an operation, commanders allocate time and resources for conducting after action reviews and assign responsibilities for them. The amount and level of detail needed during planning and preparation depends on the type of after action review and the resources available. The after action review process has four steps:

- Step 1—Plan.
- Step 2—Prepare.
- Step 3—Execute.
- Step 4—Follow-up (using after action reports).

16-15.   After action reviews during operations differ from those during training in the lack of observer controllers or observer trainers. During operations, there are no dedicated collectors of data and observations. Instead, unit generated assessments of the operation's progress form the basis of the after action review.

PLANNING AFTER ACTION REVIEWS

16-16.   An after action review plan provides the foundation for a successful after action review. Commanders develop a plan for each after action review as time allows. The plan specifies—

- Who will conduct the after action review.
- Who will provide information.
- Aspects of the operation the after action review should evaluate.
- Who will attend the after action review.
- When and where the after action review will occur.
- Aids to be used for the after action review.

16-17.   Commanders or facilitators use the after action review plan to identify critical places and events that must be covered to provide a useful after action review. Examples include the decisive operation, critical transitions, and essential tasks. The after action review plan also includes who will address each event.

16-18.   Commanders specify what they want to accomplish with the after action review and what the after action review will address. The operation order (OPORD) may provide tasks and conditions. Measures of effectiveness and some measures of performance are extracted from the order. FM 7-15 contains recommended measures of performance to develop training and evaluation outline evaluation criteria for supporting tasks. The primary source for standards for most Army units is their proponent-approved collective tasks.

16-19.   Copies of the OPORD and daily journal are given to the senior facilitator. The senior facilitator distributes these to the after action review team members to review and use them to identify critical events and times for discussion during the after action review.

Scheduling After Action Reviews

16-20.   Commanders plan for an after action review at the end of each operation whenever possible. The after action review planner should decide the scope of the after action review and allocate sufficient time. Quality after action reviews help Soldiers receive better feedback on their performance and remember lessons longer.

Determining Who Will Attend

16-21.   The after action review plan specifies who the commander wants to attend the after action review. At each echelon, an after action review has a primary set of participants. At squad and platoon levels, all Soldiers should attend and participate. At company and higher levels, it may not be practical to have everyone attend because of operations or training. In this case, unit commanders, other unit leaders, and other key players may be the only participants. Facilitators may recommend additional participants, based on their observations.

Choosing After Action Review Aids

16-22.   Appropriate aids add to an after action review's effectiveness; however, facilitators use an aid only if it makes the after action review better. Aids should promote learning and directly support discussion of the operation. Dry-erase boards, video equipment, terrain models, enlarged maps, and unit information systems are all worthwhile under the right conditions. Terrain visibility, group size, suitability for the task, and availability of electric power are all considerations when selecting after action review aids.

Preparing the After Action Review Plan

16-23.   The after action review plan is only a guide. Commanders and facilitators should review it regularly to make sure it still applies and meets the unit's needs. The plan may be adjusted as necessary, but changes take preparation and planning time away from facilitators and leaders. The after action review plan should allow facilitators and leaders as much time as possible to prepare.

PREPARING FOR AFTER ACTION REVIEWS

16-24.   Preparation is key to effectively executing any plan. Facilitators begin to prepare for an after action review before the operation and continue preparations until the actual event. Facilitators announce to unit leaders the starting time and location as soon as possible after these are set. This lets unit leaders account for personnel and equipment, perform post-operation actions, and move to the after action review site while facilitators are preparing and rehearsing.

Reviewing Objectives, Orders, Plans, and Doctrine

16-25.   Facilitators review the unit's mission before the after action review. The mission's objectives form the after action review's focus and the basis for observations. Facilitators review current doctrine, technical information, and applicable unit standard operating procedures to ensure they have the tools needed to properly guide discussion of unit and individual performance. Facilitators read and understand all warning orders (WARNORDs), OPORDs, and FRAGORDs issued before and during execution to understand what the commander wanted to happen. The detailed knowledge that facilitators display as a result of these reviews gives added credibility to their comments.

Identifying Key Events

16-26.   Facilitators identify critical events and ensure they collect data on those events or identify personnel who observed them. Examples of critical events include, but are not limited to—
- Issuance of OPORDs and FRAGORDs.
- Selected planning steps.
- Contact with opposing forces.
- Civil security attacks while conducting stability tasks.
- Passages of lines and reliefs in place.

Collecting Observations

16-27.   Facilitators need a complete picture of what happened during the operation to conduct an effective after action review. Each facilitator for subordinate, supporting, and adjacent units provides the senior facilitator with a comprehensive review of collected data on their organizations and the impact those units had on the unit accomplishing its mission.

16-28.   The senior facilitator receives input on the enemy from the G-2 (S-2). The enemy's perspective is critical to identifying why a unit succeeded or not. During formal after action reviews, the G-2 (S-2) briefs what is known of the enemy's plan and intent to set the stage for discussing what happened and why it happened. Obtaining this data after operations is extremely difficult; therefore, these observations often are treated as assumptions rather than facts.

16-29.   During their review, facilitators accurately record what they learn about events by time sequence to avoid losing valuable information and feedback. Facilitators use any recording system that is reliable (notebooks and laptops), sufficiently detailed (identifying times, places, and names), and consistent.

16-30.   Facilitators include the date-time group of each observation so that they can easily integrate it with other observations. This practice provides a comprehensive and detailed overview of what happened. When facilitators have enough time, they review their notes and fill in any details not written down earlier.

16-31.   One of the most difficult facilitator tasks is determining when and where to obtain information about the operation or the aspects of it selected for the after action review. Facilitators remain professional, courteous, and respectful at all times.

Organizing the After Action Review

16-32.   Once facilitators have gathered all available information, they organize their notes chronologically to understand the flow of events. They select and sequence key events in terms of their relevance to the unit's mission and objectives. This helps them identify key discussion and teaching points.

16-33.   An effective after action review leads participants to discover strengths and weaknesses, propose solutions, and adopt a course of action to improve future operations. Facilitators organize an after action review using one of three methods: chronological order of events, warfighting functions, or key events, themes, or issues.

Chronological Order of Events

16-34.   An after action review using the chronological order of events is logical, structured, and easy to understand. It follows the flow of the operation from start to finish. Covering actions in the order they occurred helps Soldiers and leaders better recall what happened. This method usually cannot cover all actions, only critical events.

Warfighting Functions

16-35.   This type of an after action review allows participants to discuss the operation across all its phases by warfighting function. This method identifies systemic strengths and weaknesses and is useful for staff growth and learning.

Key Events, Themes, or Issues

16-36.   An after action review using key events, themes, or issues focuses the discussion on critical operational events that directly support achieving the after action review's objectives. This works well when time is limited.

Selecting After Action Review Sites

16-37.   After action reviews should occur at or near where the operation occurred. Leaders should identify and inspect the after action review site and prepare a diagram showing placement of aids and other equipment. A good site minimizes wasted time by allowing rapid assembly of key personnel and positioning of aids. For larger units, this might not be possible for the whole operation. However, higher echelon after action reviews may include visits to selected actual sites to provide learning opportunities.

16-38.   The after action review site should let Soldiers see the terrain where the operation occurred or accurate representations of it. If this is not possible, facilitators find a location that allows Soldiers to see where the critical or most significant actions happened. Facilitators should have a map or other representation of the area of operations detailed enough to help everyone relate key events to the actual terrain. The representation may be a terrain model, enlarged map, or sketch. Facilitators also require a copy of the unit's graphics or recovered displays of the situation from the information systems databases.

16-39. Facilitators provide a comfortable setting for participants by encouraging Soldiers to remove helmets, providing shelter, and serving refreshments. These actions create an environment where participants can focus on the after action review without distractions. Participants should not face into the sun. Key leaders should have seats up front. Vehicle parking and equipment security areas should be far enough away from the after action review site to prevent distractions.

Rehearsing

16-40. After thorough preparation, the facilitator reviews the agenda and prepares to conduct the after action review. Facilitators may opt to conduct a walkthrough of the after action review site as well as review the sequence of events planned for the after action review.

EXECUTING AFTER ACTION REVIEWS

16-41. Facilitators start an after action review by reviewing its purpose and sequence: the ground rules, the objectives, and a summary of the operation that emphasizes the functions or events to be covered. This ensures that everyone present understands what the commander expects the after action review to accomplish.

Introduction and Rules

16-42. The following rules apply to all after action reviews. Facilitators emphasize them in their introduction.

- An after action review is a dynamic, candid, professional discussion that focuses on unit performance. Everyone with an insight, observation, or question participates. Total participation is necessary to maintain unit strengths and to identify and correct deficiencies.
- An after action review is not a critique. No one—regardless of rank, position, or strength of personality—has all the information or answers. After action reviews maximize learning benefits by allowing Soldiers to learn from each other.
- An after action review assesses weaknesses to improve and strengths to sustain.

16-43. Soldier participation is directly related to the atmosphere created during the introduction. Effective facilitators draw in Soldiers who seem reluctant to participate. The following ideas can help create an atmosphere conducive to maximum participation:

- Reinforce the idea that it is permissible to disagree.
- Focus on learning and encourage Soldiers to give honest opinions.
- Use open-ended and leading questions to guide the discussion.
- Facilitators enter the discussion only when necessary.

Review of Objectives and Intent

16-44. After the introduction, facilitators review the after action review's objectives. This review includes the following:

- A restatement of the events, themes, or issues being reviewed.
- The mission and commander's intent (what was supposed to happen).
- The enemy's mission and intent (how the enemy tried to defeat the force).

16-45. The commander or a facilitator restates the mission and commander's intent. Facilitators may guide the discussion to ensure that everyone present understands the plan and intent. Another method is to have subordinate leaders restate the mission and discuss the commander's intent. Automated information systems, maps, operational graphics, terrain boards, and other aids can help portray this information.

16-46. Intelligence personnel then explain as much of the enemy plan and actions as they know. The same aids the friendly force commander used can help participants understand how the plans related to each other.

Summary of Events (What Happened)

16-47.   The facilitator guides the review, using one of the methods to describe and discuss what actually happened. Facilitators avoid asking yes-or-no questions. They encourage participation and guide the discussion by using open-ended and leading questions. Open-ended questions allow those answering to reply based on what they think is significant. These questions are less likely to put Soldiers on the defensive. Open-ended questions work more effectively in finding out what happened.

16-48.   As the discussion expands and more Soldiers add their perspectives, what really happened becomes clearer. Facilitators do not tell Soldiers and leaders what was good or bad. Instead, they ensure that the discussion reveals the important issues, both positive and negative. Facilitators may want to expand this discussion and ask, "What could have been done differently?" Skillful guiding of the discussion ensures that participants do not gloss over mistakes or weaknesses.

Closing Comments (Summary)

16-49.   During the summary, facilitators review and summarize key points identified during the discussion. The after action review should end on a positive note, linking conclusions to learning and possible training. Facilitators then depart to allow unit leaders and Soldiers time to discuss the learning in private.

THE AFTER ACTION REPORT

16-50.   One of the most important collection techniques used in the Army and many other organizations is the after action report. The concept of the after action report can be easily adapted to fit any unit's lessons learned program.

16-51.   The after action report provides observations and insights from the lessons learned that allow the unit to reflect on the successes and shortcomings of the operation, and share these lessons with the Army.

16-52.   The reporting unit organizes the after action report in a logical order, usually by operational phase or warfighting function. It should be arranged chronologically when doing so facilitates the understanding and flow of the information reported. Documenting what worked well should receive as much attention as what did not.

16-53.   Table 16-1 on pages 16-8 and 16-9 is an example of what a commander and staff may elect to cover in their unit's written after action report. This approved brigade after action report template can apply across all echelons.

Table 16-1. Brigade after action report format

1. Report cover page:
     a. *Classification.*
     b. *Preparing headquarters or organization.*
     c. *Location of report preparation.*
     d. *Date of preparation.*
     e. *After action report title.*
     f. *Period covered: (date to date).*

2. Preface or foreword signed by the commander.

3. Table of contents. Keep information arranged in a logical order (by warfighting function, chronologically, or phases of operation.)

4. Executive summary and chronology of significant events:
     a. *Briefly summarize operations for all phases; include key dates for each phase starting with pre-deployment, transitioning through deployment, and ending with redeployment.*
     b. *Summarize task organization.*
     c. *Summarize key lessons learned (include level where lessons learned occurred).*
     d. *Summarize recommendations with timeline for correction to occur (makes it a historical document as leaders change units or missions).*

5. Detailed task organization. Include any significant changes and dates as appropriate:
     a. *Include organizational diagrams, including attached units, elements, and named task forces, including enablers and clearance authorities.*
     b. *Highlight any significant task organizational challenges (command and support relationships) and how they were mitigated.*
     c. *Effective dates of task organization to include all attached, operationally controlled units and individuals, including contractors.*

6. Pre-deployment phase with dates:
     a. *Unit's training focus:*
        ▪ *Describe the training plan.*
        ▪ *Did the unit have all of the elements that it would deploy with during training?*
        ▪ *Was the unit able to train as a combined arms team with all deploying assets participating?*
        ▪ *What assets outside the unit were used to support training?*
        ▪ *What were the key and essential areas trained?*
        ▪ *Describe what simulation systems (such as live, virtual, constructive, and gaming) were instrumental in training success.*
     b. *Discuss lessons learned during pre-deployment operations. What was planned but not executed?*
     c. *Discuss logistics and personnel shortages, if appropriate.*
     d. *Discuss planning for rear detachment operations.*
     e. *Describe any major shifts in personnel or manning.*
        ▪ *Coordination with non-military support agencies (unit associations, veterans groups, financial institutions, local government, law enforcement).*
        ▪ *Family readiness groups.*
     f. *What were the significant pre-deployment training lessons learned? (Use the "observation-discussion-recommendation" format).*
     g. *What were the significant gaps identified in leader development or proficiency?*

7. Deployment and reception, staging, onward movement, and integration (RSOI) with dates:
     a. *Summarize deployment and RSOI operations.*
     b. *Discuss what portions of the RSOI process went as planned and what worked.*
     c. *What were the shortcomings and delays in the RSOI? Why did these occur?*

8. Relief in place or transfer of authority with dates (if applicable):
     a. *Discuss planning and overlap.*
     b. *List or discuss key discussion topics between outgoing and incoming organizations.*
     c. *Include (either here or as an appendix) any standard operating procedures, tactics, techniques and procedures (TTP), or checklists.*

Table 16-1. Brigade after action report format (continued)

d. *Discuss relief in place or transfer of authority lessons learned in the "observation-discussion-recommendation" format.*

9. Operations phase with dates:

a. *Summarize tactical and non-tactical operations (sometimes beneficial to do this by staff element or warfighting function).*

b. *Include unit participation in named operations.*

c. *List of key operation orders (OPORDs) and fragmentary orders (FRAGORDs).*

d. *Discuss operations phases (sometimes beneficial to address by warfighting function).*

e. *Discuss any mission command challenges.*

10. Redeployment activities with dates:

a. *Summarize redeployment activities and highlight planning guidance either developed or received from higher headquarters.*

b. *Redeployment timeline.*

c. *List of critical losses, personnel, equipment, and information.*

d. *Include (either here or as an appendix) any list of instructions, TTP, or checklists developed.*

11. Post-deployment activities:

a. *Discuss combat stress planning and reintegration activities.*

b. *Discuss plans and priorities used in reconstituting and resetting the unit.*

c. *Discuss family support group operations.*

12. Provide an index or listing of all mid-tour and final unit after action report products, significant command briefings, or reports published separately:

a. *Include classification, titles, and distribution or disposition of reports.*

b. *Include a staff or section point of contact for follow-up coordination.*

c. *Include dates for scheduled umbrella week and warfighting function symposiums.*

13. Distribution (of this report).

14. Appendixes (as appropriate):

a. *List of each named operation or major event with dates.*

b. *Applicable maps.*

c. *Photographs.*

d. *Copies of key OPORDs and FRAGORDs.*

e. *Particularly useful TTP or unit products developed.*

f. *Pre-deployment site survey information.*

g. *Rear detachment operations.*

h. *Unit daily journals.*

This page intentionally left blank.

Appendix A

Operational and Mission Variables

The operational and mission variables are tools for commanders and staffs to use for analyzing an operation and organizing information. This appendix provides a description of the variables to assist commanders and staffs in applying these tools.

OPERATIONAL AND MISSION VARIABLES AND SITUATIONAL UNDERSTANDING

A-1. Commanders and staffs use the operational and mission variables to help build their situational understanding. They analyze and describe an operational environment in terms of eight interrelated operational variables: political, military, economic, social, information, infrastructure, physical environment, and time (PMESII-PT). Upon receipt of a mission, commanders filter information categorized by the operational variables into relevant information with respect to the mission. They use the mission variables, in combination with the operational variables, to refine their understanding of the situation and to visualize, describe, and direct operations. The mission variables are mission, enemy, terrain and weather, troops and support available, time available, and civil considerations (METT-TC). (See ADRP 5-0 for more information on situational understanding.)

OPERATIONAL VARIABLES

A-2. The operational variables are fundamental to developing a comprehensive understanding of the operational environment. The following is a brief description of each variable, along with examples (in parentheses) of questions a small-unit commander might need to have answered about each variable.

- *Political.* This variable describes the distribution of responsibility and power at all levels of governance—formally constituted authorities, as well as informal or covert political powers. (Who is the tribal leader in the village?)
- *Military.* This variable includes the military and paramilitary capabilities of all relevant actors (enemy, friendly, and neutral) in a given operational environment. (Does the enemy in this particular neighborhood have antitank missiles?)
- *Economic.* This variable encompasses individual and group behaviors related to producing, distributing, and consuming resources. (Does the village have a high unemployment rate?)
- *Social.* This variable includes the cultural, religious, and ethnic makeup within an operational environment and the beliefs, values, customs, and behaviors of society members. (Who are the influential people in the village—for example, religious leaders, tribal leaders, warlords, criminal bosses, or prominent families?)
- *Information.* This variable describes the nature, scope, characteristics, and effects of individuals, organizations, and systems that collect, process, disseminate, or act on information. (How much access does the local population have to news media or the Internet?)
- *Infrastructure.* This variable comprises the basic facilities, services, and installations needed for the functioning of a community or society. (Is the electrical generator in the village working?)
- *Physical Environment.* This variable includes the geography and man-made structures, as well as the climate and weather in the area of operations. (What types of terrain or weather conditions in this area of operations favor enemy operations?)
- *Time.* This variable describes the timing and duration of activities, events, or conditions within an operational environment, as well as how the timing and duration are perceived by various actors in the operational environment. (For example, at what times are people likely to congest roads or conduct activities that provide a cover for hostile operations?)

A-3. Each of the eight operational variables also has associated subvariables. Table A-1 gives examples of subvariables that might require consideration within each operational variable. The specific questions for each variable will differ, depending on the general nature of a specific operational environment.

Table A-1. Operational variables

Operational Variables		
Political Variable 　Attitude toward the United States 　Centers of political power 　Type of government 　Government effectiveness and 　legitimacy 　Influential political groups 　International relationships **Military Variable** 　Military forces 　Government paramilitary forces 　Nonstate paramilitary forces 　Unarmed combatants 　Nonmilitary armed combatants 　Military functions 　　• Command and control 　　• Maneuver 　　• Information warfare 　　• Reconnaissance, 　intelligence, surveillance, and 　target acquisition 　　• Fire support 　　• Protection 　　• Logistics **Economic Variable** 　Economic diversity 　Employment status 　Economic activity 　Illegal economic activity 　Banking and finance	**Social Variable** 　Demographic mix 　Social volatility 　Education level 　Ethnic diversity 　Religious diversity 　Population movement 　Common languages 　Criminal activity 　Human rights 　Centers of social power 　Basic cultural norms and values **Information Variable** 　Public communications media 　Information warfare 　　• Electronic warfare 　　• Computer warfare 　　• Information attack 　　• Deception 　　• Physical destruction 　　• Protection and security 　measures 　　• Perception management 　Intelligence 　Information management **Infrastructure Variable** 　Construction pattern 　Urban zones 　Urbanized building density 　Utilities present 　Utility level 　Transportation architecture	**Physical Environment Variable** 　Terrain 　　• Observation and fields of fire 　　• Avenues of approach 　　• Key terrain 　　• Obstacles 　　• Cover and concealment 　　• Landforms 　　• Vegetation 　　• Terrain complexity 　　• Mobility classification 　Natural Hazards 　Climate 　Weather 　　• Precipitation 　　• High temperature–heat index 　　• Low temperature–wind chill 　index 　　• Wind 　　• Visibility 　　• Cloud cover 　　• Relative humidity **Time Variable** 　Knowledge of the area of 　operations 　Cultural perception of time 　Information offset 　Tactical exploitation of time 　Key dates, time periods, or 　events

MISSION VARIABLES

A-4. Mission variables are fundamental in developing a course of action (COA) for a given operation. Mission variables describe characteristics of the area of operations, focusing on how they might affect a mission. Incorporating the analysis of the operational variables into METT–TC ensures Army leaders consider the best available relevant information about conditions that pertain to the mission. Using the operational variables as a source of relevant information for the mission variables allows commanders to refine their situational understanding of their operational environment and to visualize, describe, direct, lead and assess operations. (See ADRP 5-0 for more information on mission variables.)

Mission

A-5. Leaders analyze the higher headquarters' warning order (WARNORD) or operation order (OPORD) to determine how their unit contributes to the higher headquarters' mission. They examine the following information that affects their mission:

- Higher headquarters' mission and commander's intent.
- Higher headquarters' concept of operations.
- Specified, implied, and essential tasks.
- Constraints.

A-6. Leaders determine the mission and commander's intent of their higher echelon and the next higher echelon headquarters. When these are unavailable, leaders infer them based on available information. When they receive the actual mission and commander's intent, leaders revise their plan, if necessary.

A-7. Leaders examine their higher echelon's headquarters' concept of operations to determine how their unit's mission and tasks contribute to the higher mission's success. They determine details that will affect their operations, such as control measures and execution times.

A-8. Leaders extract the specified and implied tasks assigned to their unit from WARNORDs and the OPORD. They determine why each task was assigned to their unit to understand how it fits within the commander's intent and concept of operations. From the specified and implied tasks, leaders identify essential tasks. Leaders complete these tasks to accomplish the mission. Failure to complete an essential task results in mission failure.

A-9. Leaders also identify any constraints placed on their unit. Constraints can take the form of a requirement (for example, maintain a reserve of one platoon) or a prohibition on action (for example, no reconnaissance forward of Phase Line Bravo before H-hour).

A-10. The product of this part of the mission analysis is the restated mission. The restated mission is a simple, concise expression of the essential tasks the unit must accomplish and the purpose to be achieved. The mission statement states *who* (the unit), *what* (the task), *when* (either the critical time or on order), *where* (location), and *why* (the purpose of the operation).

Enemy

A-11. With the restated mission as the focus, leaders continue the analysis on the enemy. For small-unit operations, leaders need to know about the enemy's composition, disposition, strengths, recent activities, ability to reinforce, and possible COAs. Much of this information comes from higher echelon headquarters and must be refined to the level of detail required by the unit to continue with plan development. Additional information comes from adjacent units and other leaders. Some information comes from the leader's experience. Leaders determine how the available information applies to their operation. They also determine what they do not know, but should know, about the enemy. To obtain the necessary information, they identify these intelligence gaps to their higher headquarters or take action (such as sending out reconnaissance patrols).

Terrain and Weather

A-12. Leaders analyze the five military aspects of terrain expressed in the memory aid of OAKOC: observation and fields of fire, avenue of approach, key terrain, obstacles, and cover and concealment.

A-13. *Observation* is the condition of weather and terrain that permits a force to see the friendly, enemy, and neutral personnel and systems, and the key aspects of the environment (ADRP 1-02). Observation is the ability to see (or be seen by) the adversary either visually or through the use of surveillance devices. A *field of fire* is the area that a weapon or group of weapons may cover effectively from a given position (FM 3-90-1). Observation and fields of fire apply to both enemy and friendly weapons. Leaders consider direct-fire weapons and the ability of observers to mass and adjust indirect fire.

A-14. An *avenue of approach* is an air or ground route of an attacking force of a given size leading to its objective or to key terrain in its path (JP 2-01.3). Avenues of approach include overland, air, and underground routes. Underground avenues are particularly important in urban operations.

A-15. *Key terrain* is any locality, or area, the seizure or retention of which affords a marked advantage to either combatant (JP 2-01.3). *Decisive terrain* is key terrain whose seizure and retention is mandatory for successful mission accomplishment (FM 3-90-1). Terrain adjacent to the area of operations may be key if its control is necessary to accomplish the mission.

A-16. An *obstacle* is any natural or man-made obstruction designed or employed to disrupt, fix, turn, or block the movement of an opposing force, and to impose additional losses in personnel, time, and equipment on the opposing force (JP 3-15). Obstacles can exist naturally or can be man-made, or can be a combination of both. Obstacles include military reinforcing obstacles, such as minefields. (See JP 3-15 for more information on obstacles.)

A-17. *Cover* is protection from the effects of fires (ADRP 1-02). *Concealment* is protection from observation or surveillance (ADRP 1-02). Terrain that offers cover and concealment limits fields of fire. Leaders consider friendly and enemy perspectives. Although remembered as separate elements, leaders consider the military aspects of terrain together.

A-18. There are five military aspects of weather: visibility, winds, precipitation, cloud cover, and temperature and humidity. (See FM 2-01.3.) The consideration of their effects is an important part of the mission analysis. Leaders review the forecasts and considerations available from Army and Air Force weather forecast models and develop COAs based on the effects of weather on the mission. The mission analysis considers the effects on Soldiers, equipment, and supporting forces, such as air and artillery support. Leaders identify the aspects of weather that can affect the mission. They focus on factors whose effects they can mitigate. For example, leaders may modify the standard operating procedures for uniforms and carrying loads based on the temperature. Small-unit leaders include instructions on mitigating weather effects in their tentative plan. They check for compliance during preparation, especially during rehearsals.

Troops and Support Available

A-19. Perhaps the most important aspect of mission analysis is determining the combat potential of one's own force. Leaders know the status of their Soldiers' morale, their experience and training, and the strengths and weaknesses of subordinate leaders. They realistically determine all available resources. This includes troops attached to, or in direct support of, the unit. The assessment includes knowing the strength and status of their equipment. It also includes understanding the full array of assets in support of the unit. Leaders know, for example, how much indirect fire will become available, and, when it is available, they will know the type. They consider any new limitations based on the level of training or recent fighting.

Time Available

A-20. Leaders not only appreciate how much time is available, they understand the time-space aspects of preparing, moving, fighting, and sustaining. They view their own tasks and enemy actions in relation to time. They know how long it takes under such conditions to prepare for certain tasks (such as orders production, rehearsals, and subordinate element preparations). Most importantly, leaders monitor the time available. As events occur, they assess their impact on the unit timeline and update previous timelines for their subordinates. Timelines list all events that affect the unit and its subordinate elements.

Civil Considerations

A-21. *Civil considerations* are the influence of manmade infrastructure, civilian institutions, and activities of the civilian leaders, populations, and organizations within an area of operations on the conduct of military operations (ADRP 5-0). Military operations are rarely conducted in uninhabited areas. Most of the time, units are surrounded by noncombatants. These noncombatants include residents within the area of operations, local officials, and governmental and nongovernmental organizations. Based on information from higher headquarters and their own knowledge and judgment, leaders identify civil considerations that affect their mission. Commanders may analyze civil considerations using the six factors known by the memory aid ASCOPE: areas, structures, capabilities, organizations, people, and events.

Areas

A-22. Key civilian areas are localities or aspects of the terrain within an area of operations (AO) that are not normally militarily significant. This characteristic approaches terrain analysis (OAKOC) from a civilian perspective. Commanders analyze key civilian areas in terms of how they affect the missions of their individual forces and how military operations affect these areas. Failure to consider key civilian areas can seriously affect the success of an operation. Examples of key civilian areas are—

- Areas defined by political boundaries, such as districts within a city.
- Municipalities within a region.
- Locations of government centers.
- Social, political, religious, or criminal enclaves.
- Agricultural and mining regions.
- Trade routes.
- Possible sites for the temporary settlement of dislocated civilians or other civil functions.

Structures

A-23. Existing structures can play many significant roles. Some structures, such as bridges, communications towers, power plants, and dams are traditional high-payoff targets. Others, such as churches, mosques, national libraries, and hospitals are cultural sites that international law or other agreements generally protect. Still others are facilities with practical applications such as jails, warehouses, television and radio stations, and print plants that may be useful for military purposes. Some aspects of the civilian infrastructure, such as the location of toxic industrial materials, may influence operations.

A-24. Analyzing a structure involves determining how its location, functions, and capabilities can support the operation. Commanders also consider the consequences of using it. Using a structure for military purposes often competes with the civilian requirements for it. Commanders carefully weigh the expected military benefits against costs to the community that will have to be addressed in the future.

Capabilities

A-25. Commanders and staffs analyze capabilities from different levels. They view capabilities in terms of those required to save, sustain, or enhance life, in that priority. Capabilities can refer to the ability of local authorities— those of the host nation, aggressor nation, or some other body—to provide the population with key functions or services, such as public administration, public safety, emergency services, and food. Primary capabilities include those areas that the population may need help with after combat operations, such as public works and utilities, public health, economics, and commerce. Capabilities also refer to resources and services that can be contracted to support the military mission, such as interpreters, laundry services, construction materials, and equipment. The host nation or other nations might provide these resources and services.

Organizations

A-26. Organizations are nonmilitary groups or institutions in the AO. They influence and interact with the population, the force, and each other. They generally have a hierarchical structure, defined goals, established operations, fixed facilities or meeting places, and a means of financial or logistic support. Some organizations may be indigenous to the area. These may include church groups, fraternal organizations, patriotic or service organizations, labor unions, criminal organizations, and community watch groups. Other organizations may come from outside the AO. Examples of these include multinational corporations, United Nations agencies, U.S. governmental agencies, and nongovernmental organizations, such as the International Red Cross.

A-27. Operations also often require commanders to coordinate with international organizations and nongovernmental organizations. Commanders remain familiar with organizations operating in their AOs. Relevant information includes information about their activities, capabilities, and limitations. Situational understanding includes understanding how the activities of different organizations may affect military operations and how military operations may affect these organizations' activities. From this, commanders can determine how organizations and military forces can work together toward common goals when necessary.

A-28. At certain times, every echelon of command will interact with other U.S. agencies, host-nation governmental agencies, and nongovernmental organizations. However, these groups may not share the commander's objectives and point of view.

A-29. In almost every case, military forces have more resources than civilian organizations. However, civilian organizations may possess specialized capabilities that they may be willing to share with military forces. Commanders do not command civilian organizations in their AOs. However, some operations require achieving unity of effort with them and the force. These situations require commanders to influence the leaders of these organizations through persuasion.

People

A-30. People is a general term used to describe nonmilitary personnel encountered by military forces. The term includes all civilians within an AO as well as those outside the AO whose actions, opinions, or political influence can affect the mission. Individually or collectively, people can affect a military operation positively, negatively, or neutrally. In stability tasks, Army forces work closely with civilians of all types.

A-31. There can be many different kinds of people living and operating in and around an AO. As with organizations, people may be indigenous or introduced from outside the AO. An analysis of people should identify them by their various capabilities, needs, and intentions. It is useful to separate people into distinct categories. When analyzing people, commanders consider historical, cultural, ethnic, political, economic, and humanitarian factors. They also identify the key communicators and the formal and informal processes used to influence people.

Events

A-32. Events are routine, cyclical, planned, or spontaneous activities that significantly affect organizations, people, and military operations. Examples include national and religious holidays; agricultural crops, livestock, and market cycles; elections; civil disturbances; and celebrations. Other events are disasters from natural, man-made, or technological sources. These create civil hardship and require emergency responses. Examples of events precipitated by military forces include combat operations, deployments, redeployments, and paydays. Once significant events are determined, it is important to template the events and analyze them for their political, economic, psychological, environmental, and legal implications.

A-33. Technological innovation, external social influences, and natural and man-made disasters (such as hurricanes, environmental damage, and war) affect the attitudes and activities of governments and civilian populations. These changes cause stress in civilian populations and their leaders. A civilian population may or may not successfully incorporate these changes within its existing cultural value system. Addressing the problems posed by change requires considerable time and resources. The impatience of key leaders and groups, legal restrictions, and limits on resources can make resolutions difficult. However, when their resolution is necessary to accomplish the mission, commanders become concerned with them.

A-34. The existence of an independent press guarantees that U.S. military activities that do not meet America's military standards for dealing with noncombatants will be reported in U.S., host-nation, and international public forums. Commanders consider the effects of their decisions and their forces' actions on public opinion. The activities of a force, or individual members of a force, can have far-reaching effects on the legitimacy of all military operations: offense, defense, stability, or support. Commanders ensure their Soldiers understand that a tactically successful operation can also be operationally or strategically counterproductive because of the way in which they execute it or how the people perceive its execution.

A-35. Commanders have legal responsibilities to refugees and noncombatants in their AOs. These responsibilities may include providing humanitarian assistance. A commander's responsibility to protect noncombatants influences planning and preparing for operations. Commanders assess the chance that their actions may result in dislocated civilians and consider their legal obligation to respect and protect them when choosing a COA and executing an operation.

Appendix B
+Command and Support Relationships

This appendix discusses command and support relationships. Command and support relationships provide the basis for unity of command and unity of effort in operations. (See JP 3-0 for a discussion of joint command and support relationships.)

FUNDAMENTAL CONSIDERATIONS

B-1. Establishing clear command and support relationships is a key task in task organizing for any operation. (See ADRP 5-0.) These relationships establish clear responsibilities and authorities between subordinate and supporting units. Some command and support relationships limit the commander's authority to prescribe additional relationships. Knowing the inherent responsibilities of each command and support relationship allows commanders to effectively organize their forces and helps supporting commanders understand their unit's role in the organizational structure.

B-2. +Army commanders build combined arms organizations using Army command and support relationships. Command relationships define command responsibility and authority. Support relationships define the desired purpose, scope, and effect when one capability supports another.

JOINT COMMAND RELATIONSHIPS

B-3. In addition to working within Army organizations, Army commanders are often part of joint commands, or they support them. +JP 1 specifies and details four types of joint command relationships:
- Combatant command (COCOM).
- Operational control (OPCON).
- Tactical control (TACON).
- Support.

B-4. +It is important that Army leaders understand joint command relationships and how these relationships impact military operations. Paragraphs B-5 through B-14 summarize important provisions of these relationships. The glossary contains complete definitions.

COMBATANT COMMAND

B-5. +COCOM is the command authority over assigned forces vested only in commanders of combatant commands or as directed by the President or the Secretary of Defense in the Unified Command Plan and cannot be delegated or transferred. Title 10, U.S. Code, section 164 specifies this authority in law. Normally, the combatant commander exercises this authority through subordinate joint force commanders, Service component commanders, and functional component commanders. COCOM includes the directive authority for logistic matters (or the authority to delegate it to a subordinate joint force commander for common support capabilities required to accomplish the subordinate's mission).

OPERATIONAL CONTROL

B-6. +OPCON is the authority to perform those functions of command over subordinate forces involving—
- Organizing and employing commands and forces.
- Assigning tasks.
- Designating objectives.
- Giving authoritative direction necessary to accomplish missions.

B-7. +OPCON normally includes authority over all aspects of operations and joint training necessary to accomplish missions. It does not include directive authority for logistics or matters of administration, discipline, internal organization, or unit training. The combatant commander must specifically delegate these elements of COCOM. OPCON does include the authority to delineate functional responsibilities and operational areas of subordinate joint force commanders. In two instances, the Secretary of Defense may specify adjustments to accommodate authorities beyond OPCON in an establishing directive: when transferring forces between combatant commanders or when transferring members or organizations from the military departments to a combatant command. Adjustments will be coordinated with the participating combatant commanders. (JP 1 discusses operational control in detail.)

TACTICAL CONTROL

B-8. +TACON is inherent in OPCON. It may be delegated to and exercised by commanders at any echelon at or below the level of combatant command. TACON provides sufficient authority for controlling and directing the application of force or tactical use of combat support assets within the assigned mission or task. TACON does not provide organizational authority or authoritative direction for administrative and logistic support; the commander of the parent unit continues to exercise these authorities unless otherwise specified in the establishing directive. (JP 1 discusses tactical control in detail.)

SUPPORT

B-9. +Support is a command authority in joint doctrine. A supported and supporting relationship is established by a superior commander between subordinate commanders when one organization should aid, protect, complement, or sustain another force. Designating supporting relationships is important. It conveys priorities to commanders and staffs planning or executing joint operations. Designating a support relationship does not provide authority to organize and employ commands and forces, nor does it include authoritative direction for administrative and logistic support. Joint doctrine divides support into the categories listed in table B-1.

Table B-1. Joint support categories

Category	Definition
General support	That support which is given to the supported force as a whole and not to any particular subdivision thereof (JP 3-09.3).
Mutual support	That support which units render each other against an enemy, because of their assigned tasks, their position relative to each other and to the enemy, and their inherent capabilities (JP 3-31).
Direct support	A mission requiring a force to support another specific force and authorizing it to answer directly to the supported force's request for assistance (JP 3-09.3).
Close support	That action of the supporting force against targets or objectives that are sufficiently near the supported force as to require detailed integration or coordination of the supporting action (JP 3-31).

B-10. +Support is, by design, somewhat vague but very flexible. Establishing authorities ensure both supported and supporting commanders understand the authority of supported commanders. Joint force commanders often establish supported and supporting relationships among components. For example, the maritime component commander is normally the supported commander for sea control operations; the air component commander is normally the supported commander for counterair operations. An Army headquarters designated as the land component may be the supporting force during some campaign phases and the supported force in other phases.

B-11. +The joint force commander may establish a support relationship between functional and Service component commanders. Conducting operations across a large operational area often involves both the land and air component commanders. The joint task force commander places the land component in general support of the air component until the latter achieves air superiority. Conversely, within the land area of operations, the land component commander becomes the supported commander and the air component commander provides close support. A joint support relationship is not used when an Army commander task-organizes Army forces in a supporting role. When task-organized to support another Army force, Army forces use one of four Army support relationships. (See paragraphs B-35 through B-36.)

OTHER AUTHORITIES

B-12. +Although discussed in joint doctrine, coordinating authority and direct liaison authorized are directly applicable to Army forces. These relationships can assist commanders in facilitating collaboration both within and outside their respective organizations, and they can promote information sharing concerning details of military operations.

COORDINATING AUTHORITY

B-13. +Coordinating authority is the authority delegated to a commander or individual for coordinating specific functions or activities involving forces of two or more military departments, two or more joint force components, or two or more forces of the same Service. The commander or individual granted coordinating authority can require consultation between the agencies involved but does not have the authority to compel agreement. In the event that essential agreement cannot be obtained, the matter shall be referred to the appointing authority. Coordinating authority is a consultation relationship, not an authority through which command may be exercised. Coordinating authority is more applicable to planning and similar activities than to operations. (See JP 1.) For example, a joint security commander exercises coordinating authority over area security operations within the joint security area. Commanders or leaders at any echelon at or below combatant command may be delegated coordinating authority. These individuals may be assigned responsibilities established through a memorandum of agreement between military and nonmilitary organizations.

DIRECT LIAISON AUTHORIZED

B-14. +*Direct liaison authorized* is that authority granted by a commander (any level) to a subordinate to directly consult or coordinate an action with a command or agency within or outside of the granting command (JP 1). Direct liaison authorized is more applicable to planning than operations and always carries with it the requirement of keeping the commander granting direct liaison authorized informed. Direct liaison authorized is a coordination relationship, not an authority through which command may be exercised.

ARMY COMMAND AND SUPPORT RELATIONSHIPS

B-15. +Army command and support relationships are similar but not identical to joint command authorities and relationships. Differences stem from the way Army forces task-organize internally and the need for a system of support relationships between Army forces. Another important difference is the requirement for Army commanders to handle the administrative support requirements that meet the needs of Soldiers. These differences allow for flexible allocation of Army capabilities within various Army echelons. Army command and support relationships are the basis for building Army task organizations. Certain responsibilities are inherent in the Army's command and support relationships.

ARMY COMMAND RELATIONSHIPS

B-16. +Army command relationships define superior and subordinate relationships between unit commanders. By specifying a chain of command, command relationships unify effort and enable commanders to use subordinate forces with maximum flexibility. Army command relationships identify the degree of control of the gaining Army commander. The type of command relationship often relates to the expected longevity of the relationship between the headquarters involved and quickly identifies the degree of support that the gaining and losing Army commanders provide. Army command relationships include—

- Organic.
- Assigned.
- Attached.

- Operational control.
- Tactical control.

(See table B-2 on page B-5 for Army command relationships.)

Organic

B-17. *Organic* forces are those assigned to and forming an essential part of a military organization as listed in its table of organization for the Army, Air Force, and Marine Corps, and are assigned to the operating forces for the Navy (JP 1). Joint command relationships do not include organic because a joint force commander is not responsible for the organizational structure of units. That is a Service responsibility.

B-18. The Army establishes organic command relationships through organizational documents such as tables of organization and equipment and tables of distribution and allowances. If temporarily task-organized with another headquarters, organic units return to the control of their organic headquarters after completing the mission. To illustrate, within a brigade combat team, the entire brigade is organic. In contrast, within most modular support brigades, there is a "base" of organic battalions and companies and a variable mix of assigned and attached battalions and companies.

Assigned

B-19. *Assign* is to place units or personnel in an organization where such placement is relatively permanent, and/or where such organization controls and administers the units or personnel for the primary function, or greater portion of the functions, of the unit or personnel (JP 3-0). Unless specifically stated, this relationship includes administrative control.

Attached

B-20. *Attach* is the placement of units or personnel in an organization where such placement is relatively temporary (JP 3-0). A unit that is temporarily placed into an organization is attached.

Operational Control

B-21. *Operational control* is the authority to perform those functions of command over subordinate forces involving organizing and employing commands and forces, assigning tasks, designating objectives, and giving authoritative direction necessary to accomplish the mission (JP 1). +OPCON may be exercised by commanders at any echelon at or below the level of combatant command and may be delegated within the command.

Tactical Control

B-22. *Tactical control* is the authority over forces that is limited to the detailed direction and control of movements or maneuvers within the operational area necessary to accomplish missions or tasks assigned (JP 1). Tactical control allows commanders below combatant command level to apply force and direct tactical use of logistic assets but does not provide authority to change organizational structure or direct administrative and logistical support.

Table B-2. +Army command relationships

If relation-ship is:	Then inherent responsibilities:							
	Have command relation-ship with:	May be task-organized by:[1]	Unless modified, ADCON responsi-bility goes through:	Are assigned position or AO by:	Provide liaison to:	Establish/ maintain communi-cations with:	Have priorities establish-ed by:	Can impose on gained unit further command or support relation-ship of:
Organic	All organic forces organized with the HQ	Organic HQ	Army HQ specified in organizing document	Organic HQ	N/A	N/A	Organic HQ	Attached; OPCON; TACON; GS; GSR; R; DS
Assigned	Gaining unit	Gaining HQ	Gaining Army HQ	OPCON chain of command	As required by OPCON	As required by OPCON	ASCC or Service-assigned HQ	As required by OPCON HQ
Attached	Gaining unit	Gaining unit	Gaining Army HQ	Gaining unit	As required by gaining unit	Unit to which attached	Gaining unit	Attached; OPCON; TACON; GS; GSR; R; DS
OPCON	Gaining unit	Parent unit and gaining unit; gaining unit may pass OPCON to lower HQ[1]	Parent unit	Gaining unit	As required by gaining unit	As required by gaining unit and parent unit	Gaining unit	OPCON; TACON; GS; GSR; R; DS
TACON	Gaining unit	Parent unit	Parent unit	Gaining unit	As required by gaining unit	As required by gaining unit and parent unit	Gaining unit	TACON;GS GSR; R; DS

Note: [1] In NATO, the gaining unit may not task-organize a multinational force. (See TACON.)

ADCON	administrative control	HQ	headquarters
AO	area of operations	N/A	not applicable
ASCC	Army Service component command	NATO	North Atlantic Treaty Organization
DS	direct support	OPCON	operational control
GS	general support	R	reinforcing
GSR	general support–reinforcing	TACON	tactical control

ARMY SUPPORT RELATIONSHIPS

B-23. Table B-3 on page B-6 lists Army support relationships. Army support relationships are not a command authority and are more specific than joint support relationships. Commanders establish support relationships when subordination of one unit to another is inappropriate. Army support relationships are—

- Direct support.
- General support.
- Reinforcing.
- General support-reinforcing.

B-24. Commanders assign a support relationship when—

- The support is more effective if a commander with the requisite technical and tactical expertise controls the supporting unit rather than the supported commander.
- The echelon of the supporting unit is the same as or higher than that of the supported unit. For example, the supporting unit may be a brigade, and the supported unit may be a battalion. It would be inappropriate for the brigade to be subordinated to the battalion; hence, the echelon uses an Army support relationship.
- The supporting unit supports several units simultaneously. The requirement to set support priorities to allocate resources to supported units exists. Assigning support relationships is one aspect of mission command.

Table B-3. +Army support relationships

If relation-ship is:	*Then inherent responsibilities:*							
	Have command relation-ship with:	May be task-organized by:	Receive sustain-ment from:	Are assigned position or an area of operations by:	Provide liaison to:	Establish/ maintain communi-cations with:	Have priorities established by:	Can impose on gained unit further support relation-ship of:
Direct support[1]	Parent unit	Parent unit	Parent unit	Supported unit	Supported unit	Parent unit; supported unit	Supported unit	See note[1]
Reinforc-ing	Parent unit	Parent unit	Parent unit	Reinforced unit	Reinforced unit	Parent unit; reinforced unit	Reinforced unit; then parent unit	Not applicable
General support–reinforc-ing	Parent unit	Parent unit	Parent unit	Parent unit	Reinforced unit and as required by parent unit	Reinforced unit and as required by parent unit	Parent unit; then reinforced unit	Not applicable
General support	Parent unit	Parent unit	Parent unit	Parent unit	As required by parent unit	As required by parent unit	Parent unit	Not applicable

Note: [1] Commanders of units in direct support may further assign support relationships between their subordinate units and elements of the supported unit after coordination with the supported commander.

B-25. Army support relationships allow supporting commanders to employ their units' capabilities to achieve results required by supported commanders. Support relationships are graduated from an exclusive supported and supporting relationship between two units—as in direct support—to a broad level of support extended to all units under the control of the higher headquarters—as in general support (GS). Support relationships do not alter administrative control. Commanders specify and change support relationships through task organization.

B-26. +*Direct support* is a support relationship requiring a force to support another specific force and authorizing it to answer directly to the supported force's request for assistance (ADRP 5-0). A unit assigned a direct support relationship retains its command relationship with its parent unit, but is positioned by and has priorities of support established by the supported unit. (Joint doctrine considers direct support a mission rather than a support relationship.) A field artillery unit in DS of a maneuver unit is concerned primarily with the fire support needs of only that unit. The fires cell of the supported maneuver unit plans and coordinates fires to support the maneuver commander's intent. The commander of a unit in DS recommends position areas and coordinates for movement clearances where his unit can best support the maneuver commander's concept of the operation.

B-27. +*General support* is that support which is given to the supported force as a whole and not to any particular subdivision thereof (JP 3-09.3). Units assigned a GS relationship are positioned and have priorities established by their parent unit. A field artillery unit assigned in GS of a force has all of its fires under the immediate control of the supported commander or his designated force field artillery headquarters.

B-28. +*Reinforcing* is a support relationship requiring a force to support another supporting unit (ADRP 5-0). Only like units (for example, artillery to artillery) can be given a reinforcing mission. A unit assigned a reinforcing support relationship retains its command relationship with its parent unit, but is positioned by the reinforced unit. A unit that is reinforcing has priorities of support established by the reinforced unit, then the parent unit. For example, when a DS field artillery battalion requires more fires to meet maneuver force requirements, another field artillery battalion may be directed to reinforce the DS battalion.

B-29. *General support-reinforcing* is a support relationship assigned to a unit to support the force as a whole and to reinforce another similar-type unit (ADRP 5-0). A unit assigned a general support-reinforcing (GSR) support relationship is positioned and has priorities established by its parent unit and secondly by the reinforced unit. For example, an artillery unit that has a GSR mission supports the force as a whole and provides reinforcing fires for other artillery units.

ADMINISTRATIVE CONTROL

B-30. *Administrative control* is direction or exercise of authority over subordinate or other organizations in respect to administration and support (JP 1). Administrative control is not a command or support relationship; it is a Service authority. It is exercised under the authority of and is delegated by the Secretary of the Army. Administrative control (ADCON) is synonymous with the Army's Title 10 authorities and responsibilities.

B-31. ADCON responsibilities of Army forces involve the entire Army, and they are distributed between the Army generating force and operating forces. The generating force consists of those Army organizations whose primary mission is to generate and sustain the operational Army's capabilities for employment by joint force commanders. Operating forces consist of those forces whose primary missions are to participate in combat and the integral supporting elements thereof. Often, commanders in the operating force and commanders in the generating force subdivide specific responsibilities. Army generating force capabilities and organizations are linked to operating forces through co-location and reachback.

B-32. +The Army Service component commander (ASCC) is always the senior Army headquarters assigned to a combatant command. Its commander exercises command authorities as assigned by the combatant commander and ADCON as delegated by the Secretary of the Army. ADCON is the Army's authority to administer and support Army forces even while in a combatant command area of responsibility. Combatant command (command authority) is the basic authority for command and control of the same Army forces. The Army is obligated to meet the combatant commander's requirements for the operational forces. Essentially, ADCON directs the Army's support of operational force requirements. Unless modified by the Secretary of the Army, administrative responsibilities normally flow from Department of the Army through the ASCC to those Army forces assigned or attached to that combatant command. ASCCs usually "share" ADCON for at least some administrative or support functions. "Shared ADCON" refers to the internal allocation of Title 10, U.S. Code, section 3013(b) responsibilities and functions. This is especially true for Reserve Component forces. Certain administrative functions, such as pay, stay with the Reserve Component headquarters, even after unit mobilization. Shared ADCON also applies to direct reporting units of the Army that typically perform single or unique functions. The direct reporting unit, rather than the ASCC, typically manages individual and unit training for these units. The Secretary of the Army directs shared ADCON.

This page intentionally left blank.

Appendix C

Plans and Orders Formats

This appendix provides guidance for building simple, flexible plans through mission orders. It lists the different types of plans and orders, including those joint plans and orders that Army forces may receive from a joint force headquarters. Next, this appendix lists characteristics of good plans and orders and provides guidelines to ensure plans and orders are internally consistent and nested with the higher plan or order. This appendix concludes with administrative instructions for writing plans and orders. (See JP 5-0 for detailed guidance on joint operation plans and orders.)

GUIDANCE FOR PLANS

C-1. *Planning* is the art and science of understanding a situation, envisioning a desired future, and laying out effective ways of bringing that future about (ADP 5-0.) Based on this understanding and operational approach, planning continues with the development of a fully synchronized operation plan or order that arranges potential actions in time, space, and purpose to guide the force during execution.

C-2. A product of planning is a plan or order—a directive for future action. Commanders issue plans and orders to subordinates to communicate their understanding of the situation and their visualization of an operation. Plans and orders direct, coordinate, and synchronize subordinate actions and inform those outside the unit how to cooperate and provide support. To properly understand and execute the joint commander's plan, Army commanders and staffs must be familiar with joint planning processes, procedures, and orders formats. (See JP 3-33 and JP 5-0 for more information on joint planning.)

Note: An Army headquarters (battalion through Army Service component command) uses the military decisionmaking process (MDMP) and publishes plans and orders in accordance with the Army plans and orders format.

An Army headquarters that forms the base of a joint task force uses the joint operation planning process and publishes plans and orders in accordance with the joint format (see JP 5-0 and CJCSM 3122.05).

An Army headquarters (such as an Army corps) that provides the base of a joint force or coalition forces land component command headquarters will participate in joint planning and receive a joint formatted plan or order. This headquarters then has the option to use the MDMP or joint operations planning process to develop its own supporting plan or order. This headquarters may write the order in either the proper Army or joint format to distribute to subordinate commands.

BUILDING SIMPLE, FLEXIBLE PLANS

C-3. Simplicity is a principle of joint operations and is vital to effective planning. Effective plans and orders are simple and direct. Staffs prepare clear, concise, and complete plans and orders to ensure thorough understanding. They use doctrinally correct operational terms and graphics. Doing this minimizes chances of misunderstanding. Writing shorter, rather than longer, plans aids in simplicity. Shorter plans are easier to disseminate, read, and remember.

C-4. Complex plans have a greater potential to fail in execution, since they often rely on intricate coordination. Operations are always subject to friction. The more detailed the plan, the greater the chances it will no longer be applicable as friendly, enemy, and civilian actions change the situation throughout an operation.

C-5.   Simple plans require an easily understood concept of operations. Planners also promote simplicity by minimizing details where possible and by limiting the actions or tasks to what the situation requires. Subordinates can then develop specifics within the commander's intent. For example, instead of assigning a direction of attack, planners can designate an axis of advance.

C-6.   Simple plans are not simplistic plans. Simplistic refers to something made overly simple by ignoring the situation's complexity. Good plans simplify complicated situations. However, some situations require more complex plans. Commanders at all levels weigh the apparent benefits of a complex concept of operations against the risk that subordinates will be unable to understand or follow it adequately. Commanders prefer simple plans that are easy to understand and execute.

C-7.   Flexible plans help units adapt quickly to changing circumstances. Commanders and planners build opportunities for initiative into plans by anticipating events that allow them to operate inside of the enemy's decision cycle or react promptly to deteriorating situations. Identifying decision points and designing branches ahead of time—combined with a clear commander's intent—help create flexible plans. Incorporating control measures to reduce risk also makes plans more flexible. For example, a commander may hold a large, mobile reserve to compensate for the lack of information concerning an anticipated enemy attack.

MISSION ORDERS

C-8.   Commanders stress the importance of mission orders as a way of building simple, flexible plans. *Mission orders* are directives that emphasize to subordinates the results to be attained, not how they are to achieve them (ADP 6-0). Mission orders focus on what to do and the purpose of doing it without prescribing exactly how to do it. Commanders establish control measures to aid cooperation among forces without imposing needless restrictions on freedom of action. Mission orders contribute to flexibility by allowing subordinates the freedom to seize opportunities or react effectively to unforeseen enemy actions and capabilities.

C-9.   Mission orders follow the five-paragraph format (situation, mission, execution, sustainment, and command and signal) and are as brief and simple as possible. Mission orders clearly convey the unit's mission and commander's intent. Mission orders summarize the situation (current or anticipated starting conditions), provide the key tasks, describe the operation's objectives and end state (desired conditions), and provide a simple concept of operations to accomplish the unit's mission. When assigning tasks to subordinate units, mission orders include all components of a task statement: who, what, when, where, and why. However, commanders particularly emphasize the purpose (why) of the tasks to guide (along with the commander's intent) disciplined initiative. Effective plans and orders foster mission command by—

- Describing the situation to create a shared understanding.
- Conveying the commander's intent and concept of operations.
- Assigning tasks to subordinate units and stating the purposes for conducting these tasks.
- Providing the control measures necessary to synchronize the operation while retaining the maximum freedom of action for subordinates.
- Task-organizing forces and allocating resources.
- Directing preparation activities and establishing times or conditions for execution.

C-10. Mission orders contain the proper level of detail; they are neither so detailed that they stifle initiative nor so general that they provide insufficient direction. The proper level depends on each situation and is not easy to determine. Some phases of operations require tighter control over subordinate elements than others require. An air assault's air movement and landing phases, for example, require precise synchronization. Its ground maneuver plan may require less detail. As a rule, the base plan or order contains only the specific information required to provide the guidance to synchronize combat power at the decisive time and place while allowing subordinates as much freedom of action as possible. Commanders rely on disciplined initiative and coordination to act within the commander's intent and concept of operations. The attachments to the plan or order contain details regarding the situation and instructions necessary for synchronization.

TYPES OF PLANS AND ORDERS

C-11. Generally, a plan is developed well in advance of execution and is not executed until directed. A plan becomes an order when directed for execution based on a specific time or an event. Some planning results in written orders complete with attachments. Other planning results in brief fragmentary orders (FRAGORDs) issued verbally and followed in writing. Operation plans (OPLANs) and orders follow the five-paragraph format.

TYPES OF PLANS

C-12. Plans come in many forms and vary in scope, complexity, and length of planning horizons. Strategic plans establish national and multinational military objectives and include ways to achieve those objectives. Operational-level or campaign plans cover a series of related military operations aimed at accomplishing a strategic or operational objective within a given time and space. Tactical plans cover the employment of units in operations, including the ordered arrangement and maneuver of units in relation to each other and to the enemy within the framework of an operational-level or campaign plan. There are several types of plans:

- Campaign plan.
- Operation plan.
- Supporting plan.
- Concept plan.
- Branch.
- Sequel.

C-13. A *campaign plan* is a joint operation plan for a series of related major operations aimed at achieving strategic or operational objectives within a given time and space (JP 5-0). Developing and issuing a campaign plan is appropriate when the contemplated simultaneous or sequential military operations exceed the scope of a single major operation. Only joint force commanders develop campaign plans.

C-14. An *operation plan* is 1. Any plan for the conduct of military operations prepared in response to actual and potential contingencies. 2. A complete and detailed joint plan containing a full description of the concept of operations, all annexes applicable to the plan, and a time-phased force and deployment data. (JP 5-0). An OPLAN may address an extended period connecting a series of objectives and operations, or it may be developed for a single part or phase of a long-term operation. An OPLAN becomes an operation order when the commander sets an execution time or designates an event that triggers the operation.

C-15. A *supporting plan* is an operation plan prepared by a supporting commander, a subordinate commander, or an agency to satisfy the requests or requirements of the supported commander's plan (JP 5-0). For example, the Army commander develops a supporting plan for how Army forces will support the joint force commander's campaign plan or OPLAN.

C-16. In the context of joint operation planning level 3 planning detail, a *concept plan* is an operation plan in an abbreviated format that may require considerable expansion or alteration to convert it into a complete operation plan or operation order (JP 5-0). Often branches and sequels are written as concept plans. As time and the potential allow for executing a particular branch or sequel, these concept plans are developed in detail into OPLANs.

C-17. A *branch* is the contingency options built into the base plan used for changing the mission, orientation, or direction of movement of a force to aid success of the operation based on anticipated events, opportunities, or disruption caused by enemy actions and reactions (JP 5-0). Branches add flexibility to plans by anticipating situations that could alter the basic plan (See paragraph C-35 for a discussion on the basic plan). Such situations could be a result of adversary action, availability of friendly capabilities or resources, or even a change in the weather or season within the area of operations.

C-18. A *sequel* is the subsequent major operation or phase based on the possible outcomes (success, stalemate, or defeat) of the current major operation or phase (JP 5-0). For every action or major operation that does not accomplish a strategic or operational objective, there should be a sequel for each possible outcome, such as win, lose, draw, or decisive win.

TYPES OF ORDERS

C-19. An order is a communication—verbal, written, or signaled—which conveys instructions from a superior to a subordinate. Commanders issue orders verbally or in writing. The five-paragraph format (situation, mission, execution, sustainment, and command and signal) remains the standard for issuing orders. The technique used to issue orders (verbal or written) is at the discretion of the commander; each technique depends on time and the situation. Army organizations use three types of orders:

- Operation order (OPORD).
- Fragmentary order (FRAGORD).
- Warning order (WARNORD).

C-20. An *operation order* is a directive issued by a commander to subordinate commanders for the purpose of effecting the coordinated execution of an operation (JP 5-0). Commanders issue OPORDs to direct the execution of long-term operations as well as the execution of discrete short term operations within the framework of a long-range OPORD. An example of the proper naming convention for an OPORD is "OPORD 3411 (OPERATION DESERT DRAGON) (UNCLASSIFIED)."

C-21. A *fragmentary order* is an abbreviated form of an operation order issued as needed after an operation order to change or modify that order or to execute a branch or sequel to that order (JP 5-0). FRAGORDs include all five OPORD paragraph headings and differ from OPORDs only in the degree of detail provided. An example of the proper naming convention for a FRAGORD to an OPORD is "FRAGORD 11 to OPORD 3411 (OPERATION DESERT DRAGON) (UNCLASSIFIED)." If a FRAGORD contains an entire annex, then the proper naming convention for the annex would be "Annex A (Task Organization) to FRAGORD 12 to OPORD 3411 (OPERATION DESERT DRAGON) (UNCLASSIFIED)."

C-22. A *warning order* is a preliminary notice of an order or action that is to follow (JP 5-0). WARNORDs follow the five-paragraph format and help subordinate units and staffs prepare for new missions by describing the situation, providing initial planning guidance, and directing preparation activities. For example, the proper naming convention for WARNORD number 8 is "WARNORD #8."

C-23. In addition to the types of orders in paragraphs C-19 through C-22, Army forces may receive the following types of orders from a joint headquarters:

- Planning order.
- Alert order.
- Execute order.
- Prepare-to-deploy order.

(See JP 5-0 for clarification and guidance on these orders.)

VERBAL ORDERS

C-24. Commanders use verbal orders when operating in an extremely time-constrained environment. These orders offer the advantage of rapid distribution, but they risk important information being overlooked or misunderstood. Verbal orders are usually followed by written FRAGORDs.

WRITTEN ORDERS

C-25. Commanders issue written plans and orders that contain both text and graphics. Graphics convey information and instructions through military symbols. They complement the written portion of a plan or an order and promote clarity, accuracy, and brevity. Staffs often develop and disseminate written orders electronically to shorten the time needed to gather and brief the orders group. Staffs can easily edit and modify electronically produced orders. They can send the same order to multiple recipients simultaneously. Using computer programs to develop and disseminate precise, corresponding graphics adds to the efficiency and clarity of the orders process. (See ADRP 1-02 for a list of approved symbols.)

C-26. Electronic editing makes importing text and graphics into orders easy. Unfortunately, such ease can result in orders becoming unnecessarily large without added operational value. Commanders need to ensure that orders contain only that information needed to facilitate effective execution. Orders should not regurgitate unit standard operating procedures (SOPs). They should be clear, concise, and relevant to the mission.

CHARACTERISTICS OF EFFECTIVE PLANS AND ORDERS

C-27. The amount of detail provided in a plan or order depends on several factors, including the cohesion and experience of subordinate units and complexity of the operation. Effective plans and orders encourage subordinates' initiative by providing the "what" and "why" of tasks to subordinate units; they leave how to perform the tasks to subordinates. To maintain clarity and simplicity, planners keep the base plan or order as short and concise as possible. They address detailed information and instructions in attachments as required.

C-28. Effective plans and orders are simple and direct to reduce misunderstanding and confusion. The situation determines the degree of simplicity required. Simple plans executed on time are better than detailed plans executed late. Commanders at all echelons weigh potential benefits of a complex concept of operations against the risk that subordinates will fail to understand it. Multinational operations mandate simplicity due to the differences in language, doctrine, and culture. The same applies to operations involving interagency and nongovernmental organizations.

C-29. Effective plans and orders reflect authoritative and positive expression through the commander's intent. As such, the language is direct and affirmative. An example of this is, "The combat trains will remain in the assembly area" instead of "The combat trains will not accompany the unit." Effective plans and orders directly and positively state what the commander wants the unit and its subordinate units to do and why.

C-30. Effective plans and orders avoid meaningless expressions, such as "as soon as possible." Indecisive, vague, and ambiguous language leads to uncertainty and lack of confidence.

C-31. Effective plans and orders possess brevity and clarity. These plans use short words, sentences, and paragraphs. Plans use acronyms unless clarity is hindered. They do not include material covered in SOPs, but refer to those SOPs instead. Brief and clear orders use doctrinally correct terms and symbols, avoid jargon, and eliminate every opportunity for misunderstanding the commander's exact, intended meaning.

C-32. Effective plans and orders contain assumptions. This helps subordinates and others better understand the logic behind a plan or order and facilitates the preparation of branches and sequels.

C-33. Effective plans and orders incorporate flexibility. There is room built into the plan to adapt and make adjustments to counter unexpected challenges and seize opportunities. Effective plans and orders identify decision points and proposed options at those decision points to build flexibility.

C-34. Effective plans and orders exercise timeliness. Plans and orders sent to subordinates promptly allow subordinates to collaborate, plan, and prepare their own actions.

ADMINISTRATIVE INSTRUCTIONS

C-35. The following information pertains to administrative instructions for preparing all plans and orders. Unless otherwise stated, the term order refers to both plans and orders. The term base order refers to the main body of a plan or order without attachments.

C-36. Regardless of echelon, all orders adhere to the same guidance. Order writers show all paragraph headings on written orders. A paragraph heading with no text will state "None" or "See [attachment type] [attachment letter or number]." In this context, attachment is a collective term for annex, appendix, tab, and exhibit.

C-37. The base order and all attachments follow a specific template for the paragraph layout. Every order follows the five-paragraph format. Order writers underline and bold the titles of these five paragraphs: Situation, Mission, Execution, Sustainment, and Command and Signal. For example, "situation" is **Situation.** All subparagraphs and subtitles begin with capital letters and are underlined. For example, "concept of operations" is Concept of Operations.

C-38. When a paragraph is subdivided, it must have at least two subdivisions. The tabs are set at 0.25 inches and the space is doubled between paragraphs. Subsequent lines of text for each paragraph may be flush left or equally indented at the option of the chief of staff or executive officer, as long as they are consistent throughout the order. (See figure C-1.)

---

1. <u>Title</u>. Text.

    a. <u>Title</u>. Text.

    b. <u>Title</u>. Text.

        (1) <u>Title</u>. Text.

        (2) <u>Title</u>. Text.

            (a) <u>Title</u>. Text.

            (b) <u>Title</u>. Text

                1. <u>Title</u>. Text.

                2. <u>Title</u> Text.

2. <u>Title</u>. Text. (Follow the same subparagraph format as above.)

Figure C-1. Paragraph layout for plans and orders

ACRONYMS AND ABBREVIATIONS

C-39. Order writers use acronyms and abbreviations to save time and space, if these acronyms and abbreviations do not cause confusion. However, order writers do not sacrifice clarity for brevity. Order writers keep acronyms and abbreviations consistent throughout the order and its attachments. They do not use acronyms and abbreviations not found in ADRP 1-02 or JP 1-02. Before using an entire acronym or abbreviation, at its first use in the document order writers use the full form of the term and then place the acronym or abbreviation between parentheses immediately after the term. After this first use, they use the acronym or abbreviation throughout the document.

DIGITAL DISPLAY AND COMMON ACCESS TO INFORMATION

C-40. To ensure standardization and the ability to understand the common operational picture (COP), commanders must designate the standardized system to display, access, and share information. Commanders also designate which COP the command will use to gain shared understanding.

PLACE AND DIRECTION DESIGNATIONS

C-41. Order writers describe locations or points on the ground by—
- Providing the map datum used throughout the order.
- Referring to military grid reference system coordinates.
- Referring to longitude and latitude, if available maps do not have the military grid reference system.

C-42. Order writers designate directions in one of two ways:
- As a point of the compass (for example, north or northeast).
- As a magnetic, grid, or true bearing, stating the unit of measure (for example, 85 degrees [magnetic]).

C-43. When first mentioning a place or feature on a map, order writers print the name in capital letters exactly as spelled on the map and show its complete grid coordinates (grid zone designator, 100-kilometer grid square, and four-, six-, eight-, or ten-digit grid coordinates) in parentheses after it. When first using a control measure (such as a contact point), order writers print the name or designation of the point followed by its complete grid coordinates in parentheses. Thereafter, they repeat the coordinates only for clarity.

C-44. Order writers describe areas by naming the northernmost (12 o'clock) point first and the remaining points in clockwise order. They describe positions from left to right and from front to rear, facing the enemy. To avoid confusion, order writers identify flanks by compass directions, rather than right or left of the friendly force.

C-45. If the possibility of confusion exists when describing a route, order writers add a compass direction for clarity (for example, "The route is northwest along the road LAPRAIRIE–DELSON."). If a particular route already has a planning name, such as main supply route SPARTAN, order writers refer to the route using only that designator.

C-46. Order writers designate trails, roads, and railroads by the names of places along them or with grid coordinates. They precede place names with a trail, road, or railroad (for example, "road GRANT–CODY"). Order writers designate the route for a movement by listing a sequence of grids from the start point to the release point. Otherwise, they list the sequence of points from left to right or front to rear, facing the enemy.

C-47. Order writers identify riverbanks as north, south, east, or west. In wet gap-crossing operations, they identify riverbanks as either near or far.

NAMING CONVENTIONS

C-48. Unit SOPs normally designate naming conventions for graphics. Otherwise, planners select them. For clarity, order writers avoid multiword names, such as "Junction City." Simple names are better than complex ones. To ensure operations security, order writers avoid assigning names that could reveal unit identities, such as the commander's name or the unit's home station. They do not name sequential phase lines and objectives in alphabetical order. For memory aids, order writers use sets of names designated by the type of control measure or subordinate unit. For example, the division might use colors for objective names and minerals for phase line names.

CLASSIFICATION MARKINGS

C-49. AR 380-5 contains detailed information on marking documents, transmitting procedures, and other classification instructions. Each page and portions of the text on that page will be marked with the appropriate abbreviation ("TS" for TOP SECRET, "S" for SECRET, "C" for CONFIDENTIAL, or "U" for UNCLASSIFIED). Order writers place classification markings at the top and bottom of each page. All paragraphs must have the appropriate classification marking immediately following the alphanumeric designation of the paragraph (preceding the first word if the paragraph is not numbered).

C-50. The "For Official Use Only" acronym, FOUO, will be used in place of "U" when a portion is UNCLASSIFIED but contains FOUO information. FOUO is a designation that is applied to unclassified information which is exempt from mandatory release to the public under the Freedom of Information Act. AR 25-55 contains the definition and policy application of FOUO markings. (See chapter 3 in this manual for more details on information protection.)

C-51. Leaders may have to handle Department of State information. Sensitive But Unclassified (SBU) information is information originating from within the Department of State. This information requires protection and administrative control. It meets the criteria for exemption from mandatory public disclosure under the Freedom of Information Act. The Department of State does not require that SBU information be specifically marked, but does require that holders of this information are aware of the need for control measures and protection. When including SBU information in Department of Defense documents, these documents will be marked as if the information were FOUO. (See AR 380-5 for more information.)

C-52. The Army continues its involvement in numerous multinational commitments and operations. This involves an understanding of how commanders may release to or withhold information from select unified action partners. Intelligence information previously marked "Not Releasable To Foreign Nationals" (NOFORN) continues to be non-releasable to foreigners and must be referred to the originator. NOFORN is not authorized for new classification decisions. A limited amount of information will contain the marking

"U.S. ONLY". This information cannot be shared with any foreign government. (See AR 380-5 for more details. See local SOPs for classification and dissemination guidance.)

EXPRESSING UNNAMED DATES AND HOURS

C-53. Order writers use specific letters to designate unnamed dates and times in plans and orders. (See table C-1.)

Table C-1. Designated letters for dates and times

Term	Designates
C-Day	The unnamed day on which a deployment operation commences or is to commence (JP 5-0).
D-day	The unnamed day on which a particular operation commences or is to commence (JP 3-02).
F-hour	The effective time of announcement by the Secretary of Defense to the Military Departments of a decision to mobilize reserve units (JP 3-02).
H-hour	The specific hour on D-Day at which a particular operation commences (JP 5-0).
L-hour	The specific hour on C-day at which a deployment operation commences or is to commence (JP 5-0).
M-day	The term used to designate the unnamed day on which full mobilization commences or is due to commence (JP 3-02).
N-day	The unnamed day an active duty unit is notified for deployment or redeployment (JP 3-02).
P-hour	**The specific hour on D-day at which a parachute assault commences with the exit of the first Soldier from an aircraft over a designated drop zone. P-hour may or may not coincide with H-hour.**
R-day	Redeployment day. The day on which redeployment of major combat, combat support, and combat service support forces begins in an operation (JP 3-02).
S-day	The day the President authorizes Selective reserve callup (not more than 200,000) (JP 3-02).
T-day	The effective day coincident with Presidential declaration of national emergency and authorization of partial mobilization (not more than 1,000,000 personnel exclusive of the 200,000 callup) (JP 3-02).
W-day	Declared by the President, W-day is associated with an adversary decision to prepare for war (unambiguous strategic warning) (JP 3-02).

EXPRESSING TIME

C-54. The effective time for implementing the plan or order is the same as the date-time group of the order. Order writers express the date and time as a six-digit date-time group. The first two digits indicate the day of the month; the next four digits indicate the time. The letter at the end of the time indicates the time zone. Staffs add the month and year to the date-time group to avoid confusion. For example, a complete date-time group for 6 August 20XX at 1145 appears as 061145Z August 20XX.

C-55. If the effective time of any portion of the order differs from that of the order, staffs identify those portions at the beginning of the coordinating instructions (in paragraph 3). For example, order writers may use "Effective only for planning on receipt" or "Task organization effective 261300Z May 20XX."

C-56. Order writers express all times in a plan or order in terms of one time zone, for example ZULU (Z) or LOCAL. (Order writers do not abbreviate local time as [L]. The abbreviation for the LIMA time is L.)

Staffs include the appropriate time zone indicator in the heading data and mission statement. For example, the time zone indicator for Central Standard Time in the continental United States is SIERRA. When daylight savings time is in effect, the time zone indicator for Central Standard Time is ROMEO. The relationship of local time to ZULU time, not the geographic location, determines the time zone indicator to use.

C-57. When using inclusive dates, staffs express them by writing both dates separated by a dash (6–9 August 20XX or 6 August–6 September 20XX). They express times in the 24-hour clock system by means of four-digit Arabic numbers, including the time zone indicator.

IDENTIFYING PAGES

C-58. Staffs identify pages following the first page of plans and orders with a short title identification heading located two spaces under the classification marking. They include the number (or letter) designation of the plan, and the issuing headquarters. For example, OPLAN 09-15–23d AD (U) or Annex B (Intelligence) to OPLAN 09-15–23rd AD (U).

NUMBERING PAGES

C-59. Order writers use the following convention to indicate page numbers:

- Order writers number the pages of the base order and each attachment separately beginning on the first page of each attachment. They use a combination of alphanumeric designations to identify each attachment.
- Order writers use Arabic numbers only to indicate page numbers. They place page numbers after the alphanumeric designation that identifies the attachment. (Use Arabic numbers without any proceeding alphanumeric designation for base order page numbers.) For example, the designation of the third page to Annex C is C-3. Order writers assign each attachment either a letter or Arabic number that corresponds to the letter or number in the attachment's short title. They assign letters to annexes, Arabic numbers to appendixes, letters to tabs, and Arabic numbers to exhibits. For example, the designation of the third page to Appendix 5 to Annex C is C-5-3.
- Order writers separate elements of the alphanumeric designation with hyphens. For example, the designation of the third page of exhibit 2 to Tab B to Appendix 5 to Annex C is C-5-B-2-3.

ATTACHMENTS (ANNEXES, APPENDIXES, TABS, AND EXHIBITS)

C-60. Attachments (annexes, appendixes, tabs, and exhibits) are information management tools. They simplify orders by providing a structure for organizing information. However, even when attachments are used, an effective base order contains enough information to be executed without them. The organizational structure for attachments to Army OPLANs and OPORDs is in table C-2 on page C-17 through C-21.

C-61. Attachments are part of an order. Using them increases the base order's clarity and usefulness by keeping the base order or plan short. Attachments include information (such as sustainment), administrative support details, and instructions that expand upon the base order.

C-62. Commanders and staffs are not required to develop all attachments listed in table C-2on pages C-17 through C-21. The number and type of attachments depend on the commander, level of command, and complexity or needs of a particular operation. Minimizing the number of attachments keeps the order consistent with completeness and clarity. If the information relating to an attachment's subject is brief, the order writer places the information in the base order and "omits" the attachment. (See paragraph C-64 for information on omitting attachments.)

C-63. Staffs list attachments under an appropriate heading at the end of the document they expand. For example, they list annexes at the end of the base order, appendixes at the end of annexes, and so forth. Paragraph C-68 and table C-2 (on pages C-17 through C-21) provide the required sequence of attachments at the end of the base plan or order.

C-64. Army OPLANs or OPORDs do not use Annexes I and O as attachments. Army orders label these annexes "Not Used." Annexes T, X, and Y are available for use and are labeled as "Spare." If the commander needs to use any of these three spare annexes (T, X, Y), orders writers use the same attachment format as described as figure C-3 on page C-22 and C-23. When an attachment required by doctrine or an SOP is unnecessary, staffs indicate this by stating, "Type of attachment and its alphanumeric identifier omitted." For example, the order writer would state, "Annex R (Reports) omitted." If the situation requires an additional attachment not provided in table C-2 leaders can add to this structure. For example, if there is a requirement to add an additional tab to Appendix 1 (Intelligence Estimate) to Annex B (Intelligence), the order writer would label that additional attachment as Tab E (Attachment name) to Appendix 1 (Intelligence Estimate) to Annex B (Intelligence).

C-65. Staffs refer to attachments by letter or number and title. They use the following naming conventions:

- Annexes. Staffs designate annexes with capital letters, for example, Annex D (Fires) to OPORD 09 06—1 ID.
- Appendixes. Staffs designate appendixes with Arabic numbers, for example, Appendix 1 (Intelligence Estimate) to Annex B (Intelligence) to OPORD 09-06—1 ID.
- Tabs. Staffs designate tabs with capital letters, for example, Tab B (Target Synchronization Matrix) to Appendix 3 (Targeting) to Annex D (Fires) to OPORD 09-06—1 ID.
- Exhibits. Staffs designate exhibits with Arabic numbers, for example, Exhibit 1 (Traffic Circulation and Control) to Tab C (Transportation) to Appendix 1 (Logistics) to Annex F (Sustainment) to OPORD 09-06—1 ID.

C-66. If an attachment has wider distribution than the base order or is issued separately, the attachment requires a complete heading and acknowledgment instructions. When staffs distribute attachments with the base order, these elements are not required.

EXAMPLES AND PROCEDURES FOR CREATING PLANS, ORDERS, AND ANNEXES

C-67. All plans and orders follow the five-paragraph order format. Attachments also follow the five-paragraph format except matrixes, overlays, and lists. The example in figure C-2 provides the format and instructions for developing the base OPLAN or OPORD.

[CLASSIFICATION]

Place the classification at the top and bottom of every page of the OPLAN or OPORD. Place the classification marking at the front of each paragraph and subparagraph in parentheses. Refer to AR 380-5 for classification and release marking instructions.

Copy ## of ## copies
Issuing headquarters
Place of issue
Date-time group of signature
Message reference number

The first line of the heading is the copy number assigned by the issuing headquarters. Maintain a log of specific copies issued to addressees. The second line is the official designation of the issuing headquarters (for example, 1st Infantry Division). The third line is the place of issue. It may be a code name, postal designation, or geographic location. The fourth line is the date or date-time group that the plan or order was signed or issued and becomes effective unless specified otherwise in the coordinating instructions. The fifth line is a headquarters internal control number assigned to all plans and orders in accordance with unit standard operating procedures (SOPs).

OPERATION PLAN/ORDER [number] [(code name)] [(classification of title)]
Example: **OPORD 3411 (OPERATION DESERT DRAGON) (UNCLASSIFIED)**

Number plans and orders consecutively by calendar year. Include code name, if any.

(U) References: *List documents essential to understanding the OPLAN or OPORD. List references concerning a specific function in the appropriate attachments.*

    (a) *List maps and charts first. Map entries include series number, country, sheet names, or numbers, edition, and scale.*

    (b) *List other references in subparagraphs.*

(U) Time Zone Used Throughout the OPLAN/OPORD: *State the time zone used in the area of operations during execution. When the OPLAN or OPORD applies to units in different time zones, use Greenwich Mean (ZULU) Time.*

(U) Task Organization: *Describe the organization of forces available to the issuing headquarters and their command and support relationships. Refer to Annex A (Task Organization) if long or complicated.*

1. (U) Situation. *The situation paragraph describes the conditions of the operational environment that impact operations in the following subparagraphs:*

    a. (U) Area of Interest. *Describe the area of interest. Refer to Annex B (Intelligence) as required.*

    b. (U) Area of Operations. *Describe the area of operations. Refer to the appropriate map by its subparagraph under references, for example, "Map, reference (b)." See Appendix 2 (Operation Overlay) to Annex C (Operations) as required.*

        (1) (U) Terrain. *Describe the aspects of terrain that impact operations. Refer to Annex B (Intelligence) as required.*

        (2) (U) Weather. *Describe the aspects of weather that impact operations. Refer to Annex B (Intelligence) as required.*

[page number]
[CLASSIFICATION]

Figure C-2. Operation plan or operation order format

CLASSIFICATION]

OPLAN/OPORD [number] [(code name)] [issuing headquarters] [(classification of title)]

Place the classification and title of the OPLAN or OPORD and the issuing headquarters at the top of the second and any subsequent pages of the base plan or order.

    c. (U) <u>Enemy Forces</u>. *Identify enemy forces and appraise their general capabilities. Describe the enemy's composition, disposition, location, strength, and probable courses of action. Identify adversaries and known or potential terrorist threats within the area of operations. Refer to Annex B (Intelligence) as required.*

    d. (U) <u>Friendly Forces</u>. *Briefly identify the missions of friendly forces and the objectives, goals, and missions of civilian organizations that impact the issuing headquarters in the following subparagraphs:*

        (1) (U) <u>Higher Headquarters Mission and Intent</u>. *Identify and state the mission and commander's intent for headquarters two levels up and one level up from the issuing headquarters.*

            (a) (U) <u>Higher Headquarters Two Levels Up.</u> *Identify the higher headquarters two echelons above (for example, Joint Task Force-18).*

                1. (U) <u>Mission</u>.

                2. (U) <u>Commander's Intent</u>.

            (b) (U) <u>Higher Headquarters</u>. *Identify the higher headquarters one echelon above (for example, 1st [U.S.] Armored Division).*

                1. (U) <u>Mission</u>.

                2. (U) <u>Commander's Intent</u>.

        (2) (U) <u>Missions of Adjacent Units</u>. *Identify and state the missions of adjacent units and other units whose actions have a significant impact on the issuing headquarters.*

    e. (U) <u>Interagency, Intergovernmental, and Nongovernmental Organizations</u>. *Identify and state the objective or goals and primary tasks of those non-Department of Defense organizations that have a significant role within the area of operations. Refer to Annex V (Interagency Coordination) as required.*

    f. (U) <u>Civil Considerations</u>. *Describe the critical aspects of the civil situation that impact operations. Refer to Appendix 1 (Intelligence Estimate) to Annex B (Intelligence) as required.*

    g. (U) <u>Attachments and Detachments</u>. *List units attached to or detached from the issuing headquarters. State when each attachment or detachment is effective (for example, on order, on commitment of the reserve) if different from the effective time of the OPLAN or OPORD. Do not repeat information already listed in Annex A (Task Organization).*

    h. (U) <u>Assumptions</u>. *List assumptions used in the development of the OPLAN or OPORD.*

2. (U) <u>Mission</u>. *State the unit's mission—a short description of the who, what (task), when, where, and why (purpose) that clearly indicates the action to be taken and the reason for doing so.*

3. (U) <u>Execution</u>. *Describe how the commander intends to accomplish the mission in terms of the commander's intent, an overarching concept of operations, schemes of employment for each warfighting function, assessment, specified tasks to subordinate units, and key coordinating instructions in the subparagraphs below.*

[page number]

[CLASSIFICATION]

Figure C-2. Operation plan or operation order format (continued)

[CLASSIFICATION]

OPLAN/OPORD [number] [(code name)]—[issuing headquarters] [(classification of title)]

a. (U) <u>Commander's Intent</u>. *Commanders develop their intent statement personally. The commander's intent is a clear, concise statement of what the force must do and conditions the force must establish with respect to the enemy, terrain, and civil considerations that represent the desired end state. It succinctly describes what constitutes the success of an operation and provides the purpose and conditions that define that desired end state. The commander's intent must be easy to remember and clearly understood two echelons down. The commander's intent includes:*
***Purpose**–an expanded description of the operation's purpose beyond the "why" of the mission statement.*
***Key tasks**–those significant activities the force as a whole must perform to achieve the desired end state.*
***End state**–a description of the desired future conditions that represent success.*

b. (U) <u>Concept of Operations</u>. *The concept of operations is a statement that directs the manner in which subordinate units cooperate to accomplish the mission and establishes the sequence of actions the force will use to achieve the end state. It is normally expressed in terms of the commander's desired operational framework as discussed in ADRP 3-0. It states the principal tasks required, the responsible subordinate units, and how the principal tasks complement one another. Normally, the concept of operations projects the status of the force at the end of the operation. If the mission dictates a significant change in tasks during the operation, the commander may phase the operation. The concept of operations may be a single paragraph, divided into two or more subparagraphs, or if unusually lengthy, summarize here with details located in Annex C (Operations). If the concept of operations is phased, describe each phase in a subparagraph. Label these subparagraphs as "Phase" followed by the appropriate Roman numeral, for example, "Phase I." If the operation is phased, all paragraphs and subparagraphs of the base order and all annexes must mirror the phasing established in the concept of operations. The operation overlay and graphic depictions of lines of effort help portray the concept of operations and are located in Annex C (Operations).*

c. (U) <u>Scheme of Movement and Maneuver</u>. *Describe the employment of maneuver units in accordance with the concept of operations. Provide the primary tasks of maneuver units conducting the decisive operation and the purpose of each. Next, state the primary tasks of maneuver units conducting shaping operations, including security operations, and the purpose of each. For offensive tasks, identify the form of maneuver. For defensive tasks, identify the type of defense. For stability tasks, describe the role of maneuver units by primary stability tasks. If the operation is phased, identify the main effort by phase. Identify and include priorities for the reserve. Refer to Annex C (Operations) as required.*

(1) (U) <u>Scheme of Mobility/Countermobility</u>. *State the scheme of mobility/countermobility including priorities by unit or area. Refer to Annex G (Engineer) as required.*

(2) (U) <u>Scheme of Battlefield Obscuration</u>. *State the scheme of battlefield obscuration, including priorities by unit or area. Refer to Appendix 9 (Battlefield Obscuration) to Annex C (Operations) as required.*

(3) (U) <u>Scheme of Information Collection</u>. *Describe how the commander intends to use reconnaissance missions and surveillance tasks to support the concept of operations. Include the primary reconnaissance objectives. Refer to Annex L (Information Collection) as required.*

*(**Note:** Army forces do not conduct reconnaissance missions and surveillance within the United States and its territories. For domestic operations, this paragraph is titled "Information Awareness and Assessment" and the contents of this paragraph comply with Executive Order 12333.)*

[page number]

[CLASSIFICATION]

Figure C-2. Operation plan or operation order format (continued)

[CLASSIFICATION]

OPLAN/OPORD [number] [(code name)]—[issuing headquarters] [(classification of title)]

d. (U) <u>Scheme of Intelligence</u>. *Describe how the commander envisions intelligence supporting the concept of operations. Include the priority of effort for situation development, targeting, and assessment. State the priority of intelligence support to units and areas. Refer to Annex B (Intelligence) as required.*

e. (U) <u>Scheme of Fires</u>. *Describe how the commander intends to use fires to support the concept of operations with emphasis on the scheme of maneuver. State the fire support tasks and the purpose of each task. State the priorities for, allocation of, and restrictions on fires. Refer to Annex D (Fires) as required.*

f. (U) <u>Scheme of Protection</u>. *Describe how the commander envisions protection supporting the concept of operations. Include the priorities of protection by unit and area. Include survivability. Address the scheme of operational area security, including security for routes, bases, and critical infrastructure. Identify tactical operating forces and other reaction forces. Use subparagraphs for protection categories (for example, air and missile defense and explosive ordnance disposal) based on the situation. Refer to Annex E (Protection) as required.*

g. (U) <u>Cyber Electromagnetic Activities.</u> *Describe how cyber electromagnetic activities (including cyberspace operations, electronic warfare and spectrum management operations), supports the concept of operations. Refer to Appendix 12 (Cyber Electromagnetic Activities) to Annex C (Operations) as required. Refer to Annex H (Signal) for defensive cyberspace operations, network operations and spectrum management operations as required.*

h. (U) <u>Stability Tasks</u>. *Describe how stability tasks support the concept of operations. Describe how the commander envisions the conduct of stability tasks in coordination with other organizations. (See ADRP 3-07.) If other organizations or the host nation cannot provide for civil security, restoration of essential services, and civil control, then commanders with an assigned area of operations must do so with available resources, request additional resources, or request relief for these requirements from higher headquarters. Commanders assign specific responsibilities for stability tasks to subordinate units in paragraph 3j (Tasks to Subordinate Units) and paragraph 3k (Coordinating Instructions). Refer to Annex C (Operations) and Annex K (Civil Affairs Operations) as required.*

i. (U) <u>Assessment</u>. *Describe the priorities for assessment and identify the measures of effectiveness used to assess end state conditions and objectives. Refer to Annex M (Assessment) as required.*

j. (U) <u>Tasks to Subordinate Units</u>. *State the task assigned to each unit that reports directly to the headquarters issuing the order. Each task must include who (the subordinate unit assigned the task), what (the task itself), when, where, and why (purpose). Use a separate subparagraph for each unit. List units in task organization sequence. Place tasks that affect two or more units in paragraph 3k (Coordinating Instructions).*

k. (U) <u>Coordinating Instructions</u>. *List only instructions and tasks applicable to two or more units not covered in unit SOPs.*

(1) (U) <u>Time or condition when the OPORD becomes effective</u>.

(2) (U) <u>Commander's Critical Information Requirements</u>. *List commander's critical information requirements (CCIRs).*

(3) (U) <u>Essential Elements of Friendly Information</u>. *List essential elements of friendly information (EEFIs).*

(4) (U) <u>Fire Support Coordination Measures</u>. *List critical fire support coordination or control measures.*

[page number]

[CLASSIFICATION]

Figure C-2. Operation plan or operation order format (continued)

[CLASSIFICATION]

OPLAN/OPORD [number] [(code name)]—[issuing headquarters] [(classification of title)]

(5) (U) Airspace Coordinating Measures. *List critical airspace coordinating or control measures.*

(6) (U) Rules of Engagement. *List rules of engagement. Refer to Appendix 11 (Rules of Engagement) to Annex C (Operations) as required.*

(**Note:** *For operations within the United States and its territories, title this paragraph "Rules for the Use of Force").*

(7) (U) Risk Reduction Control Measures. *State measures specific to this operation not included in unit SOPs. They may include mission-oriented protective posture, operational exposure guidance, troop-safety criteria, and fratricide avoidance measures. Refer to Annex E (Protection) as required.*

(8) (U) Personnel Recovery Coordination Measures. *Refer to Appendix 13 (Personnel Recovery) to Annex E (Protection) as required.*

(9) (U) Environmental Considerations. *Refer to Appendix 5 (Environmental Considerations) to Annex G (Engineer) as required.*

(10) (U) Soldier and Leader Engagement. *State commander's guidance for target audiences and reporting requirements.*

(11) (U) Other Coordinating Instructions. *List in subparagraphs any additional coordinating instructions and tasks that apply to two or more units, such as the operational timeline and any other critical timing or events.*

4. (U) Sustainment. *Describe the concept of sustainment, including priorities of sustainment by unit or area. Include instructions for administrative movements, deployments, and transportation—or references to applicable appendixes—if appropriate. Use the following subparagraphs to provide the broad concept of support for logistics, personnel, and health service support. Provide detailed instructions for each sustainment subfunction in the appendixes to Annex F (Sustainment).*

a. (U) Logistics. *Refer to Annex F (Sustainment) as required.*

b. (U) Personnel. *Refer to Annex F (Sustainment) as required.*

c. (U) Health Service Support. *Refer to Annex F (Sustainment) as required.*

5. (U) Command and Signal.

a. (U) Command.

(1) (U) Location of Commander and Key Leaders. *State where the commander and key leaders intend to be during the operation, by phase if the operation is phased.*

(2) (U) Succession of Command. *State the succession of command if not covered in the unit's SOPs.*

(3) (U) Liaison Requirements. *State liaison requirements not covered in the unit's SOPs.*

b. (U) Control.

(1) (U) Command Posts. *Describe the employment of command posts (CPs), including the location of each CP and its time of opening and closing, as appropriate. State the primary controlling CP for specific tasks or phases of the operation (for example, "The division tactical command post will control the air assault").*

[page number]

[CLASSIFICATION]

Figure C-2. Operation plan or operation order format (continued)

[CLASSIFICATION]

OPLAN/OPORD [number] [(code name)]—[issuing headquarters] [(classification of title)]

(2) (U) <u>Reports</u>. *List reports not covered in SOPs. Refer to Annex R (Reports) as required.*

c. (U) <u>Signal</u>. *Describe the concept of signal support, including location and movement of key signal nodes and critical electromagnetic spectrum considerations throughout the operation. Refer to Annex H (Signal) as required.*

ACKNOWLEDGE: *Provide instructions for how the addressees acknowledge receipt of the OPLAN or OPORD. The word "acknowledge" may suffice. Refer to the message reference number if necessary. Acknowledgement of an OPLAN or OPORD means that it has been received and understood.*

[Commander's last name]
[Commander's rank]

The commander or authorized representative signs the original copy. If the representative signs the original, add the phrase "For the Commander." The signed copy is the historical copy and remains in the headquarters' files.

OFFICIAL:

[Authenticator's name]
[Authenticator's position]

Use only if the commander does not sign the original order. If the commander signs the original, no further authentication is required. If the commander does not sign, the signature of the preparing staff officer requires authentication and only the last name and rank of the commander appear in the signature block.

ANNEXES: *List annexes by letter and title. Army and joint OPLANs or OPORDs do not use Annexes I and O as attachments and in Army orders label these annexes "Not Used." Annexes T, X, and Y are available for use in Army OPLANs or OPORDs and are labeled as "Spare." When an attachment required by doctrine or an SOP is unnecessary, label it "Omitted."*

Annex A–Task Organization
Annex B–Intelligence
Annex C–Operations
Annex D–Fires
Annex E–Protection
Annex F–Sustainment
Annex G–Engineer
Annex H–Signal
Annex I–Not Used
Annex J–Public Affairs
Annex K–Civil Affairs Operations
Annex L–Information Collection
Annex M–Assessment
Annex N–Space Operations
Annex O–Not Used
Annex P–Host-Nation Support
Annex Q–Knowledge Management
Annex R–Reports
Annex S–Special Technical Operations

[page number]

[CLASSIFICATION]

Figure C-2. Operation plan or operation order format (continued)

```
[CLASSIFICATION]

OPLAN/OPORD [number] [(code name)]—[issuing headquarters] [(classification of title)]

Annex T–Spare
Annex U–Inspector General
Annex V–Interagency Coordination
Annex W–Operational Contract Support
Annex X–Spare
Annex Y–Spare
Annex Z–Distribution

DISTRIBUTION: Furnish distribution copies either for action or for information. List in detail those
who are to receive the plan or order. Refer to Annex Z (Distribution) if lengthy.

[page number]

[CLASSIFICATION
```

Figure C-2. Operation plan or operation order format (continued)

C-68. Table C-2 (on pages C-17 through C-21) lists the attachments (annexes, appendixes, tabs, and exhibits) to the base OPLAN or OPORD and identifies the staff officers responsible for developing each attachment.

Table C-2. List of attachments and responsible staff officers

ANNEX A–TASK ORGANIZATION (G-5 or G-3 [S-3])
ANNEX B–INTELLIGENCE (G-2 [S-2])
Appendix 1–Intelligence Estimate
Tab A–Terrain (Engineer Officer)
Tab B–Weather (Staff Weather Officer)
Tab C–Civil Considerations
Tab D–Intelligence Preparation of the Battlefield Products
Appendix 2–Counterintelligence
Appendix 3–Signals Intelligence
Appendix 4–Human Intelligence
Appendix 5–Geospatial Intelligence
Appendix 6–Measurement and Signature Intelligence
Appendix 7–Open-Source Intelligence

Table C-2. List of attachments and responsible staff officers (continued)

ANNEX C–OPERATIONS (G-5 or G-3 [S-3])
Appendix 1–Army Design Methodology Products
Appendix 2–Operation Overlay
Appendix 3–Decision Support Products
Tab A–Execution Matrix
Tab B–Decision Support Template and Matrix
Appendix 4–Gap Crossing Operations
Tab A–Traffic Control Overlay
Appendix 5–Air Assault Operations
Tab A–Pickup Zone Diagram
Tab B–Air Movement Table
Tab C–Landing Zone Diagram
Appendix 6–Airborne Operations
Tab A–Marshalling Plan
Tab B–Air Movement Plan
Tab C–Drop Zone/Extraction Zone Diagram
Appendix 7–Amphibious Operations
Tab A–Advance Force Operations
Tab B–Embarkation Plan
Tab C–Landing Plan
Tab D–Rehearsal Plan
Appendix 8–Special Operations (G-3 [S-3])
Appendix 9–Battlefield Obscuration (CBRN Officer)
Appendix 10–Airspace Control (G-3 [S-3] or Airspace Control Officer)
Tab A–Air Traffic Services
Appendix 11–Rules of Engagement (Staff Judge Advocate)
Tab A–No Strike List
Tab B–Restricted Target List (G-3 [S-3] with Staff Judge Advocate)
Appendix 12–Cyber Electromagnetic Activities (Electronic Warfare Officer)
Tab A–Offensive Cyberspace Operations
Tab B–Defensive Cyberspace Operations–Response Actions
Tab C–Electronic Attack
Tab D–Electronic Protection
Tab E–Electronic Warfare Support
Appendix 13–Military Information Support Operations (Military Information Support Officer)
Appendix 14–Military Deception (Military Deception Officer)
Appendix 15–Information Operations (Information Operations Officer)

Table C-2. List of attachments and responsible staff officers (continued)

ANNEX D–FIRES (Chief of Fires/Fire Support Officer)
Appendix 1–Fire Support Overlay
Appendix 2–Fire Support Execution Matrix
Appendix 3–Targeting
Tab A–Target Selection Standards
Tab B–Target Synchronization Matrix
Tab C–Attack Guidance Matrix
Tab D–Target List Work Sheets
Tab E–Battle Damage Assessment (G-2 [S-2])
Appendix 4–Field Artillery Support
Appendix 5–Air Support
Appendix 6–Naval Fire Support
Appendix 7–Air and Missile Defense (Air and Missile Defense Officer)
Tab A–Enemy Air Avenues of Approach
Tab B–Enemy Air Order of Battle
Tab C–Enemy Theater Ballistic Missile Overlay
Tab D–Air and Missile Defense Protection Overlay

ANNEX E–PROTECTION (Chief of Protection/Protection Officer as designated by the commander)
Appendix 1–Operational Area Security
Appendix 2–Safety (Safety Officer)
Appendix 3–Operations Security
Appendix 4–Intelligence Support to Protection
Appendix 5–Physical Security
Appendix 6–Antiterrorism
Appendix 7–Police Operations (Provost Marshal)
Appendix 8–Survivability Operations
Appendix 9–Force Health Protection (Surgeon)
Appendix 10–Chemical, Biological, Radiological, and Nuclear Defense (CBRN Officer)
Appendix 11–Explosive Ordnance Disposal (EOD Officer)
Appendix 12–Coordinate Air and Missile Defense (Air Defense Officer)
Appendix 13–Personnel Recovery (Personnel Recovery Officer)
Appendix 14–Detainee and Resettlement

Table C-2. List of attachments and responsible staff officers (continued)

ANNEX F–SUSTAINMENT (Chief of Sustainment [S-4])
Appendix 1–Logistics (G-4 [S-4])
Tab A–Sustainment Overlay
Tab B–Maintenance
Tab C–Transportation
Exhibit 1–Traffic Circulation and Control (Provost Marshal)
Exhibit 2–Traffic Circulation Overlay
Exhibit 3–Road Movement Table
Exhibit 4–Highway Regulation (Provost Marshal)
Tab D–Supply
Tab E–Field Services
Tab F–Distribution
Tab G–Contract Support Integration
Tab H–Mortuary Affairs
Appendix 2–Personnel Services Support (G-1 [S-1])
Tab A–Human Resources Support (G-1 [S-1])
Tab B–Financial Management (G-8)
Tab C–Legal Support (Staff Judge Advocate)
Tab D–Religious Support (Chaplain)
Tab E–Band Operations (G-1 [S-1])
Appendix 3–Health Service Support (Surgeon)
ANNEX G–ENGINEER (Engineer Officer)
Appendix 1–Mobility/Countermobility
Tab A–Obstacle Overlay
Appendix 2–Survivability
Appendix 3–General Engineering
Appendix 4–Geospatial Engineering
Appendix 5–Environmental Considerations
Tab A–Environmental Assessments
Tab B–Environmental Assessment Exemptions
Tab C–Environmental Baseline Survey
ANNEX H–SIGNAL (G-6 [S-6])
Appendix 1–Defensive Cyberspace Operations
Appendix 2–Information Network Operations
Appendix 3–Voice, Video, and Data Network Diagrams
Appendix 4–Satellite Communications
Appendix 5–Foreign Data Exchanges
Appendix 6–Spectrum Management Operations
Appendix 7–Information Services
ANNEX I–Not Used

Table C-2. List of attachments and responsible staff officers (continued)

ANNEX J–PUBLIC AFFAIRS
Appendix 1–Public Affairs Running Estimate   Appendix 2–Public Affairs Guidance
ANNEX K–CIVIL AFFAIRS OPERATIONS (G-9 [S-9])

Appendix 1–Execution Matrix
    Appendix 2–Populace and Resources Control Plan
    Appendix 3–Civil Information Management Plan

ANNEX L–INFORMATION COLLECTION (G-3 [S-3])
Appendix 1–Information Collection Plan   Appendix 2–Information Collection Overlay
ANNEX M–ASSESSMENT (G-5 [S-5] or G-3 [S-3])

Appendix 1–Nesting of Assessment Efforts
    Appendix 2–Assessment Framework
    Appendix 3–Assessment Working Group

ANNEX N–SPACE OPERATIONS (Space Operations Officer)
ANNEX O–Not Used
ANNEX P–HOST-NATION SUPPORT (G-4 [S-4])
ANNEX Q–KNOWLEDGE MANAGEMENT (Knowledge Management Officer)

Appendix 1–Knowledge Management Decision Support Matrix
    Appendix 2–Common Operational Picture Configuration Matrix
    Appendix 3–Mission Command Information Systems Integration Matrix
    Appendix 4–Content Management
    Appendix 5–Battle Rhythm

ANNEX R–REPORTS (G-3 [S-3], G-5 [S-5], and Knowledge Management Officer)
ANNEX S–SPECIAL TECHNICAL OPERATIONS (Special Technical Operations Officer)

Appendix 1–Special Technical Operations Capabilities Integration Matrix
    Appendix 2–Functional Area I Program and Objectives
    Appendix 3–Functional Area II Program and Objectives

ANNEX T–Spare
ANNEX U–INSPECTOR GENERAL (Inspector General)
ANNEX V–INTERAGENCY COORDINATION (G-3 [S-3] and G-9 [S-9])
ANNEX W–OPERATIONAL CONTRACT SUPPORT (G-4 [S-4])
ANNEX X–Spare
ANNEX Y–Spare
ANNEX Z–DISTRIBUTION (G-3 [S-3] and Knowledge Management Officer)

C-69. The example in figure C-3 (on pages C-22 through C-23) provides the format and instructions for developing an attachment to an OPORD or OPLAN: an annex, appendix, tab, or exhibit. The reference to functional area in this attachment format refers to the subject of this attachment. If this attachment is Appendix 6 (Airborne Operations) to Annex C (Operations) the orders writer will substitute airborne operations for functional area. For example, paragraph 1.b (2) would read, "Describe aspects of weather that impact airborne operations."

[CLASSIFICATION]

(Change from verbal orders, if any)

Copy ## of ## copies
Issuing headquarters
Place of issue
Date-time group of signature
Message reference number

Include heading if attachment is distributed separately from the base order or higher-level attachment.
[Attachment type and number/letter] [(attachment title)] TO [higher-level attachment type and number/letter, if applicable] [(higher-level attachment title, if applicable)] TO OPERATION PLAN/ORDER [number] [(code name)] [(classification of title)]
Example: **EXHIBIT 1 (TRAFFIC CIRCULATION AND CONTROL) TO TAB C (TRANSPORTATION) TO APPENDIX 1 (LOGISTICS) TO ANNEX F (SUSTAINMENT) TO OPORD 3411 (OPERATION DESERT DRAGON) (UNCLASSIFIED)**

References: *Refer to higher headquarters' OPLAN or OPORD and identify map sheets for operation (Optional).*

Time Zone Used Throughout the Order:

1. (U) <u>Situation</u>. *Include information affecting the functional area that paragraph 1 of the OPLAN or OPORD does not cover or that needs expansion.*

    a. (U) <u>Area of Interest</u>. *Refer to Annex B (Intelligence) as required.*

    b. (U) <u>Area of Operations</u>. *Refer to Appendix 2 (Operation Overlay) to Annex C (Operations).*

        (1) (U) <u>Terrain</u>. *Describe aspects of terrain that impact functional area operations. Refer to Annex B (Intelligence) as required.*

        (2) (U) <u>Weather</u>. *Describe aspects of weather that impact functional area operations. Refer to Annex B (Intelligence) as required.*

    c. (U) <u>Enemy Forces</u>. *List known and templated locations and activities of enemy functional area units for one echelon up and two echelons down. List enemy maneuver and other area capabilities that will impact friendly operations. State expected enemy courses of action and employment of enemy functional area assets. Refer to Annex B (Intelligence) as required.*

    d. (U) <u>Friendly Forces</u>. *Outline the higher headquarters' plan as it pertains to the functional area. List designation, location, and outline of plan of higher, adjacent, and other functional area assets that support or impact the issuing headquarters or require coordination and additional support.*

    e. (U) <u>Interagency, Intergovernmental, and Nongovernmental Organizations</u>. *Identify and describe other organizations in the area of operations that may impact the conduct of functional area operations or implementation of functional area-specific equipment and tactics.*

    f. (U) <u>Civil Considerations</u>. *Describe critical aspects of the civil situation that impact functional area operations. Refer to Annex K (Civil Affairs Operations) as required.*

    g. (U) <u>Attachments and Detachments</u>. *List units attached or detached only as necessary to clarify task organization. Refer to Annex A (Task Organization) as required.*

    h. (U) <u>Assumptions</u>. *List any functional area-specific assumptions that support the development of this attachment.*

[page number]

[CLASSIFICATION

Figure C-3. Operation order or operation plan attachment format

---

[CLASSIFICATION]

[Attachment type and number/letter] [(attachment title)] TO [higher-level attachment type and number/letter, if applicable] [(higher-level attachment title, if applicable)] TO OPERATION PLAN/ORDER [number] [(code name)] [(classification of title)]

2. (U) Mission. *State the mission of the functional area in support of the base plan or order.*

3. (U) Execution.

    a. (U) Scheme of Support. *Describe how the functional area supports the commander's intent and concept of operations. Establish the priorities of support to units for each phase of the operation. Refer to Annex C (Operations) as required.*

    b. (U) Tasks to Subordinate Units. *List functional area tasks assigned to specific subordinate units not contained in the base order.*

    c. (U) Coordinating Instructions. *List only instructions applicable to two or more subordinate units not covered in the base order.*

4. (U) Sustainment. *Identify priorities of sustainment for functional area key tasks and specify additional instructions as required. Refer to Annex F (Sustainment) as required.*

5. (U) Command and Signal.

    a. (U) Command. *State the location of commander and key leaders.*

    b. (U) Control. *State the functional area liaison requirements not covered in the base order.*

    c. (U) Signal. *Address any functional area-specific communications requirements or reports. Refer to Annex H (Signal) as required.*

ACKNOWLEDGE: *Include only if attachment is distributed separately from the base order.*

[Commander's last name]
[Commander's rank]

OFFICIAL:

[Authenticator's name]
[Authenticator's position]

Either the commander or principal staff officer responsible for the functional area will sign attachments.

ATTACHMENT: *List lower-level attachments as required.*

DISTRIBUTION: *Show only if distributed separately from the base order or higher-level attachments.*

[page number]

[CLASSIFICATION]

Figure C-3. Operation order or operation plan attachment format (continued)

C-70. The example in figure C-4 on page C-24 provides the format and instructions for developing a WARNORD. The example in figure C-5 on page C-25 provides the format and instructions for developing a FRAGORD.

|CLASSIFICATION|
(Change from verbal orders, if any) (Optional)

Copy ## of ## copies
Issuing headquarters
Place of issue
Date-time group of signature
Message reference number

WARNING ORDER [number] Example: WARNING ORDER #8

(U) References: *Refer to higher headquarters' OPLAN or OPORD and identify map sheets for operation (Optional).*

(U) Time Zone Used Throughout the OPLAN/OPORD: *(Optional).*

(U) Task Organization: *(Optional).*

1. (U) <u>Situation</u>. *The situation paragraph describes the conditions and circumstances of the operational environment that impact operations in the following subparagraphs:*

    a. (U) <u>Area of Interest</u>.

    b. (U) <u>Area of Operations</u>.

    c. (U) <u>Enemy Forces</u>.

    d. (U) <u>Friendly Forces</u>.

    e. (U) <u>Interagency, Intergovernmental, and Nongovernmental Organizations</u>.

    f. (U) <u>Civil Considerations</u>.

    g. (U) <u>Attachments and Detachments</u>. *Provide initial task organization.*

    h. (U) <u>Assumptions</u>. *List any significant assumptions for order development.*

2. (U) <u>Mission</u>. *State the issuing headquarters' mission.*

3. (U) <u>Execution</u>.

    a. (U) <u>Initial Commander's Intent</u>. *Provide brief commander's intent statement.*

    b. (U) <u>Concept of Operations</u>. *This may be "to be determined" for an initial WARNORD.*

    c. (U) <u>Tasks to Subordinate Units</u>. *Include any known tasks at time of issuance of WARNORD.*

    d. (U) <u>Coordinating Instructions</u>.

4. (U) <u>Sustainment</u>. *Include any known logistics, personnel, or health service support preparation tasks.*

5. (U) <u>Command and Signal</u>. *Include any changes to the existing order or state "No change."*

ACKNOWLEDGE:

               [Commander's last name]
               [Commander's rank]

OFFICIAL:

[Authenticator's name]
[Authenticator's position]

ANNEXES: *List annexes by letter and title.*

DISTRIBUTION: *List recipients.*

[page number]

|CLASSIFICATION|

Figure C-4. Warning order format

[CLASSIFICATION]
(Change from verbal orders, if any) (Optional)

Copy ## of ## copies
Issuing headquarters
Place of issue
Date-time group of signature
Message reference number

FRAGMENTARY ORDER [number] to OPERATION PLAN/ORDER [number] [(code name)]— [(classification of title)]

Example: **FRAGORD #1 TO OPORD 3411 (OPERATION DESERT DRAGON) (UNCLASSIFIED)**

(U) **References:** *Refer to the higher order being modified.*

(U) **Time Zone Used Throughout the OPLAN/OPORD:** *(Optional)*

1. (U) <u>Situation</u>. *Include any changes to this paragraph or state "No change."*

2. (U) <u>Mission</u>. *Include any changes to this paragraph or state "No change."*

3. (U) <u>Execution</u>. *Include any changes to this paragraph or state "No change."*

    a. (U) <u>Commander's Intent</u>. *Include any changes or state "No change."*

    b. (U) <u>Concept of Operations</u>. *Include any changes or state "No change."*

    c. (U) <u>Scheme of Movement and Maneuver</u>. *Include any changes or state "No change."*

    d. (U) <u>Scheme of Intelligence</u>. *Include any changes or state "No change."*

    e. (U) <u>Scheme of Fires</u>. *Include any changes or state "No change."*

    f. (U) <u>Scheme of Protection</u>. *Include any changes or state "No change."*

    g. (U) <u>Cyber Electromagnetic Activities</u>. *Include any changes or state "No change."*

    h. (U) <u>Stability Tasks</u>. *Include any changes or state "No change."*

    i. (U) <u>Assessment</u>. *Include any changes or state "No change."*

    j. (U) <u>Tasks to Subordinate Units</u>. *Include any changes or state "No change."*

    k. (U) <u>Coordinating Instructions</u>. *Include any changes or state "No change"*

4. (U) <u>Sustainment</u>. *Include any changes to this paragraph or state "No change."*

5. (U) <u>Command and Signal</u>. *Include any changes to this paragraph or state "No change."*

ACKNOWLEDGE:

                        [Commander's last name]
                        [Commander's rank]

OFFICIAL:

[Authenticator's name]
[Authenticator's position]

ANNEXES: *List annexes by letter and title. Army and joint OPLANs or OPORDs do not use Annexes I and O as attachments and in Army orders label these annexes "Not Used." Annexes T, X, and Y are available for use in Army OPLANs or OPORDs and are labeled as "Spare." When an attachment required by doctrine or an SOP is unnecessary, label it "Omitted."*

DISTRIBUTION:

[page number]

[CLASSIFICATION]

Figure C-5. Fragmentary order format

C-71. If, on occasion, a FRAGORD has an annex as an attachment, order writers use the naming convention for that attachment, for example "ANNEX A (TASK ORGANIZATION) to FRAGMENTARY ORDER #1 to OPERATION ORDER 3411 (Operation Desert Dragon) (Unclassified)." (See figure C-6 for a sample overlay order graphic with text.)

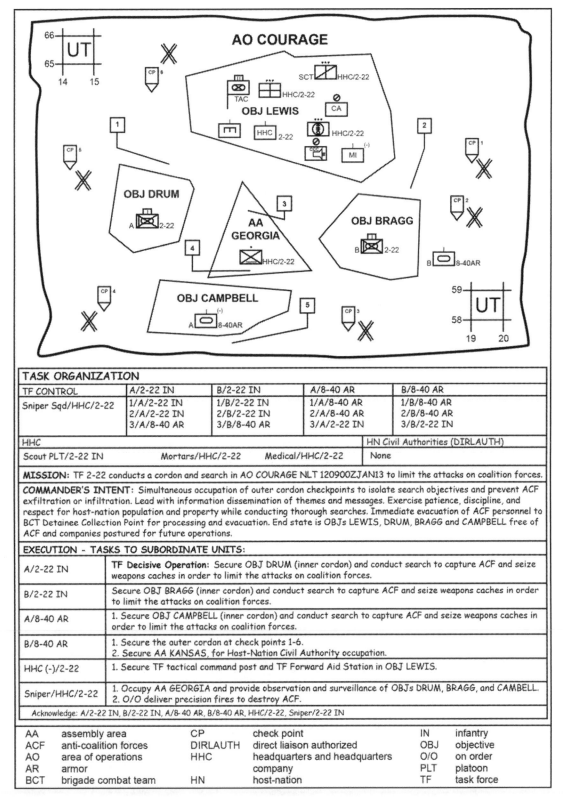

Figure C-6. Example of overlay order graphic

This page intentionally left blank.

Appendix D

Annex Formats

This appendix provides commanders and staffs guidance and formats to build annexes for plans and orders. This appendix lists 26 annexes and provides formats for 21 annexes. Two annexes are not used, annexes I and O. There are three annexes designated as spares: annexes T, X, and Y. Commanders and staffs use these annexes as required.

ANNEX A (TASK ORGANIZATION) FORMAT AND INSTRUCTIONS

D-1. This annex discusses the fundamentals of task organization and provides the format and instructions for developing Annex A (Task Organization) to the base plan or order. This annex does not follow the five-paragraph attachment format. Unit standard operating procedures (SOPs) will dictate development and format for this annex.

FUNDAMENTAL CONSIDERATIONS

D-2. *Task-organizing* is the act of designing an operating force, support staff, or sustainment package of specific size and composition to meet a unique task or mission (ADRP 3-0). Characteristics to examine when task-organizing the force include, but are not limited to, training, experience, equipment, sustainability, operational environment, (including enemy threat), and mobility. For Army forces, it includes allocating available assets to subordinate commanders and establishing their command and support relationships. Command and support relationships provide the basis for unity of command in operations. The assistant chief of staff, plans (G-5) or assistant chief of staff, operations (G-3 [S-3]) develops Annex A (Task Organization).

> *Note:* Army command relationships are similar but not identical to joint command authorities and relationships. Differences stem from the way Army forces task-organize internally and the need for a system of support relationships between Army forces. Another important difference is the requirement for Army commanders to handle the administrative control requirements.

D-3. Military units consist of organic components. Organic parts of a unit are those forming an essential part of the unit and are listed in its table of organization and equipment (TOE). Commanders can alter organizations' organic unit relationships to better allocate assets to subordinate commanders. They also can establish temporary command and support relationships to facilitate exercising mission command.

D-4. Establishing clear command and support relationships is fundamental to organizing any operation. These relationships establish clear responsibilities and authorities between subordinate and supporting units. Some command and support relationships (for example, tactical control) limit the commander's authority to prescribe additional relationships. Knowing the inherent responsibilities of each command and support relationship allows commanders to effectively organize their forces and helps supporting commanders understand their unit's role in the organizational structure.

D-5. Commanders designate command and support relationships to weight the decisive operation and support the concept of operations. Task organization also helps subordinate and supporting commanders support the commander's intent. These relationships carry with them varying responsibilities to the subordinate unit by the parent and gaining units as discussed in paragraphs B-3 and B-10. Commanders consider two organizational principles when task-organizing forces:

- Maintain cohesive mission teams.
- Do not exceed subordinates' span of control capabilities.

D-6. When possible, commanders maintain cohesive mission teams. They organize forces based on standing headquarters, their assigned forces, and habitual associations, when possible. When this is not feasible, and commanders create ad hoc organizations, commanders arrange time for training and establishing functional working relationships and procedures. Once commanders have organized and committed a force, they keep its task organization unless the benefits of a change clearly outweigh the disadvantages. Reorganizations may result in a loss of time, effort, and tempo. Sustainment considerations may also preclude quick reorganization.

D-7. Commanders carefully avoid exceeding the span of control capabilities of subordinates. Span of control refers to the number of subordinate units under a single commander. This number depends on the situation and may vary. As a rule, commanders can effectively command two to six subordinate units. Allocating subordinate commanders more units gives them greater flexibility and increases options and combinations. However, increasing the number of subordinate units increases the number of decisions commanders have to make. This slows down the reaction time among decisionmakers.

D-8. Running estimates and course of action (COA) analysis of the military decisionmaking process provide information that helps commanders determine the best task organization. An effective task organization—

- Facilitates the commander's intent and concept of operations.
- Retains flexibility within the concept of operations.
- Adapts to conditions imposed by mission variables.
- Accounts for the requirements to conduct essential stability tasks for populations within an area of operations.
- Creates effective combined arms teams.
- Provides mutual support among units.
- Ensures flexibility to meet unforeseen events and to support future operations.
- Allocates resources with minimum restrictions on their employment.
- Promotes unity of command.
- Offsets limitations and maximizes the potential of all forces available.
- Exploits enemy vulnerabilities.

D-9. Creating an appropriate task organization requires understanding—

- The mission, including the higher commander's intent and concept of operations.
- The fundamentals of offense, defense, stability, and defense support of civil authorities tasks (see ADRP 3-0) and basic tactical concepts (see ADRP 3-90).
- The roles and relationships among the warfighting functions.
- The status of available forces, including morale, training, and equipment capabilities.
- Specific unit capabilities, limitations, strengths, and weaknesses.
- The risks inherent in the plan.

D-10. During COA analysis, commanders identify what resources they need, and where, when, and how frequently they will need them. Formal task organization and the change from generic to specific units begin after COA analysis when commanders assign tasks to subordinate commanders. Staffs assign tasks to subordinate headquarters, determine if subordinate headquarters have enough combat power, and re-allocate combat power as necessary. They then refine command and support relationships for subordinate units and decide the priorities of support. Commanders approve or modify the staff's recommended task organization based on their evaluation of the factors listed in paragraphs D-8 and D-9 and information from running estimates and COA analysis as part of the military decisionmaking process. In allocating assets, the commander and staff consider—

- The task organization for the ongoing operation.
- Potentially adverse effects of breaking up cohesive teams by changing the task organization.
- Time necessary to re-align the organization after receipt of the task organization.
- Limits on control over supporting units provided by higher headquarters.

FORMAT AND INSTRUCTIONS

D-11. Annex A (Task Organization) of the operation plan (OPLAN) and operation order (OPORD) is one of the annexes that does not follow the standard five-paragraph attachment format. (Refer to appendix C for more information on OPLAN and OPORD attachments.) Task organization is typically displayed in a list or an outline format following the unit listing convention shown in table D-1. (See table D-1 on pages D-3 through D-5.)

Table D-1. Army unit listing convention

	Corps	Division	Brigade	Battalion	Company
Movement and Maneuver	Divisions Separate maneuver brigades or battalions Combat aviation brigades or battalions Special operations forces - Ranger - Special forces **MISO**	Brigade-size ground units in alpha-numerical order - Infantry - Armor - Stryker **Battalion TF** - Named TFs in alphabetical order - Numbered TFs in numerical order **MISO** **Combat aviation brigade** **Special operations forces** - Ranger - Special forces	Battalion TFs Battalions or squadrons - Combined arms - Infantry - Reconnaissance Company teams Companies Air cavalry squadrons MISO	Company teams - Named teams in alphabetical order - Letter designated teams in alphabetical order Companies or troops (in alphabetical order) - Infantry - Armor - Stryker MISO	Platoons - Organic platoons - Attached platoons - Weapons squads
Fires	Fires brigade USAF air support unit - Air defense	Fires brigade USAF air support unit - Air defense	Fires battalion USAF air support unit - Air defense	FA batteries Fire support team Mortar platoon USAF air support unit - Air defense	FA firing platoons Fire support team Mortar section - Air defense
Intelligence	Battlefield surveillance brigade - MI - Recon squads - Human terrain team	Battlefield surveillance brigade - MI - Recon squads - Human terrain team	CI teams Ground sensor teams Human terrain team HUMINT teams Scout platoon TUAS platoon	CI teams Ground sensor teams HUMINT teams Scout platoon TUAS platoon	CI teams Ground sensor teams HUMINT teams

Table D-1. Army unit listing convention (continued)

	Corps	Division	Brigade	Battalion	Company
Protection	**MEB** **Functional brigades** - Air defense - CBRN - Engineer - EOD - Military police	**MEB** **Functional brigades** - Air defense - CBRN - Engineer - EOD - Military police	**Functional battalions or companies or batteries and detachments** - Air defense - CBRN - Engineer - EOD - Military police	**Functional companies or batteries and detachments** - Air defense - CBRN - Engineer - EOD - Military police	**Functional platoons and detachments** - Air defense - CBRN - Engineer - EOD - Military police
Sustainment	**Sustainment brigade (attached functional units are listed in alpha-numerical order)** - Contracting - Finance - Ordnance - Personnel services - Transportation - Quartermaster **Medical brigade (support)**	**Sustainment brigade (attached functional units are listed in alpha-numerical order)** - Contracting - Finance - Ordnance - Personnel services - Transportation - Quartermaster **Medical brigade (support)**	**Brigade support battalion (attached or supporting functional units are listed first by branch in alphabetical order and then in numerical order)**	**Forward support company (attached or supporting functional units are listed first by branch in alphabetical order and then in numerical order)**	**Attached or supporting functional platoons and teams listed in alpha-numerical order**

Table D-1. Army unit listing convention (continued)

	Corps	Division	Brigade	Battalion	Company
Mission Command	Signal	Signal	Signal	Signal	
	Public Affairs	Public affairs	Public affairs	Public affairs	
	Civil Affairs	Civil affairs	Civil affairs	Civil affairs	
	PRT	PRT	PRT	PRT	
	MISO	MISO	MISO	MISO	
	Space	Space	Airfield Operations Battalion (AOB)	OGA (listed in alphabetical order with reference to any applicable nonstandard command and support relationship)	
	Theater Airfield Operations Group (TAOG)	Theater Airfield Operations Group (TAOG)	OGA (listed in alphabetical order with reference to any applicable nonstandard command and support relationship)		
	OGA, such as an FBI forensics team (listed in alphabetical order with reference to any applicable nonstandard command and support relationship)	OGA (listed in alphabetical order with reference to any applicable nonstandard command and support relationship)			

AOB	airfield operations battalion		MISO	military information support operations
CBRN	chemical, biological, radiological, and nuclear		OGA	other governmental agencies
CI	counterintelligence		PRT	provincial reconstruction team
EOD	explosive ordnance disposal		TAOG	theater airfield operations group
FA	field artillery		TF	task force
HUMINT	human intelligence		TUAS	tactical unmanned aerial system
MEB	maneuver enhancement brigade		USAF	United States Air Force
MI	military intelligence			

D-12. Order writers group units by headquarters. They list major subordinate maneuver units first (for example, 2d ABCT; 1-77th IN; A/4-52d CAV). Order writers place them by size in numerical order. They list brigade combat teams (BCTs) ahead of combat aviation brigades. In cases where two BCTs are numbered the same, order writers use the division number (by type). For example, 1st ABCT (armored brigade combat team) 1st Infantry Division (Mechanized) is listed before the 1st ABCT 1st Armored Division (AD). In turn, the 1st ABCT 1st Armored Division is listed before the 1st ABCT 1st Cavalry Division. Combined arms battalions are listed before battalions, and company teams before companies. Order writers follow maneuver units with multifunctional supporting units in the following order: fires, battlefield surveillance, maneuver enhancement, and sustainment. Supporting units (in alpha-numerical order) follow multifunctional supporting units. For example, a medical brigade (support) is listed after a functional engineer brigade but before a functional military police brigade. The last listing should be any special troops units under the command of the force headquarters.

D-13. Order writers use a plus (+) symbol when attaching one or more subordinate elements of a similar function to a headquarters. They use a minus symbol (–) when deleting one or more subordinate elements of a similar function to a headquarters. Order writers always show the symbols in parenthesis. They do not use a plus symbol when the receiving headquarters is a combined arms task force or company team. Order writers do not use plus and minus symbols together (as when a headquarters detaches one element and receives attachment of another); they use the symbol that portrays the element's combat power with respect to other similar elements. Order writers do not use either symbol when two units swap subordinate elements and their combat power is unchanged. Here are some examples:

- Within the 3-68th Combined Arms Battalion, C Company loses one platoon to A Company; the battalion task organization will show A Co. (+) and C Co. (–).
- Within the 3-68th Combined Arms Battalion, C Company swaps one platoon with A Company; the battalion task organization will show Team A and Team C. (The teams can also be named for their commanders, their unit nickname, or some other naming scheme.)
- 4-77th Infantry receives a tank company from 1-30th Armor; the BCT task organization will show Task Force 4-77 IN and 1-30 AR (–).
- Division and corps headquarters are always task organized. Therefore, order writers do not show these headquarters with either the plus (+) or minus (–) symbol.

D-14. If applicable, order writers list task organizations according to phases of the operation. When the effective attachment time of a nonorganic unit to another unit differs from the effective time of the plan or order, order writers add the effective attachment time in parentheses after the attached unit—for example, 1-80 IN (OPCON 2 ABCT Phase II). They list this information either in the task organization (preferred) or in paragraph 1c of the plan or order, but not both. For clarity, order writers list subsequent command or support relationships under the task organization in parentheses following the affected unit—for example, "...on order, OPCON (operational control) to 2 ABCT" is written (O/O OPCON 2 ABCT).

D-15. Long or complex task organizations are displayed in outline format in Annex A (Task Organization) of the OPLAN or OPORD in lieu of being placed in the base plan or order. Units are listed under the headquarters to which they are allocated or that they support in accordance with the organizational taxonomy previously provided in this chapter. The complete unit task organization for each major subordinate unit should be shown on the same page. Order writers only show command or support relationships if they are other than organic or attached. Other Services and multinational forces recognize and understand this format. Planners should use it during joint and multinational operations.

D-16. Order writers list subordinate units under the higher headquarters to which they are assigned, attached, or in support. They place direct support (DS) units below the units they support. Order writers indent subordinate and supporting units two spaces. They identify relationships other than attached with parenthetical terms—for example, (GS) or (DS).

D-17. Order writers provide the numerical designations of units as Arabic numerals, unless they are shown as Roman numerals. For example, an Army corps is numbered in series beginning with Roman numeral "I"—for example, I Corps or XVIII Airborne Corps.

D-18. During multinational operations, order writers insert the country code between the numeric designation and the unit name—for example, 3d (DE) Corps. (Here, DE designates that the corps is German. ADRP 1-02 contains authorized country codes.)

D-19. Order writers use abbreviated designations for organic units. They use the full designation for nonorganic units—for example, 1-52 FA (MLRS) (GS), rather than 1-52 FA. They specify a unit's command or support relationship only if it differs from that of its higher headquarters.

D-20. Order writers designate task forces with the last name of the task force (TF) commander (for example, TF WILLIAMS), a code name (for example, TF DESERT DRAGON), or a number (for example, TF 47 or TF 1-77 IN).

D-21. For unit designation at theater army level, order writers list major subordinate maneuver units first, placing them in alpha-numerical order, followed by multifunctional brigades in the following order: fires, intelligence, maneuver enhancement, sustainment, then followed by functional brigades in alpha-numerical order, and any units under the command of the force headquarters. For each function following maneuver, they list headquarters in the order of commands, brigades, groups, battalions, squadrons, companies, detachments, and teams.

D-22. Figure D-1 (on pages D-7 through D-9) illustrates a sample Annex A (Task Organization) format and provides a sample acronym list (on page D-9) for task organization.

[CLASSIFICATION]

Place the classification at the top and bottom of every page of the attachments. Place the classification marking at the front of each paragraph and subparagraph in parentheses. Refer to AR 380-5 for classification and release marking instructions.

Copy ## of ## copies
Issuing headquarters
Place of issue
Date-time group of signature
Message reference number

Include the full heading if attachment is distributed separately from the base order or higher-level attachment.

ANNEX A (TASK ORGANIZATION) TO OPERATION PLAN/ORDER [number] [(code name)]—[issuing headquarters] [(classification of title)]

(U) References: *List documents essential to understanding Annex A (Task Organization).*

   a. *List maps and charts first. Map entries include series number, country, sheet names or numbers, edition, and scale.*

   b. *List other references in subparagraphs labeled as shown.*

   c. *Doctrinal references for task organization include ADRP 3-0, ADRP 5-0, ADRP 6-0, FM 6-0, JP 1, and JP 5-0.*

(U) Time Zone Used Throughout the OPLAN/OPORD: *Write the time zone established in the base plan or order.*

(U) Task Organization: *Use the outline format for listing units as shown in the example below. (The acronym list in this annex is helpful if attached units are unfamiliar with Army acronyms.) If applicable, list task organization according to the phases of the operation during which it applies.*

[page number]

[CLASSIFICATION

Figure D-1. Sample Annex A (Task Organization) format

```
[CLASSIFICATION]
ANNEX A (TASK ORGANIZATION) TO OPERATION PLAN/ORDER [number] [(code
name)]—[issuing headquarters] [(classification of title)]
                        (sample task organization)
```

2/52 ABCT	2/54 ABCT	116 ABCT (+)
1-31 IN (-)	4-77 IN	3-116 AR
1-30 AR (-)	8-40 AR	1-163 IN
1-20 CAV	3-20 CAV	2-116 AR
A/4-52 CAV (ARS) (DS)	2/C/4-52 CAV (ARS) (DS)	1-148 FA
2-606 FA (2x8)	2-607 FA	145 BSB
TACP/52 ASOS (USAF)	TACP/52 ASOS (USAF)	4/B/2-52 AV (GSAB) (TACON)
521 BSB	105 BSB	4/2/311 QM CO (MA)
2/2/311 QM CO (MA)	3/2/311 QM CO (MA)	4/577 MED CO (GRD AMB)
1/B/2-52 AV (GSAB) (TACON)	2/B/2-52 AV (GSAB) (TACON)	844 FST
2/577 MED CO (GRD AMB)	843 FST	116 BSTB
(attached)	3/577 MED CO (GRD AMB)	366 EN CO (SAPPER) (DS)
842 FST	3 BSTB	1/401 EN CO (ESC) (DS)
2 BSTB	A 388 CA BN	2/244 EN CO (RTE CL) (DS)
31 EN CO (MRBC) (DS)	1/244 EN CO (RTE CL) (DS)	52 EOD
63 EOD	763 EOD	1/301 MP CO
2/244 EN CO (RTE CL) (DS)	2/2/1/55 SIG CO (COMCAM)	1/3/1/55 SIG CO (COMCAM)
1/2/1/55 SIG CO (COMCAM)	3D MP PLT	1/467 CM CO (MX) (S)
2D MP PLT	**52 CAB AASLT**	C/388 CA BN
RTS TM 1/A/52 BSTB	HHC/52 CAB	116 MP PLT
RTS TM 2/A/52 BSTB	1/B/1-31 IN (DIV QRF) (OPCON)	**52 SUST BDE**
RTS TM 3/A/52 BSTB	1-52 AV (ARB) (-)	52 BTB
RTS TM	4-52 CAV (ARS) (-)	520 CSSB
87 IBCT	3-52 AV (ASLT) (-)	521 CSSB
1-80 IN	2-52 AV (GSAB)	10 CSH
2-80 IN	1 (TUAS)/B/52 BSTB (-) (GS)	168 MMB
3-13 CAV	2/694 EN CO (HORIZ) (DS)	
A/3-52 AV (ASLT) (DS)		**52 HHB**
B/1-52 AV (ARB) (DS)	**52 FIRES BDE**	A/1-30 AR (DIV RES)
C/4-52 CAV (ARS) (-) (DS)	HHB	35 SIG CO (-) (DS)
2-636 FA	TAB (-)	154 LTF
A/3-52 FA (+)	1-52 FA (MLRS)	2/1/55 SIG CO (-)
TACP/52 ASOS (USAF)	3-52 FA (-) (M109A6)	14 PAD
Q37 52 FA BDE (GS)	1/694 EN CO (HORIZ) (DS)	388 CA BN (-) (DS)
99 BSB		
845 FST	**17 MEB 52 ID**	
1/577 MED CO (GRD AMB)	25 CM BN (-)	
3/B/2-52 AV (GSAB) (TACON)	700 MP BN	
1/2/311 QM CO (MA)	7 EN BN	
87 BSTB	2/2/1/55 SIG CO (COMCAM)	
53 EOD	11 ASOS (USAF)	
3/2/1/55 SIG CO (COMCAM)		
B/420 CA BN		
2 HCT/3/B/52 BSTB		
745 EN CO (MAC) (DS)		
1/1/52 CM CO (R/D) (R)		
2/467 CM CO (MX) (S)		
1/1102 MP CO (DS)		
4/A/52 BSTB		

```
                            [page number]
                          [CLASSIFICATION]
```

Figure D-1. Sample Annex A (Task Organization) format (continued)

[CLASSIFICATION] ANNEX A (TASK ORGANIZATION) TO OPERATION PLAN/ORDER [number] [(code name)]—[issuing headquarters] [(classification of title)]		
(sample acronym list)		
AASLT air assault	EOD explosive ordnance disposal	MLRS multiple launch rocket system
ABCT armored brigade combat team	ESC expeditionary sustainment command	MMB multifunctional medical battalion
AR armor	FA field artillery	MP military police
ARB attack reconnaissance battalion	FST forward surgical team	MRBC multi-role bridge company
	GRD AMB ground ambulance	MX mechanized
	GS general support	OPCON operational control
ARS attack reconnaissance squadron	GSAB general support aviation battalion	PAD public affairs detachment
		PLT platoon
ASLT assault	HCT human intelligence collection team	QM quartermaster
ASOS air support operations squadron		QRF quick reaction force
	HHB headquarters and headquarters battalion	R reinforcing
AV aviation		R/D reconnaissance/ decontamination
BDE brigade	HHC headquarters and headquarters company	
BN battalion		RES reserve
BSB brigade support battalion	HORIZ horizontal	RTE CL route clearance
BSTB brigade special troops battalion	IBCT infantry brigade combat team	RTS retransmission
		S smoke
BTB brigade troop battalion	ID infantry division	SIG signal
CA civil affairs	IN infantry	SUST sustainment
CAB combat aviation brigade	LTF logistics task force	TAB target acquisition battery
CAV cavalry	MA mortuary affairs	TACON tactical control
CM chemical	MAC mobility augmentation company	TACP tactical air control party
CO company		TM team
COMCAM combat camera	MEB maneuver enhancement brigade	TUAS tactical unmanned aircraft system
CSH combat support hospital		
CSSB combat sustainment support battalion	MED medical	USAF United States Air Force
DIV division		
DS direct support		
EN engineer		
[page number] [CLASSIFICATION]		

Figure D-1. Sample Annex A (Task Organization) format (continued)

ANNEX B (INTELLIGENCE) FORMAT AND INSTRUCTIONS

D-23. This annex provides fundamental considerations, formats, and instructions for developing Annex B (Intelligence) to the base plan or order. This annex follows the five-paragraph attachment format (see figure D-2).

D-24. Commanders and staffs use Annex B (Intelligence) to describe how intelligence supports the concept of operations described in the base plan or order. The assistant chief of staff, intelligence (G-2 [S-2]) develops Annex B (Intelligence).

D-25. The purpose of Annex B (Intelligence) is to provide detailed information and intelligence on the characteristics of the operational environment and to direct intelligence and counterintelligence activities. Staffs use appendixes to provide detailed analysis of the operational environment and instructions from the various intelligence disciplines. (See figure D-2 on pages D-10 through D-14.)

[CLASSIFICATION]

Place the classification at the top and bottom of every page of the attachments. Place the classification marking at the front of each paragraph and subparagraph in parentheses. Refer to AR 380-5 for classification and release marking instructions.

Copy ## of ## copies
Issuing headquarters
Place of issue
Date-time group of signature
Message reference number

Include the full heading if attachment is distributed separately from the base order or higher-level attachment.

ANNEX B (INTELLIGENCE) TO OPERATION PLAN/ORDER [number] [(code name)]— [issuing headquarters] [(classification of title)]

(U) References: *List documents essential to understanding the attachment.*

   a. *List maps and charts first. Map entries include series number, country, sheet names or numbers, edition, and scale.*

   b. *List other references in subparagraphs labeled as shown.*

   c. *Doctrinal references for this annex include ADRP 2-0 and FM 6-0.*

(U) Time Zone Used Throughout the Plan/Order: *Write the time zone established in the base plan or order.*

1. (U) Situation. *Include information affecting intelligence that paragraph 1 of the OPLAN or OPORD does not cover or that needs expansion.*

   a. (U) <u>Area of Interest</u>. *Describe the area of interest as it relates to intelligence.*

   b. (U) <u>Area of Operations</u>. *Refer to Appendix 2 (Operation Overlay) to Annex C (Operations) as required.*

[page number]
[CLASSIFICATION]

Figure D-2. Sample Annex B (Intelligence) format

[CLASSIFICATION]

ANNEX B (INTELLIGENCE) TO OPERATION PLAN/ORDER [number] [(code name)]— [issuing headquarters] [(classification of title)]

   (1) (U) <u>Terrain</u>. *Describe the aspects of terrain that impact intelligence operations. Refer to Tab A (Terrain) to Appendix 1 (Intelligence Estimate) to Annex B (Intelligence) as required.*

   (2) (U) <u>Weather</u>. *Describe the aspects of weather that impact intelligence operations. Refer to Tab B (Weather) to Appendix 1 (Intelligence Estimate) to Annex B (Intelligence) as required.*

  c. (U) <u>Enemy Forces</u>. *List known and templated locations and activities of enemy intelligence units for one echelon up and two echelons down. List enemy maneuver and other area capabilities that will impact friendly intelligence operations. State expected enemy courses of action and employment of enemy intelligence assets.*

  d. (U) <u>Friendly Forces</u>. *Outline the higher headquarters' intelligence plan. List designation, location, and outline the plan of higher, adjacent, and other intelligence organizations and assets that support or impact the issuing headquarters or require coordination and additional support.*

  e. (U) <u>Interagency, Intergovernmental, and Nongovernmental Organizations</u>. *Identify and describe other organizations in the area of operations that may impact the conduct of intelligence operations or implementation of intelligence-specific equipment and tactics. Refer to Annex V (Interagency Coordination) as required.*

  f. (U) <u>Civil Considerations</u>. *Describe the aspects of the civil situation that impact intelligence operations. Refer to Tab C (Civil Considerations) to Appendix 1 (Intelligence Estimate) to Annex B (Intelligence) and Annex K (Civil Affairs Operations) as required.*

  g. (U) <u>Attachments and Detachments</u>. *List units attached or detached only as necessary to clarify task organization. Refer to Annex A (Task Organization) as required.*

  h. (U) <u>Assumptions</u>. *List any intelligence-specific assumptions that support the annex development.*

2. (U) <u>Mission</u>. *State the mission of intelligence in support of the base plan or order.*

3. (U) <u>Execution</u>.

  a. (U) <u>Scheme of Intelligence Support</u>. *Outline the purpose of intelligence operations and summarize the means and agencies used in planning, directing, collecting, processing, exploiting, producing, disseminating, and evaluating intelligence in support of the concept of operations. When available and appropriate, integrate the resources of other Services and multinational forces. Refer to the base plan or order and Annex C (Operations) as required.*

  b. (U) <u>Tasks to Subordinate Units</u>. *List intelligence tasks assigned to specific subordinate units not contained in the base plan or order. Use subparagraphs to list detailed instructions for each unit performing intelligence functions.*

  c. (U) <u>Counterintelligence</u>. *Refer to Appendix 2 (Counterintelligence) to Annex B (Intelligence).*

  d. (U) <u>Coordinating Instructions</u>. *List only instructions applicable to two or more subordinate units not covered in the base plan or order.*

   (1) (U) <u>Requirements</u>. *Provide guidance for determining intelligence requirements (including those of subordinate commanders), issuing orders, and issuing requests to information collection agencies.*

[page number]
[CLASSIFICATION]

Figure D-2. Sample Annex B (Intelligence) format (continued)

[CLASSIFICATION]

ANNEX B (INTELLIGENCE) TO OPERATION PLAN/ORDER [number] [(code name)]—[issuing headquarters] [(classification of title)]

(a) (U) <u>Priority Intelligence Requirements</u>. *List the priority intelligence requirements (PIRs) along with the latest time intelligence of value for each PIR.*

(b) (U) <u>Friendly Force Information Requirements</u>. *List the friendly force information requirements.*

(c) (U) <u>Requests for Information</u>. *Provide separate, numbered subparagraphs applicable to each unit not organic or attached and from which intelligence support is requested, including multinational forces.*

(2) (U) <u>Measures for Handling Personnel, Documents, and Material</u>. *Describe in the following subparagraphs procedures for handling captured or detained personnel, captured documents, and materiel.*

(a) (U) <u>Prisoners of War, Deserters, Repatriates, Inhabitants, and Other Persons</u>. *State special handling, segregation instructions, and locations of the command's and next higher headquarters' personnel collection points.*

(b) (U) <u>Captured Documents</u>. *List instructions for handling and processing captured documents from time of capture to receipt by specified intelligence personnel.*

(c) (U) <u>Captured Materiel</u>. *Designate items or categories of enemy materiel required for examination. Include any specific instructions for processing and disposition (such as the effects of the Geneva Conventions on the disposition of captured medical materiel). Give locations of the command's and next higher headquarters' captured materiel collection points.*

(d) (U) <u>Documents or Equipment Required</u>. *List in each category the conditions under which units can obtain or request certain documents or equipment. Items may include aerial photographs and maps, charts, and geodesy (satellite) products.*

(3) (U) <u>Distribution of Intelligence Products</u>. *Identify and list in the following subparagraphs any special request procedures for intelligence products in support of this operation. List in each category the conditions under which units can obtain or request certain documents or equipment.*

(a) (U) <u>Special Request for Reports</u>. *Identify, list, or describe the following: periods that routine reports and distribution address; updates to the threat and environment portions of the common operational picture; formats and methods for push and pull intelligence support; and distribution of special intelligence studies, such as defense overprints, photo intelligence reports, and order of battle overlays.*

(b) (U) <u>Special Request Liaison Requirements</u>. *Identify, list, or describe the following liaison requirements: periodic or special intelligence meetings and conferences and special intelligence liaison, when indicated.*

(4) (U) <u>Other Instructions</u>. *Identify, list, or describe any other instructions not covered in the above paragraphs.*

4. (U) <u>Sustainment</u>. *Identify and list sustainment priorities for intelligence key tasks and specify additional sustainment instructions as necessary, to include contractor support. Refer to Annex F (Sustainment) as required.*

[page number]
[CLASSIFICATION]

Figure D-2. Sample Annex B (Intelligence) format (continued)

[CLASSIFICATION]

ANNEX B (INTELLIGENCE) TO OPERATION PLAN/ORDER [number] [(code name)]— [issuing headquarters] [(classification of title)]

a. (U) <u>Logistics</u>. *Identify unique sustainment requirements, procedures, and guidance to support intelligence teams and operations. Specify procedures for specialized technical logistics support from external organizations as necessary. Use subparagraphs to identify priorities and specific instructions for logistics support for intelligence. Refer to Annex F (Sustainment) and Annex P (Host-Nation Support) as required.*

b. (U) <u>Personnel</u>. *Identify intelligence unique personnel requirements and concerns, including global sourcing support and contracted linguist requirements. Use subparagraphs to identify priorities and specific instructions for human resources support, financial management, legal support, and religious support. Refer to Annex F (Sustainment) as required.*

c. (U) <u>Health Service Support</u>. *Identify medical intelligence requirements of the area of operations from the National Center for Medical Intelligence on health hazards to include endemic and epidemic diseases, toxic industrial materials, and known disease vectors. Identify availability, priorities, and instructions for medical care. Refer to Annex F (Sustainment) as required.*

5. (U) <u>Command and Signal</u>.

a. (U) <u>Command</u>.

(1) (U) <u>Location of the Commander and Key Leaders</u>. State the location of the commander and key intelligence leaders.

(2) (U) <u>Succession of Command</u>. *State the succession of command if not covered in the unit's SOPs.*

(3) (U) <u>Liaison Requirements</u>. *State the intelligence liaison requirements not covered in the base order or unit standard operating procedures (SOPs).*

b. (U) <u>Control</u>.

(1) (U) <u>Command Posts</u>. *Describe the employment of intelligence-specific command posts (CPs), including the location of each CP and its time of opening and closing.*

(2) (U) <u>Intelligence Coordination Line</u>. *Identify the intelligence coordination line.*

(3) (U) <u>Special Security</u>. *Identify special security office arrangements and coordination.*

(4) (U) <u>Reports</u>. *List intelligence-specific reports not covered in SOPs. Refer to Annex R (Reports) as required.*

c. (U) <u>Signal</u>. *Address any intelligence-specific communications requirements. Refer to Annex H (Signal) as required.*

[page number]
[CLASSIFICATION]

Figure D-2. Sample Annex B (Intelligence) format (continued)

[CLASSIFICATION]

ANNEX B (INTELLIGENCE) TO OPERATION PLAN/ORDER [number] [(code name)]—[issuing headquarters] [(classification of title)]

ACKNOWLEDGE: *Include only if attachment is distributed separately from the base order.*

[Commander's last name]
[Commander's rank]

The commander or authorized representative signs the original copy of the attachment. If the representative signs the original, add the phrase "For the Commander." The signed copy is the historical copy and remains in the headquarters' files.

OFFICIAL:

[Authenticator's name]
[Authenticator's position]

Use only if the commander does not sign the original attachment. If the commander signs the original, no further authentication is required. If the commander does not sign, the signature of the preparing staff officer requires authentication and only the last name and rank of the commander appear in the signature block.

ATTACHMENTS: *List lower-level attachment (appendixes, tabs, and exhibits).*

Appendix 1–Intelligence Estimate
Appendix 2–Counterintelligence
Appendix 3–Signals Intelligence
Appendix 4–Human Intelligence
Appendix 5–Geospatial Intelligence
Appendix 6–Measurement and Signature Intelligence
Appendix 7–Open-Source Intelligence

DISTRIBUTION: *Show only if distributed separately from the base order or higher-level attachments.*

[page number]
[CLASSIFICATION]

Figure D-2. Sample Annex B (Intelligence) format (continued)

ANNEX C (OPERATIONS) FORMAT AND INSTRUCTIONS

D-26. This annex provides fundamental considerations, formats, and instructions for developing Annex C (Operations) to the base plan or order. This annex follows the five-paragraph attachment format.

D-27. Commanders and staffs use Annex C (Operations) to describe and outline how this annex supports the concept of operations described in the base plan or order. The G-5 or G-3 (S-3) develops Annex C (Operations).

D-28. This annex describes the operation's objectives. A complex operation's concept of support may require a schematic to show the operations objectives and task relationships. It includes a discussion of the overall operations concept of support with specific details in element subparagraphs and attachments. It refers to the execution matrix to clarify timing relationships among various operations tasks. This annex also contains the information needed to synchronize timing relationships of each element related to operations. It includes operations-related constraints, if appropriate. (See figure D-3 on pages D-15 through D-20.)

[CLASSIFICATION]

Place the classification at the top and bottom of every page of the attachments. Place the classification marking at the front of each paragraph and subparagraph in parentheses. Refer to AR 380-5 for classification and release marking instructions.

Copy ## of ## copies
Issuing headquarters
Place of issue
Date-time group of signature
Message reference number

Include the full heading if attachment is distributed separately from the base order or higher-level attachment.

ANNEX C (OPERATIONS) TO [OPERATION PLAN/ORDER [number] [(code name)]—[(classification of title)]

(U) References: *List documents essential to understanding the attachment.*

a. *List maps and charts first. Map entries include series number, country, sheet names or numbers, edition, and scale.*

b. *List other references in subparagraphs labeled as shown.*

c. *Doctrinal references for this annex are ADRP 3-0, ADRP 5-0, ADRP 6-0, and FM 6-0.*

(U) Time Zone Used Throughout the Order: *Write the time zone established in the base plan or order.*

1. (U) Situation. *Include information affecting operations that paragraph 1 of the OPLAN or OPORD does not cover or that needs expansion. If there is no new information from what is contained in the base order then indicate this by stating "See base order."*

a. (U) Area of Interest. *Describe the area of interest as it relates to operations. Reference the digital overlay(s) within systems such as command post of the future. Refer to Annex B (Intelligence) as required.*

[page number]
[CLASSIFICATION]

Figure D-3. Sample Annex C (Operations) format

[CLASSIFICATION]

ANNEX C (OPERATIONS) TO [OPERATION PLAN/ORDER [number] [(code name)]—[(classification of title)]

b. (U) Area of Operations. *Refer to Appendix 2 (Operation Overlay) to Annex C (Operations).*

(1) (U) Terrain. *Describe the aspects of terrain that impact operations. Refer to Annex B (Intelligence) as required.*

(2) (U) Weather. *Describe the aspects of weather that impact operations. Refer to Annex B (Intelligence) as required.*

c. (U) Enemy Forces. *Identify and reference enemy overlays. First, list known and templated locations and activities of enemy units for two echelons down. For example, a U.S. division would address enemy battalions; a U.S. battalion would address enemy platoons. Second, list enemy maneuver and other capabilities that will impact friendly operations. Third, state the enemy most likely and most dangerous courses of action and employment of enemy assets. A staff more easily understands these enemy courses of action when they are depicted in sketches. Fourth, include an assessment of terrorist or criminal activities directed against U.S. government interests in the area of operations. Refer to Annex B (Intelligence) and other sources as required.*

(Note: *If conducting operations focused on stability or defense support of civil authorities, change the title of this subparagraph to "Terrorist/Criminal Threats.")*

d. (U) Friendly Forces. *Subparagraphs outline the mission, commander's intent, and concept of operations for headquarters one and two command echelons above the unit. Subparagraphs also provide the missions and concept of operations of flank units, supported units, supporting units, and other units and organizations, such as special operations forces, whose actions have a significant effect on the issuing headquarters or require coordination. This subparagraph uses the same format as the base order and can be shortened by using the phrase "See Base Order" if there is no change.*

e. (U) Interagency, Intergovernmental, and Nongovernmental Organizations. *Identify and describe other organizations in the area of operations that may impact the conduct of the unit's operations or require support not identified in the base order. Also identify nongovernmental organizations in the area of operations that want nothing to do with the U.S. military and are not identified in the base order. Refer to Annex V (Interagency Coordination) as required.*

f. (U) Civil Considerations. *List all critical civil considerations that impact on the unit's operations, such as cultural or religious sensitivities to male Soldiers searching female civilians, searching civilian homes at night, or resolving injury or damage claims not established in the base order. Refer to Annex B (Intelligence) and Annex K (Civil Affairs Operations) as required.*

[page number]
[CLASSIFICATION]

Figure D-3. Sample Annex C (Operations) format (continued)

[CLASSIFICATION]

ANNEX C (OPERATIONS) TO [OPERATION PLAN/ORDER [number] [(code name)]—[(classification of title)]

g. (U) <u>Attachments and Detachments</u>. *List units attached or detached only as necessary to clarify task organization. Do not repeat information already listed under Task Organization in the base order or in Annex A (Task Organization). Try to put all information in the task organization annex and state "See Annex A (Task Organization)." Otherwise, list units that are attached or detached to the issuing headquarters. State when attachment or detachment is effective, if different from the effective time of the operation plan or order, such as on-order, or commitment of reserve forces. Use the term "remains attached" when units will be or have been attached for some time. Refer to Annex A (Task Organization) as required.*

h. (U) <u>Assumptions</u>. *List any operations-specific assumptions that support the annex development.*

2. (U) <u>Mission</u>. *Enter the unit's restated mission only if this annex is distributed separately from the base order. It should contain a short description of the who, what (task), when, where, and why (purpose) that clearly indicates the action to be taken and the reason for doing so. A mission statement contains no subparagraphs. The mission statement covers on-order missions, otherwise state "See base order."*

3. (U) <u>Execution</u>. *Describe how the commander intends to accomplish the mission in terms of the commander's intent, an overarching concept of operations, scheme of employing maneuver, assessment, specified tasks to subordinate units, and key coordinating instructions in the subparagraphs below only if this annex is distributed separately from the base order.*

Commanders ensure that their scheme of maneuver is consistent with their intent and that of the next two higher echelon commanders. This paragraph and the operation overlay are complementary, each adding clarity to, rather than duplicating, the other. Do not duplicate information in unit subparagraphs and coordinating instructions contained in the base order. Provide the primary tasks of maneuver units conducting the decisive operation and the purpose of each. Next, state the primary tasks of maneuver units conducting shaping operations, including security operations, and the purpose of each. For offensive-focused operations, identify the form of maneuver. For defensive-focused operations, identify the type of defense. For stability-focused operations, describe the role of maneuver units by primary stability tasks. For defense support of civil authorities-focused operations, describe the role of maneuver units by primary defense support of civil authorities support tasks. If the operation is phased, identify the main effort by phase. Identify and include priorities for the reserves. Refer to attached appendixes as required.

a. (U) <u>Scheme of Movement and Maneuver</u>. *State the scheme of movement and maneuver by describing the employment of maneuver units, such as divisions, brigade combat teams, and combat aviation brigades in accordance with the concept of operations. Ensure that this paragraph is consistent with the operation overlay in Appendix 2 (Operation Overlay) to Annex C (Operations). Describe how the actions of subordinate maneuver units fit together to accomplish the mission. The scheme of maneuver expands the commander's selected course of action and expresses how each maneuver element of the force will cooperate. As the commander's intent focuses on the end state, the scheme of maneuver focuses on the maneuver tactics and techniques employed during the operation as well as synchronizes the actions of each maneuver element.*

[page number]
[CLASSIFICATION]

Figure D-3. Sample Annex C (Operations) format (continued)

[CLASSIFICATION]
ANNEX C (OPERATIONS) TO [OPERATION PLAN/ORDER [number] [(code name)]—[(classification of title)]

(1) (U) <u>Scheme of Mobility/Countermobility</u>. *State the scheme of mobility/countermobility including priorities by unit or area. Refer to Annex G (Engineer) as required.*

(2) (U) <u>Scheme of Battlefield Obscuration</u>. *State the scheme of battlefield obscuration, including priorities by unit or area. Refer to Appendix 9 (Battlefield Obscuration) to Annex C (Operations) as required.*

(3) (U) <u>Scheme of Information Collection</u>. *Describe how the commander intends to use information collection to support the concept of operations. Include the primary objectives. Refer to Annex L (Information Collection) as required.*

b. (U) <u>Assessment</u>. *Describe the priorities for assessment and identify the measures of performance and effectiveness used to assess end state conditions and objectives. Refer to Annex M (Assessment) as required.*

c. (U) <u>Tasks to Subordinate Units</u>. *List movement and maneuver tasks assigned to specific subordinate units not contained in the base order. Each task must include* who *(the subordinate unit assigned the task),* what *(the task itself),* when, where, *and* why *(purpose). Use a separate subparagraph for each unit. List units in sequence of task organization. Place tasks that affect two or more units in paragraph 3d of this annex.*

d. (U) <u>Coordinating Instructions</u>. *List only instructions applicable to two or more subordinate units not covered in the base plan or order.*

4. (U) <u>Sustainment</u>. *Describe priorities of sustainment by unit or area. Highlight subordinate allocations of command-regulated classes of supply that impact movement and maneuver, such as controlled supply rates. Include instructions for deployment or redeployment. Identify priorities of sustainment for operations key tasks and specify additional instructions as required. Refer to Annex F (Sustainment) as required.*

a. (U) <u>Logistics</u>. *Use subparagraphs to identify priorities and specific instructions for logistics support. Refer to Annex F (Sustainment) and Annex P (Host-Nation Support) as required.*

b. (U) <u>Personnel</u>. *Use subparagraphs to identify priorities and specific instructions for human resources support, financial management, legal support, and religious support. Refer to Annex F (Sustainment) as required.*

c. (U) <u>Health Service Support</u>. *Identify availability, priorities, and instructions for medical care. Refer to Annex F (Sustainment) as required.*

[page number]
[CLASSIFICATION]

Figure D-3. Sample Annex C (Operations) format (continued)

[CLASSIFICATION]

ANNEX C (OPERATIONS) TO [OPERATION PLAN/ORDER [number] [(code name)]—[(classification of title)]

5. (U) Command and Signal. *List information in this paragraph and its subparagraphs only if annex distributed separately from base order, otherwise state "Same as base order."*

 a. (U) Command.

   (1) (U) Location of Commander and Key Leaders. *State the location of the commander and key leaders.*

   (2) (U) Succession of Command. *State the succession of command if not covered in the unit's SOPs.*

   (3) (U) Liaison Requirements. *State the liaison requirements not covered in the base order.*

 b. (U) Control.

   (1) (U) Command Posts. *Describe the employment of command posts (CPs), including the location of each CP and its time of opening and closing.*

   (2) (U) Reports. *List reports not covered in standard operating procedures (SOPs). Refer to Annex R (Reports) as required.*

 c. (U) Signal. *Address any communications requirements. Refer to Annex H (Signal) as required.*

ACKNOWLEDGE: *Include only if attachment is distributed separately from the base order.*

[Commander's last name]

[Commander's rank]

The commander or authorized representative signs the original copy of the attachment. If the representative signs the original, add the phrase "For the Commander." The signed copy is the historical copy and remains in the headquarters' files.

OFFICIAL:

[Authenticator's name]

[Authenticator's position]

Use only if the commander does not sign the original attachment. If the commander signs the original, no further authentication is required. If the commander does not sign, the signature of the preparing staff officer requires authentication and only the last name and rank of the commander appear in the signature block.

[page number]
[CLASSIFICATION]

Figure D-3. Sample Annex C (Operations) format (continued)

[CLASSIFICATION]

ANNEX C (OPERATIONS) TO [OPERATION PLAN/ORDER [number] [(code name)]—[(classification of title)]

ATTACHMENT: *List lower-level attachment (appendixes, tabs, and exhibits).*
Appendix 1–Army Design Methodology Products
Appendix 2–Operation Overlay
Appendix 3–Decision Support Products
Appendix 4–Gap Crossing Operations
Appendix 5–Air Assault Operations
Appendix 6–Airborne Operations
Appendix 7–Amphibious Operations
Appendix 8–Special Operations
Appendix 9–Battlefield Obscuration
Appendix 10 –Airspace Control
Appendix 11 –Rules of Engagement
Appendix 12–Cyber Electromagnetic Activities
Appendix 13–Military Information Support Operations
Appendix 14–Military Deception
Appendix 15–Information Operations

DISTRIBUTION: *Show only if distributed separately from the base order or higher-level attachments.*

[page number]
[CLASSIFICATION]

Figure D-3. Sample Annex C (Operations) format (continued)

ANNEX D (FIRES) FORMAT AND INSTRUCTIONS

D-29. This annex provides fundamental considerations, format, and instructions for developing Annex D (Fires) to the base plan or order. This annex follows the five-paragraph attachment format.

D-30. Commanders and staffs use Annex D (Fires) to describe how fires support the concept of operations described in the base plan or order. The chief of fires (fire support officer) develops Annex D (Fires).

D-31. This annex describes the fires concept of support objectives. A complex fires concept of support may require a schematic to show the fires objectives and task relationships. It includes a discussion of the overall fires concept of support with the specific details in element subparagraphs and attachments. It refers to the execution matrix to clarify timing relationships among various fires tasks. This annex also contains the information needed to synchronize timing relationships of each element related to fires. It includes fires-related constraints, if appropriate. (See figure D-4 on pages D-21 through D-26.)

[CLASSIFICATION]

Place the classification at the top and bottom of every page of the attachments. Place the classification marking at the front of each paragraph and subparagraph in parentheses. Refer to AR 380-5 for classification and release marking instructions.

Copy ## of ## copies
Issuing headquarters
Place of issue
Date-time group of signature
Message reference number

Include the full heading if attachment is distributed separately from the base order or higher-level attachment.

ANNEX D (FIRES) TO OPERATION PLAN/ORDER [number] [(code name)]—[issuing headquarters] [(classification of title)]

(U) References: *List documents essential to understanding the attachment.*

   a. *List maps and charts first. Map entries include series number, country, sheet names or numbers, edition, and scale.*

   b. *List other references in subparagraphs labeled as shown.*

   c. *Doctrinal references for this annex include the ADRP 3-09, FM 3-09, FM 3-36, FM 3-60, FM 6-0, FM 6-20-40, and FM 6-20-50.*

(U) Time Zone Used Throughout the Plan/Order: *Write the time zone established in the base plan or order.*

1. (U) Situation. *Include information affecting fires that paragraph 1 of the OPLAN or OPORD does not cover or that needs expansion.*

   a. (U) Area of Interest. *Describe the area of interest as it relates to fires. Refer to Annex B (Intelligence) as required.*

   b. (U) Area of Operations. *Refer to Annex C (Operations) as required.*

[page number]
[CLASSIFICATION]

Figure D-4. Sample Annex D (Fires) format

[CLASSIFICATION]

ANNEX D (FIRES) TO OPERATION PLAN/ORDER [number] [(code name)]—[issuing headquarters] [(classification of title)]

(1) (U) <u>Terrain</u>. *Describe the aspects of terrain that impact fires. Refer to Annex B (Intelligence) as required.*

(2) (U) <u>Weather</u>. *Describe the aspects of weather that impact fires. Refer to Annex B (Intelligence) as required.*

c. (U) <u>Enemy Forces</u>. *List known and templated locations and activities of enemy fires units for one echelon above and two echelons below the unit. List enemy maneuver, indirect fire/counterfire, air, and electronic warfare threats and other capabilities that will impact friendly fires operations. State expected enemy courses of action and employment of enemy fires assets. Refer to Annex B (Intelligence) as required.*

d. (U) <u>Friendly Forces</u>. *Outline the higher headquarters' fires plan. List designation, location, and outline the plan of higher, adjacent, and other fires organizations and assets that support or impact the issuing headquarters or require coordination and additional support.*

e. (U) <u>Interagency, Intergovernmental, and Nongovernmental Organizations</u>. *Identify and describe other organizations in the area of operations that may impact the conduct of fires or implementation of fires-specific equipment and tactics. Refer to Annex V (Interagency Coordination) as required.*

f. (U) <u>Civil Considerations</u>. *Describe the aspects of the civil situation that impact fires. Refer to Annex B (Intelligence) and Annex K (Civil Affairs Operations) as required.*

g. (U) <u>Attachments and Detachments</u>. *List fires resources attached or under operational control to the unit by higher headquarters and any units detached or under operational control to other headquarters. Refer to Annex A (Task Organization) as required.*

h. (U) <u>Assumptions</u>. *List any fires-specific assumptions that support the annex development.*

2. (U) <u>Mission</u>. *State the mission of fires in support of the base plan or order.*

3. (U) <u>Execution</u>.

a. (U) <u>Scheme of Fires</u>. *Describe how fires support the commander's intent and concept of operations. Establish the priorities of fires to units for each phase of the operation. The scheme of fires must be concise but specific enough to clearly state what fires are to accomplish in the operation. The scheme of fires must answer the "who, what, when, where, and why" of the fires to be provided, but provide enough flexibility to allow subordinate commanders to determine the "how" to the maximum extent possible by ensuring necessary procedural and positive control. The scheme of fires may include a general narrative for the entire operation that should address the fire support task and purpose, allocation of assets, positioning guidance for fire support assets and observers, and attack guidance to include the entire scalable range of effects (lethal to nonlethal effects). Add subparagraphs addressing fire support tasks for each phase of the operation use the following format: task, purpose, execution, and assessment in matrix form. Refer to the base plan or order and Annex C (Operations) as required.*

(1) <u>Task, Purpose, Execution, and Assessment</u>: *The example below provides a sample matrix for task, purpose, execution, and assessment, to be used at the discretion of the commander. See local SOPs for additional guidance and details.*

[page number]
[CLASSIFICATION]

Figure D-4. Sample Annex D (Fires) format (continued)

[CLASSIFICATION]   **ANNEX D (FIRES) TO OPERATION PLAN/ORDER [number] [(code name)]—[issuing headquarters] [(classification of title)]**
Sample matrix for each fire support task
PHASE: *State the phase of the operation.*
TASK (what): *State the supported/maneuver commander task and the type(s) of effects the fires unit must provide for that phase of the operation (suppress, neutralize, interdict, divert, exploit, deny, delay, deceive, disrupt, degrade, destroy, obscuration, or screening).*
PURPOSE (why): *State the supported/maneuver commander purpose and the desired end state for the targeted enemy formation/function/capability. (There may be more that one task/purpose per phase.)*
Priority of fire: *State the priority of fire to subordinate units for all fires assets under the unit's command or control.*
Allocations: *List any additional assets assigned to subordinates for planning. Some examples are primary targets, radar zones, and attack aviation.*
Positioning Guidance: *Provide positioning guidance to assets such as mortars or observers required for execution.*
Restrictions: *List all restrictions for the phase.*
Target Information: *Target number, trigger, location, observer, delivery, attack guidance, and communications (TTLODAC) refined by executer.*

Fire Support Task	Target Number	Trigger	Location	Observer	Delivery	Attack Guidance	Communications
List the task number the target supports.	*List the target number or type of target.*	*State the trigger (tactical/technical) for the target.*	*Give the location of the target.*	*State the observer of the target (primary and alternate).*	*State the delivery system for the target (primary and alternate).*	*State the attack guidance/ method of engagement for the target.*	*State the frequency and communications net the target will be called in on (primary, alternate, contingency, or emergency).*

    b. (U) <u>Scheme of Field Artillery Support</u>*. Describe the scheme of cannon, rocket, and missile fires in support of operations. Include specific tasks to subordinate field artillery headquarters. Address any potential requirements for massing fires that may affect organic, direct support, or reinforcing fires units. Identify the timing and duration of specific identified fire plans, such as counterfire, preparations, suppression of enemy air defenses, or joint suppression of enemy air defenses. Refer to Appendix 4 (Field Artillery Support) to Annex D (Fires) as required.*

[page number]
[CLASSIFICATION]

Figure D-4. Sample Annex D (Fires) format (continued)

[CLASSIFICATION]

ANNEX D (FIRES) TO OPERATION PLAN/ORDER [number] [(code name)]—[issuing headquarters] [(classification of title)]

(1) (U) <u>Organization for Combat</u>. *Provide direction for the proper organization for combat, including the unit designation, nomenclature, and tactical task.*

(2) (U) <u>Miscellaneous</u>. *Provide any other information necessary for planning not already mentioned. Other information in this subparagraph may include changes to the targeting numbering system, the use of pulse repetition frequency codes, positioning restrictions, and a position area overlay.*

c. (U) <u>Scheme of Air Support</u>. *Briefly describe the maneuver commander's guidance for the use of air power. Refer to Appendix 5 (Air Support) to Annex D (Fires) as required.*

(1) (U) <u>Organization for Combat</u>. *Provide direction for the proper organization for combat, including the unit designation, nomenclature, and tactical task.*

(2) (U) <u>Air Interdiction Operations</u>. *Briefly describe the joint force air component commander's intent for air interdiction. Describe the maneuver commander's air interdiction concept and priorities for target attack within the area of operations.*

(3) (U) <u>Close Air Support Operations</u>. *Provide the allocation and distribution of close air support sorties by subordinate unit. Provide the desired method for planning close air support (immediate or pre-planned) or any special control arrangements.*

(4) (U) <u>Air Reconnaissance Operations</u>. *Provide the concept for use of reconnaissance aircraft if resources are provided by the joint force air component commander. Refer to Annex L (Information Collection).*

(5) (U) <u>Miscellaneous</u>. *Provide any other information necessary for planning not already mentioned, including the following:*

(a) *The air tasking order's effective time.*

(b) *Deadlines for submission of air interdiction, close air support, reconnaissance aircraft, and electronic warfare aircraft requests.*

(c) *The mission request numbering system as it relates to the target numbering system.*

(d) *The joint suppression of enemy air defenses tasking from the joint force land component commander.*

(e) *Reference to essential airspace control measures (coordinating altitude, target areas, low level transit route requirements, and so on) identified in Annex C (Operations).*

d. (U) <u>Scheme of Naval Fire Support</u>. *Describe the concept for use of naval fire support. Include specific tasks to supporting units. Include trajectory limitations or minimum safe distances. Refer to Appendix 6 (Naval Fire Support) to Annex D (Fires) as required.*

(1) (U) <u>Organization for Combat</u>. *List the grouping or organization for combat, including the following:*

(a) (U) *Identify and list the allocation of observers and spotters.*

(b) (U) *Identify and list the allocation of ships to units.*

[page number]
[CLASSIFICATION]

Figure D-4. Sample Annex D (Fires) format (continued)

ANNEX D (FIRES) TO OPERATION PLAN/ORDER [number] [(code name)]—[issuing headquarters] [(classification of title)]

(2) (U) <u>Miscellaneous</u>. *Provide any other information necessary for planning not already mentioned.*

e. (U) <u>Battlefield Obscuration Support</u>. *Describe the concept for use of artillery smoke and battlefield obscuration. Refer to Annex C (Operations) as required.*

f. (U) <u>Target Acquisition</u>. *Provide information pertaining to the employment and allocation of fires target acquisition systems and assets. Refer to Appendix 3 (Targeting) and Appendix 4 (Field Artillery Support) to Annex D (Fires) as required.*

g. (U) <u>Tasks to Subordinate Units</u>. *List fires tasks assigned to specific subordinate units not contained in the base order.*

h. (U) <u>Coordinating Instructions</u>. *List only instructions applicable to two or more subordinate units not covered in the base plan or order. Provide subordinates and adjacent units the following information to coordinate fires:*

(1) *A clear definition of the boundary of the operational area if not specified in the basic plan. This area may be identified by phase if it is a phased operation.*

(2) *Targeting products.*

(3) *Fire support coordination measures.*

(4) *The time of execution of program of fires relative to H-hour (counterfire, preparations or counterpreparations, joint suppression of enemy air defenses), if needed.*

(5) *Rules of engagement specific to fires.*

4. (U) <u>Sustainment</u>. *Identify sustainment priorities for fires key tasks and specify additional sustainment instructions as necessary. Describe critical or unusual sustainment actions that might occur before, during, and after the battle to support the commander's scheme of fires. Refer to Annex F (Sustainment) as required.*

a. (U) <u>Logistics</u>. *Use subparagraphs to identify priorities and specific instructions for fires logistics support. Refer to Annex F (Sustainment) and Annex P (Host-Nation Support) as required.*

(1) (U) <u>Supply</u>. *Identify the location of ammunition transfer holding points and ammunition supply points. Refer to Annex F (Sustainment) as required.*

(2) (U) <u>Allocation of Ammunition</u>. *List the allocation of cannon, rocket, and missile ammunition for each phase of the operation based on the amount of Class V available. Refer to Annex F (Sustainment) as required.*

b. (U) <u>Personnel</u>. *Use subparagraphs to identify priorities and specific instructions for human resources support, financial management, legal support, and religious support. Refer to Annex F (Sustainment) as required.*

c. (U) <u>Health Service Support</u>. *Identify ground and air medical evacuation requirements and the availability, priorities, and instructions for medical care. Refer to Annex F (Sustainment) as required.*

[page number]
[CLASSIFICATION]

Figure D-4. Sample Annex D (Fires) format (continued)

[CLASSIFICATION

ANNEX D (FIRES) TO OPERATION PLAN/ORDER [number] [(code name)]—[issuing headquarters] [(classification of title)]

5. (U) **Command and Signal.**

    a. (U) <u>Command</u>.

        (1) (U) <u>Location of the Commander and Key Leaders</u>. *State the location of the commander and key fires leaders.*

        (2) (U) <u>Succession of Command</u>. *State the succession of command if not covered in the unit's SOPs.*

        (3) (U) <u>Liaison Requirements</u>. *State the fires liaison requirements not covered in the base order.*

    b. (U) <u>Control</u>.

        (1) (U) <u>Command Posts</u>. *Describe the employment of maneuver units and fires-specific command posts, including the location of each command post and its time of opening and closing.*

        (2) (U) <u>Reports</u>. *List fires-specific reports not covered in standard operating instructions. Refer to Annex R (Reports) as required.*

    c. (U) <u>Signal</u>. *Address any fires-specific communications requirements. Identify the current standard operating instructions edition. Refer to Annex H (Signal) as required.*

[Commander's last name]
[Commander's rank]

ACKNOWLEDGE: *Include only if attachment is distributed separately from the base order.*

The commander or authorized representative signs the original copy of the attachment. If the representative signs the original, add the phrase "For the Commander." The signed copy is the historical copy and remains in the headquarters' files.

OFFICIAL:

[Authenticator's name]
[Authenticator's position]

Use only if the commander does not sign the original attachment. If the commander signs the original, no further authentication is required. If the commander does not sign, the signature of the preparing staff officer requires authentication and only the last name and rank of the commander appear in the signature block.

ATTACHMENTS: *List lower-level attachment (appendixes, tabs, and exhibits).*

Appendix 1–Fire Support Overlay
Appendix 2–Fire Support Execution Matrix
Appendix 3–Targeting
Appendix 4–Field Artillery Support
Appendix 5–Air Support
Appendix 6–Naval Fire Support
Appendix 7–Air and Missile Defense

DISTRIBUTION: *Show only if distributed separately from the base order or higher-level attachments.*

[page number]
[CLASSIFICATION]

Figure D-4. Sample Annex D (Fires) format (continued)

ANNEX E (PROTECTION) FORMAT AND INSTRUCTIONS

D-32. This annex provides fundamental considerations, formats, and instructions for developing Annex E (Protection) to the base plan or order. This annex follows the five-paragraph attachment format.

D-33. Commanders and staffs use Annex E (Protection) to describe how protection supports the concept of operations described in the base plan or order. This annex describes how the commander intends to preserve the force through the protection tasks (listed in this annex's appendixes). The chief of protection or a designated staff officer (engineer; chemical, biological, radiological, and nuclear; air and missile defense; or provost marshal) develops Annex E (Protection). The surgeon provides the chief of protection with input for Appendix 10 (Force Health Protection).

D-34. This annex describes the protection concept of support objectives. A complex protection concept of support may require a schematic to show the protection objectives and task relationships. This annex includes a discussion of the overall protection concept of support, with the specific details in element subparagraphs and attachments. It refers to the execution matrix to clarify timing relationships among various protection tasks. This annex also contains information needed to synchronize timing relationships of each element related to protection. It includes protection-related constraints, if appropriate. (See figure D-5 on pages D-27 through D-33.)

[CLASSIFICATION]

Place the classification at the top and bottom of every page of the attachments. Place the classification marking at the front of each paragraph and subparagraph in parentheses. Refer to AR 380-5 for classification and release marking instructions.

Copy ## of ## copies
Issuing headquarters
Place of issue
Date-time group of signature
Message reference number

Include the full heading if attachment is distributed separately from the base order or higher-level attachment.

ANNEX E (PROTECTION) TO OPERATION PLAN/ORDER [number] [(code name)]—[issuing headquarters] [(classification of title)]

(U) References: *List documents essential to understanding the attachment.*

    a. List maps and charts first. Map entries include series number, country, sheet names or numbers, edition, and scale.

    b. List other references in subparagraphs labeled as shown.

    c. Doctrinal references for protection include ADRP 3-0, ADRP 3-37, ADRP 3-90, AR 385-10, AR 525-13, AR 525-28, AR 530-1, ATP 3-37.34, ATP 4-32, DA Pamphlet 385-10, FM 3-01, FM 3-11, FM 3-13, FM 3-50.1, FM 4-02.7, FM 4-02.17, FM 4-02.18, FM 4-02.19, FM 4-02.51, FM 5-19, and FM 6-0.

[page number]
[CLASSIFICATION]

Figure D-5. Sample Annex E (Protection) format

|CLASSIFICATION|

ANNEX E (PROTECTION) TO OPERATION PLAN/ORDER [number] [(code name)]—[issuing headquarters] [(classification of title)]

(U) Time Zone Used Throughout the Plan/Order: *Write the time zone established in the base plan or order.*

1. (U) <u>Situation</u>. *Provide situational information affecting the protection tasks and systems that paragraph 1 of the OPLAN or OPORD does not cover or that needs expansion.*

    a. (U) <u>Area of Interest</u>. *Describe the area of interest as it impacts protection. Identify area of interest characteristics and hazards (including health hazards) that require coordinated protection actions to preserve the force. Refer to Annex B (Intelligence) as required.*

    b. (U) <u>Area of Operations</u>. *Describe the area of operations as it impacts protection. Identify and describe the area of operation's characteristics and hazards that require coordinated protection actions to preserve the force. Refer to Annex C (Operations) as required.*

        (1) (U) <u>Terrain</u>. *Describe the aspects of terrain that impact protection operations. Identify terrain features in the area of interest and area of operations that create a hazard or enhance the threat. Specify protection actions that may be required as a result of the terrain. Identify terrain that may benefit protection capabilities. Refer to Annex B (Intelligence) as required.*

        (2) (U) <u>Weather</u>. *Describe the aspects of weather that impact protection operations, tasks, and systems. Refer to Appendix 2 (Safety) to Annex E (Protection) and Annex B (Intelligence) as required.*

    c. (U) <u>Enemy Forces</u>. *List known and templated locations and activities of enemy protection units for one echelon up and two echelons down. List enemy maneuver and other area capabilities that will impact friendly operations. State expected enemy courses of action and employment of enemy protection assets. Include consideration of civil disturbances and criminal acts. Narrow the focus to offensive-minded threats that require planning, resources, and actions to protect the force. Refer to Annex B (Intelligence) as required.*

    d. (U) <u>Friendly Forces</u>. *Outline the higher headquarters' protection plan. List designation, location, and outline of plan of higher, adjacent, and other protection assets that support or impact the issuing headquarters or require coordination and additional support. List areas of the operation most vulnerable to enemy attack or adverse influence.*

    e. (U) <u>Interagency, Intergovernmental, and Nongovernmental Organizations</u>. *Identify and describe other organizations in the area of operations that may impact the conduct of protection operations or impact protection specific equipment and tactics. Outline the results of the risk management process to mitigate the risk of fratricide. Enhance continual situational understanding by frequently updating data of friendly forces. Describe the method and timing of the data updates. Refer to Annex V (Interagency Coordination) as required.*

    f. (U) <u>Civil Considerations</u>. *Describe the aspects of the civil situation that impact protection operations. Refer to Annex B (Intelligence) and Annex K (Civil Affairs Operations) as required.*

    g. (U) <u>Attachments and Detachments</u>. *List units attached or detached only as necessary to clarify task organization. Refer to Annex A (Task Organization) as required.*

    h. (U) <u>Assumptions</u>. *List any protection-specific assumptions that support the annex development.*

2. (U) <u>Mission</u>. *State the protection mission in support of the base plan or order.*

3. (U) <u>Execution</u>.

[page number]
|CLASSIFICATION|

Figure D-5. Sample Annex E (Protection) format (continued)

|CLASSIFICATION|

ANNEX E (PROTECTION) TO OPERATION PLAN/ORDER |number| |(code name)|—|issuing headquarters| |(classification of title)|

a. (U) <u>Scheme of Protection</u>. *Describe how the protection tasks and systems support the commander's intent and concept of operations. Establish the priorities of support to units for each phase of the operation. If required information for a specific protection task or system is brief, include it in this paragraph and eliminate the associated appendix. Refer to the base order and Annex C (Operations) as required.*

(1) (U) <u>Operational Area Security</u>. *State the scheme of operational area security and overall area security objective. Describe how operational area security supports the commander's intent, the maneuver plan, and protection priorities. Direct how each element of the force will cooperate to accomplish operational area security and tie that to support of the operation with the task and purpose statement. Discuss how operational area security orients on the force, installation, route, area, or asset to be protected. Discuss how operational area security is often an economy of force role assigned in some manner to many organizations. Discuss how operational area security is often designed to ensure the continued conduct of sustainment operations and to support decisive and shaping operations. Describe how forces engaged in area security operations saturate an area or position on key terrain to provide protection through early warning, reconnaissance, or surveillance and guard against unexpected enemy attack with an active response. Discuss the role of response forces in the operational area security scheme. Refer to Appendix 1 (Operational Area Security) to Annex E (Protection) as required.*

(2) (U) <u>Safety</u>. *Describe how mission-dictated safety program requirements support the commander's intent and concept of operations. Describe how safety tasks and functions are prioritized to eliminate or mitigate hazards on a greatest risk first basis to support the unit. Refer to Appendix 2 (Safety) to Annex E (Protection) as required.*

(3) (U) <u>Operations Security</u>. *Describe how operations security applies to all operations. All units conduct operations security to preserve essential secrecy from threat exploitation, and support the commander's intent and concept of operations. Describe the general concept and any additional operations security measures with other staff and command elements, and synchronize with adjacent units. Refer to Appendix 3 (Operations Security) to Annex E (Protection) as required.*

a. (U) *Identify actions that can be observed by threat intelligence systems.*

b. (U) *Determine indicators of threat intelligence that systems might obtain which could be interpreted or pieced together to derive critical information in time to be useful to the threat.*

c. (U) *Describe how to execute measures that eliminate or reduce (to an acceptable level) the vulnerabilities of friendly actions.*

(4) (U) <u>Intelligence Support to Protection</u>. *Describe how providing intelligence supports measures that the command takes to remain viable and functional by protecting itself from the effects of threat activities. Describe how it also provides intelligence that supports recovery from threat actions. It includes analyzing the threats, hazards, and other aspects of an operational environment and using the intelligence preparation of the battlefield process to describe the operational environment and identify threats and hazards that may impact protection. Refer to Appendix 4 (Intelligence Support to Protection) to Annex E (Protection) as required.*

(5) (U) <u>Physical Security</u>. *Describe how physical security consists of physical measures that are designed to safeguard personnel; to prevent unauthorized access to equipment, installations, material, and documents; and to safeguard them against espionage, sabotage, damage, and theft. Refer to Appendix 5 (Physical Security Procedures) to Annex E (Protection) as required.*

|page number|
|CLASSIFICATION|

Figure D-5. Sample Annex E (Protection) format (continued)

[CLASSIFICATION]

ANNEX E (PROTECTION) TO OPERATION PLAN/ORDER [number] [(code name)]—[issuing headquarters] [(classification of title)]

(6) (U) <u>Antiterrorism</u>. *State the overall antiterrorism objective. Describe how the commander envisions antiterrorism measures in support of the scheme of protection that supports the concept of operations in the base order. It should stress detection, deterrence, and mitigation of the terrorist threat in the applicable environment (in-transit, on a base, during operations, and in protection of host-nation and local civilians). Refer to Appendix 6 (Antiterrorism) to Annex E (Protection) as required.*

(7) (U) <u>Police Operations</u>. *Describe how police operations encompass policing and the associated law enforcement activities to control and protect populations and resources and to facilitate the existence of a lawful and orderly environment. Describe how police operations are conducted across the range of military operations. As the operation transitions and the operational environment stabilizes, civil control efforts are implemented and the rule of law is established. Refer to Appendix 7 (Police Operations) to Annex E (Protection) as required.*

(8) (U) <u>Survivability Operations</u>. *Describe how personnel and physical assets have inherent survivability qualities or capabilities that can be enhanced through various means and methods. When existing terrain features offer insufficient cover and concealment, survivability can be enhanced by altering the physical environment to provide or improve cover and concealment. Describe how natural or artificial materials may be used as camouflage to confuse, mislead, or evade the enemy or adversary. Refer to Appendix 8 (Survivability Operations) to Annex E (Protection) as required.*

(9) (U) <u>Force Health Protection</u>. *Describe how force health protection supports the commander's intent and concept of operations. Establish the priorities of support to units for each phase of the operation. Identify and describe medical defensive measures to be taken (chemoprophylaxis, pretreatments, and barrier creams) in the event of chemical, biological, radiological, and nuclear operations. Identify and describe any chemoprophylaxis requirements for endemic diseases (such as malaria). Describe medical and occupational and environmental health surveillance activities which will be established. Identify and describe food safety and food defense activities to include inspection of Class I rations. Refer to Appendix 9 (Force Health Protection) to Annex E (Protection) as required.*

(10) (U) <u>Chemical, Biological, Radiological and Nuclear Defense</u>. *Describe how the chemical, biological, radiological, and nuclear defense unit supports the commander's intent and concept of operations. Establish the priorities of support to units or the concept for employing chemical, biological, radiological, and nuclear defense units for each phase of the operation. Detail the priority of chemical, biological, radiological, and nuclear defense reconnaissance support to the maneuver forces based on the mission and chemical, biological, radiological, and nuclear defense threat. Focus on the commander's guidance, mission, and intent. Emphasize how chemical, biological, radiological, and nuclear defense operations affect readiness and warfighting capability. Refer to Appendix 10 (Chemical, Biological, Radiological, and Nuclear Defense) to Annex E (Protection) as required.*

(11) (U) <u>Explosive Ordnance Disposal</u>. *Describe how explosive ordnance disposal supports the commander's intent and concept of operations. Establish the priorities of explosive ordnance disposal support to units for each phase of the operation. Refer to Appendix 11 (Explosive Ordnance Disposal) to Annex E (Protection) as required.*

(12) (U) <u>Coordinate Air and Missile Defense</u>. *Describe how air and missile defense protects the force from missile attack, air attack, and aerial surveillance by ballistic missiles, cruise missiles, conventional fixed- and rotary-wing aircraft, and unmanned aircraft systems. Indirect-fire protection systems protect forces from threats that are largely immune to air defense artillery systems. Describe how protection cell planners coordinate with the air defense airspace management cell for air and missile defense for the protection of the critical asset list and defended asset list and for other air and missile defense protection as required. Refer to Appendix 12 (Coordinate Air and Missile Defense) to Annex E (Protection) as required.*

[page number]
[CLASSIFICATION]

Figure D-5. Sample Annex E (Protection) format (continued)

[CLASSIFICATION]

ANNEX E (PROTECTION) TO OPERATION PLAN/ORDER [number] [(code name)]—[issuing headquarters] [(classification of title)]

(13) (U) <u>Personnel Recovery</u>. *Describe the manner in which subordinate units execute personnel recovery operations in support of the mission, including phasing and the principal tasks to accomplish. This narrative of how the operation will proceed includes support from host-nation, coalition, and multinational forces and capabilities. Discuss the roles of specialized personnel recovery assets from other Services and special operations forces (for unconventional assisted recovery). Refer to Appendix 13 (Personnel Recovery) to Annex E (Protection) as required.*

(14) (U) <u>Detainee and Resettlement</u>. *Describe how detainee and resettlement operations are conducted by military police to shelter, sustain, guard, protect, and account for populations (detainees, dislocated civilians, and U.S. military prisoners) as a result of military or civil conflict and natural or man-made disasters or to facilitate criminal prosecution. Refer to Appendix 14 (Detainee and Resettlement) to Annex E (Protection) as required.*

b. (U) <u>Tasks to Subordinate Units</u>. *List protection tasks assigned to specific subordinate units not contained in the base order.*

c. (U) <u>Coordinating Instructions</u>. *List only instructions applicable to two or more subordinate units not covered in the base plan or order. Identify any nonstandard operating procedure type of information that will enhance protection by coordinated actions. Examples include personnel identification, vehicle identification, and control measures. Provide additional coordinating instructions for the following:*

(1) (U) <u>Critical Asset List</u>. *Identify, assess, and prioritize all critical assets and develop a critical asset list for a given area of operations.*

(2) (U) <u>Defended Asset List</u>. *Develop measures (forces and procedures) to mitigate threats against critical assets. The defended asset list is a listing of those assets from the critical asset list prioritized by the joint force commander to be defended with the resources available.*

(3) (U) <u>Criticality Assessment</u>. *The criticality assessment identifies key assets that are required to accomplish a mission. It addresses the impact of a temporary or permanent loss of key assets or the unit ability to conduct a mission. It should also include high-population facilities (recreational centers, theaters, sports venues) which may not be mission-essential. It examines the costs of recovery and reconstitution, including time, expense, capability, and infrastructure support. The staff gauges how quickly a lost capability can be replaced before giving an accurate status to the commander.*

(4) (U) <u>Vulnerability Assessment</u>. *The vulnerability assessment identifies physical characteristics or procedures that render critical assets, areas, infrastructures, or special events vulnerable to known or potential threats and hazards. The staff addresses "who" or "what" is vulnerable and "how" it is vulnerable.*

(5) (U) <u>Capability Assessment</u>. *Capability assessment of an organization determines its current capacity to perform protection tasks based on the integrated material and nonmaterial readiness of the assets. A capability assessment considers the mitigating effects of existing manpower, procedures, and equipment. It is especially important in identifying capability gaps, which may be addressed to reduce the consequences of a specific threat or hazard.*

(6) (U) <u>Essential Elements of Friendly Information</u>.

(a) (U) *Date-time group, location, size, disposition, and flight path of aviation units in the area of operations.*

[page number]
[CLASSIFICATION]

Figure D-5. Sample Annex E (Protection) format (continued)

[CLASSIFICATION]

ANNEX E (PROTECTION) TO OPERATION PLAN/ORDER [number] [(code name)]—[issuing headquarters] [(classification of title)]

(b) (U) *Date-time group, location, size, disposition, and mobility of units in the area of operations.*

(c) (U) *Location and disposition of command nodes.*

(d) (U) *Sustainment plans and sustainment operations.*

(e) (U) *Methods of locating and neutralizing enemy weapons of mass destruction and tactical ballistic missile capabilities.*

(f) (U) *Sustainment, operational, intelligence, command, control, and communication limitations and vulnerabilities.*

(g) (U) *Vulnerabilities that could be exploited to recue or eliminate international support of ongoing operations.*

(6) (U) Risk Reduction Control Measures. *Provide the required information in the blank spaces.*

(a) (U) *Air and Missile Defense Warning:* _____

(b) (U) *Air and Missile Defense Weapon Control Status:* _____

(c) (U) *Operational Exposure Guidance:* _____

(d) (U) *Mission-Oriented Protective Posture:* _____

(e) (U) *Force Protection Level:* _____

(f) (U) *Information Operations Condition Level:* _____

(g) (U) *Operations Security:* _____

4. (U) Sustainment. *Identify priorities of sustainment for key protection tasks and specify additional instructions as required. Refer to Annex F (Sustainment) as required.*

a. (U) Logistics. *Use subparagraphs to identify priorities and specific instructions for protection logistics support. Refer to Annex F (Sustainment) and Annex P (Host-Nation Support) as required.*

b. (U) Personnel. *Use subparagraphs to identify priorities and specific instructions for human resources support, financial management, legal support, and religious support. Refer to Annex F (Sustainment) as required.*

c. (U) Health Service Support. *Identify availability, priorities, and instructions for medical care. Address treatment and medical evacuation issues affecting protection forces and synchronize health threat reporting and statistics (such as the disease and nonbattle injury rate). Refer to Annex F (Sustainment) as required.*

5. (U) Command and Signal.

a. (U) Command.

(1) (U) Location of the Commander and Key Leaders. *State the location of the commander and key protection leaders.*

(2) (U) Succession of Command. *State the succession of command if not covered in the unit's standard operating procedures.*

[page number]
[CLASSIFICATION]

Figure D-5. Sample Annex E (Protection) format (continued)

[CLASSIFICATION]

ANNEX E (PROTECTION) TO OPERATION PLAN/ORDER [number] [(code name)]—[issuing headquarters] [(classification of title)]

(3) (U) <u>Liaison Requirements</u>. *State the protection liaison requirements not covered in the unit's standard operating procedures.*

b. (U) <u>Control</u>.

(1) (U) <u>Command Posts</u>. *Describe the employment of protection-specific command posts, including the location of each command post and its time of opening and closing.*

(2) (U) <u>Reports</u>. *List protection-specific reports not covered in standard operating procedures. Refer to Annex R (Reports) as required.*

c. (U) <u>Signal</u>. *Address any protection-specific communications requirements. Refer to Annex H (Signal) as required.*

[Commander's last name]
[Commander's rank]

ACKNOWLEDGE: *Include only if attachment is distributed separately from the base order.*

The commander or authorized representative signs the original copy of the attachment. If the representative signs the original, add the phrase "For the Commander." The signed copy is the historical copy and remains in the headquarters' files.

OFFICIAL:

[Authenticator's name]
[Authenticator's position]

Use only if the commander does not sign the original attachment. If the commander signs the original, no further authentication is required. If the commander does not sign, the signature of the preparing staff officer requires authentication and only the last name and rank of the commander appear in the signature block.

ATTACHMENTS: *List lower-level attachment (appendixes, tabs, and exhibits). When an attachment required by doctrine or a standard operating procedure is unnecessary, label it "Omitted." Unit standard operating procedures will dictate attachment development and format. Common attachments include the following:*

Appendix 1–Operational Area Security
Appendix 2–Safety
Appendix 3–Operations Security
Appendix 4–Intelligence Support to Protection
Appendix 5–Physical Security
Appendix 6–Antiterrorism
Appendix 7–Police Operations
Appendix 8–Survivability Operations
Appendix 9–Force Health Protection
Appendix 10–Chemical, Biological, Radiological, and Nuclear Defense
Appendix 11–Explosive Ordnance Disposal
Appendix 12–Coordinate Air and Missile Defense
Appendix 13–Personnel Recovery
Appendix 14–Detainee and Resettlement

DISTRIBUTION: *Show only if distributed separately from the base order or higher-level attachments.*

[page number]
[CLASSIFICATION]

Figure D-5. Sample Annex E (Protection) format (continued)

ANNEX F (SUSTAINMENT) FORMAT AND INSTRUCTIONS

D-35. This annex provides fundamental considerations, formats, and instructions for developing Annex F (Sustainment) to the base plan or order. This annex follows the five-paragraph attachment format.

D-36. Commanders and staffs use Annex F (Sustainment) to describe how sustainment operations support the concept of operations described in the base plan or order. The chief of sustainment (G-4[S-4]) develops Annex F (Sustainment). (See figure D-6 on pages D-34 through D-40.)

[CLASSIFICATION]

Place the classification at the top and bottom of every page of the attachments. Place the classification marking at the front of each paragraph and subparagraph in parentheses. Refer to AR 380-5 for classification and release marking instructions.

Copy ## of ## copies
Issuing headquarters
Place of issue
Date-time group of signature
Message reference number

Include the full heading if attachment is distributed separately from the base order or higher-level attachment.

ANNEX F (SUSTAINMENT) TO OPERATION PLAN/ORDER [number] [(code name)]— [issuing headquarters] [(classification of title)]

(U) References: List documents essential to understanding the attachment.

 a. *List maps and charts first. Map entries include series number, country, sheet names or numbers, edition, and scale.*

 b. *List other references in subparagraphs labeled as shown.*

 c. *Doctrinal references for sustainment include ADRP 4-0, FM 3-09, FM 4-02, and FM 6-0.*

(U) Time Zone Used Throughout the Order: *Write the time zone established in the base plan or order.*

1. (U) Situation. *Include information affecting the sustainment operations that paragraph 1 of the OPLAN or OPORD does not cover or that needs expansion.*

 a. (U) Area of Interest. *Describe the area of interest as it relates to the sustainment. Refer to Annex B (Intelligence) as required.*

 b. (U) Area of Operations. *Refer to Appendix 2 (Operation Overlay) to Annex C (Operations) as required.*

  (1) (U) Terrain. *Describe the aspects of terrain that impact sustainment operations. Refer to Annex B (Intelligence) as required.*

  (2) (U) Weather. *Describe the aspects of weather that impact sustainment operations. Refer to Annex B (Intelligence) as required.*

[page number]
[CLASSIFICATION]

Figure D-6. Sample Annex F (Sustainment) format

|CLASSIFICATION|

ANNEX F (SUSTAINMENT) TO OPERATION PLAN/ORDER |number| |(code name)|— |issuing headquarters| |(classification of title)|

c. (U) <u>Enemy Forces</u>. *List known and templated locations and activities of enemy sustainment units for one echelon up and two echelons down. List enemy maneuver and other capabilities that will impact friendly sustainment operations. State expected enemy sustainment courses of action and employment of enemy sustainment assets. Refer to Annex B (Intelligence) as required.*

d. (U) <u>Friendly Forces</u>. *Outline the higher headquarters' sustainment plan. List designation, location, and outline of plan of higher, adjacent, and other sustainment assets that support or impact the issuing headquarters or require coordination and additional support.*

e. (U) <u>Interagency, Intergovernmental, and Nongovernmental Organizations</u>. *Identify and describe other organizations in the area of operations that may impact the conduct of sustainment operations or implementation of sustainment-specific equipment and tactics. Refer to Annex V (Interagency Coordination) as required.*

f. (U) <u>Civil Considerations</u>. *Describe the aspects of the civil situation that impact sustainment operations. Refer to Annex B (Intelligence) and Annex K (Civil Affairs Operations) as required.*

g. (U) <u>Attachments and Detachments</u>. *List units attached or detached only as necessary to clarify task organization. Refer to Annex A (Task Organization) as required.*

h. (U) <u>Assumptions</u>. *List any sustainment-specific assumptions that support the annex development.*

2. (U) <u>Mission</u>. *State the mission of sustainment in support of the base plan or order.*

3. (U) <u>Execution</u>.

a. (U) <u>Scheme of Sustainment Support</u>. *Describe how sustainment supports the commander's intent and concept of operations. Establish the priorities of sustainment support to units for each phase of the operation. Refer to Annex C (Operations) as required.*

b. (U) <u>Tasks to Subordinate Units</u>. *List sustainment tasks assigned to specific subordinate units not contained in the base order.*

c. (U) <u>Coordinating Instructions</u>. *List only instructions applicable to two or more subordinate units not covered in the base plan or order.*

4. (U) <u>Sustainment</u>. *Identify priorities of sustainment for key tasks and specify additional instructions as required.*

a. (U) <u>Materiel and Services</u>. *Provide materiel and services information in the following subparagraphs.*

(1) (U) <u>Maintenance</u>. *Provide maintenance information for each subparagraph, including priority of maintenance, location of facilities and collection points, repair time limits at each level of maintenance, and evacuation procedures. Post maintenance collection points and command posts to the sustainment overlay at Tab A (Sustainment Overlay) to Appendix 1 (Logistics) to Annex F (Sustainment). Refer to Tab B (Maintenance) to Appendix 1 (Logistics) to Annex F (Sustainment) as required.*

(a) (U) <u>Ground</u>. *Identify the proper procedures to request ground recovery and maintenance.*

(b) (U) <u>Watercraft</u>. *Identify the proper procedures to request watercraft recovery and maintenance.*

(c) (U) <u>Aircraft</u>. *Identify the proper procedures to request aircraft recovery and maintenance.*

|page number|
|CLASSIFICATION|

Figure D-6. Sample Annex F (Sustainment) format (continued)

[CLASSIFICATION]

ANNEX F (SUSTAINMENT) TO OPERATION PLAN/ORDER [number] [(code name)]— [issuing headquarters] [(classification of title)]

(d) (U) *Field Maintenance. Identify, list, and describe the recovery plan and types of recovery vehicles available; Class IX parts support; the locations of maintenance collection points; logistics civil augmentation program capabilities and availability; and field maintenance support relationships at each phase of the operation.*

(e) (U) *Sustainment Maintenance. Identify, list, and describe the location of sustainment maintenance units and services; the locations of maintenance collection points; the logistics civil augmentation program capabilities and availability; and sustainment maintenance support relationships at each phase of the operation.*

(2) (U) *Transportation. Provide transportation information for each subparagraph. Identify facility locations, traffic control, regulation measures, main supply routes, alternate supply routes, transportation critical shortages, and other essential transportation data not provided elsewhere. Post main supply routes, alternate supply routes, and transportation facilities to the logistics synchronization matrix and the overlay at Tab A (Sustainment Overlay) to Appendix 1 (Logistics) to Annex F (Sustainment). Identify and list transportation request procedures. Refer to Tab C (Transportation) to Appendix 1 (Logistics) to Annex F (Sustainment) as required.*

(a) (U) *Ground. Identify the proper procedures to request ground transportation.*

(b) (U) *Sea/River/Water. Identify the proper procedures to request sea, river, and water transportation.*

(c) (U) *Air. Identify the proper procedures to request air transportation.*

(d) (U) *Container Management. Describe the container management plan.*

(3) (U) *Supply. Provide information by class of supply in each subparagraph. Identify and list maps, water, special supplies, and excess and salvage materiel, as applicable. For each subparagraph, list supply point locations and state supply plan and procedures. Post supply points and facilities to the logistics synchronization matrix and the overlay at Tab A (Sustainment Overlay) to Appendix 1 (Logistics) to Annex F (Sustainment). Refer to Tab D (Supply) to Appendix 1 (Logistics) to Annex F (Sustainment) as required. Coordinate with the surgeon for information for subparagraph 4.a(3)(h) in this annex on medical supplies. Refer to Appendix 3 (Health Service Support) to Annex F (Sustainment) for additional information on medical logistics.*

(a) (U) *Class I Rations. Identify and list the issue and ration cycle, ration stockage objectives, and the bulk water locations.*

(b) (U) *Class II Organizational Clothing and Individual Equipment and Maps. Identify and list organizational clothing and individual equipment available for this operation. Submit classified map requests through G-2 (S-2) channels.*

(c) (U) *Class III Bulk Fuel; Class III Package Petroleum, Oils, and Lubricants. Identify and list quantities of petroleum, oil, and lubricant; locations of the retail and bulk fuel points; and types of products available at each site available to support the operation.*

(d) (U) *Class IV Construction and Fortification Material. Identify and list construction and fortification or barrier material available for this operation including command-controlled items.*

(e) (U) *Class V Munitions. Identify and list available ammunition and the controlled supply rates. List the procedures to request explosive ordnance disposal support. Refer to Annex E (Protection) as required for explosive ordnance disposal support.*

(f) (U) *Class VI Personal Demand Items. Describe the Class VI plan. Identify and list items available.*

(g) (U) *Class VII Major End Items. Identify and list major end items available for this operation.*

[page number]
[CLASSIFICATION]

Figure D-6. Sample Annex F (Sustainment) format (continued)

[CLASSIFICATION]

ANNEX F (SUSTAINMENT) TO OPERATION PLAN/ORDER [number] [(code name)]—[issuing headquarters] [(classification of title)]

(h) (U) <u>Class VIII Medical Supply</u>. *Identify and list medical supplies available for this operation.*

(i) (U) <u>Class IX Repair Parts</u>. *Identify and list all critical shortage repair part and command-controlled items available for this operation. State the approving authority for controlled exchange of parts.*

(j) (U) <u>Class X Material for Nonmilitary or Civil Affairs Operations</u>. *Identify and list material available for this operation.*

(k) (U) <u>Miscellaneous</u>. *Identify and list any other available materiel and supplies not mentioned in the above subparagraphs available for this operation.*

(4) (U) <u>Field Services</u>. *Identify and list key field services available during this operation. At a minimum, this paragraph and subparagraphs must contain the location and the responsible unit for each separate field service activity. Identify and list locations and operating hours for laundry facilities, shower facilities, clothing repair facilities, food services facilities, billeting facilities, and field sanitation facilities. Highlight field sanitation requirements for each service, such as water purification and trash removal. Post field service facilities to the logistics synchronization matrix and the overlay at Tab A (Sustainment Overlay) to Appendix 1 (Logistics) to Annex F (Sustainment). Refer to Tab E (Field Services) to Appendix 1 (Logistics) to Annex F (Sustainment) as required.*

(a) (U) <u>Construction</u>. *Identify and list available construction material. Provide essential information as appropriate.*

(b) (U) <u>Light Textile Repair and Showers, Laundry, and Clothing Repair</u>. *Identify and list locations of showers, laundry, and clothing repair available for this operation.*

(c) (U) <u>Food Preparation</u>. *Identify and list food preparation available for this operation.*

(d) (U) <u>Water Purification</u>. *Identify and list water purification locations and units available for this operation.*

(e) (U) <u>Aerial Delivery</u>. *Identify and list aerial delivery available for this operation.*

(f) (U) <u>Installation Services</u>. *Identify and list installation services available for this operation.*

(5) (U) <u>Distribution</u>. *Provide information about distribution support. Refer to Tab F (Distribution) to Appendix 1 (Logistics) to Annex F (Sustainment) as required.*

(a) (U) <u>Distribution Nodes' Locations</u>. *Identify and list the location of distribution nodes (seaport of debarkation and arrival/departure airfield control group).*

(b) (U) <u>Tracking Procedures</u>. *Identify and discuss the tracking procedures.*

(c) (U) <u>Distribution Modes</u>. *Identify and list the various distribution modes: land, sea, or air.*

(d) (U) <u>Movement Request Format</u>. *Discuss the movement request format and processing requirements.*

(e) (U) <u>Container Operations</u>. *Discuss container management and operations.*

(f) (U) <u>Movement Control Responsibility</u>. *Identify units at each level responsible for movement control.*

[page number]
[CLASSIFICATION]

Figure D-6. Sample Annex F (Sustainment) format (continued)

[CLASSIFICATION]

ANNEX F (SUSTAINMENT) TO OPERATION PLAN/ORDER [number] [(code name)]— [issuing headquarters] [(classification of title)]

(6) (U) <u>Contract Support Integration</u>. *Identify and list key contract support integration functions for this operation. Identify the location and contract support unit responsible at each level. Identify contract support capabilities, limitations, and priority of support. Refer to Annex W (Operational Contract Support) as required.*

(7) (U) <u>Mortuary Affairs</u>. *Provide information about mortuary affairs support. Refer to Tab H (Mortuary Affairs) to Appendix 1 (Logistics) to Annex F (Sustainment) as required.*

(8) (U) <u>Labor</u>. *Provide information about contract labor. Refer to Appendix 1 (Logistics) to Annex F (Sustainment) and Annex P (Host-Nation Support) as required.*

b. (U) <u>Personnel</u>. *Provide personnel information. Outline plans for unit-strength maintenance; personnel management; morale development and maintenance; discipline, law, and order; headquarters management; force provider; religious support; and legal and finance support. Post personnel services unit locations to the logistics synchronization matrix and the overlay at Tab A (Sustainment Overlay) to Appendix 1 (Logistics) to Annex F (Sustainment). Refer to Appendix 2 (Personnel Services Support) to Annex F (Sustainment) as required.*

(1) (U) <u>Human Resources Support</u>. *Provide human resources support information. Refer to Tab A (Human Resources Support) to Appendix 2 (Personnel Services Support) to Annex F (Sustainment) as required.*

(2) (U) <u>Financial Management</u>. *Provide financial management support information. Refer to Tab B (Financial Management) to Appendix 2 (Personnel Services Support) to Annex F (Sustainment) as required.*

(3) (U) <u>Legal Support</u>. *Provide legal support information. Refer to Tab C (Legal Support) to Appendix 2 (Personnel Services Support) to Annex F (Sustainment) as required.*

(4) (U) <u>Religious Support</u>. *Provide religious support information. Refer to Tab D (Religious Support) to Appendix 2 (Personnel Services Support) to Annex F (Sustainment) as required.*

(5) (U) <u>Band Operations</u>. *Provide band operations support information. Refer to Tab E (Band Operations) to Appendix 2 (Personnel Services Support) to Annex F (Sustainment) as required.*

c. (U) <u>Health Service Support</u>. *Provide health service support information. Identify availability, priorities, and instructions for medical care. Describe the plan for collection and medical treatment of sick, injured, or wounded U.S., multinational, and joint force Soldiers, enemy prisoners of war, detainees, and, when authorized, civilians. Describe support requirements for medical logistics (including blood management), combat and operational stress control, preventive medicine, dental services, medical laboratory support, and veterinary services. Post hospital and medical treatment facility locations to the logistics synchronization matrix and the overlay at Tab A (Sustainment Overlay) to Appendix 1 (Logistics) to Annex F (Sustainment). Refer to Appendix 3 (Health Service Support) to Annex F (Sustainment) as required.*

(1) (U) <u>Medical Evacuation</u>. *Provide medical evacuation information. Address the theater evacuation policy, en route care, medical regulating (if appropriate), casualty evacuation, and the medical evacuation of casualties contaminated with chemical, biological, radiological, and nuclear ordnance.*

[page number]
[CLASSIFICATION]

Figure D-6. Sample Annex F (Sustainment) format (continued)

[CLASSIFICATION]

ANNEX F (SUSTAINMENT) TO OPERATION PLAN/ORDER [number] [(code name)]—[issuing headquarters] [(classification of title)]

(2) (U) <u>Hospitalization</u>. *Provide hospitalization information and guidelines. List the locations of medical treatment facilities. Identify and list area units without organic medical resources requiring support and describe how to support these units. Describe the procedures for mass casualty operations and patient decontamination operations. Identify and list roles of medical care (1, 2, and 3) by treatment facility and location. Refer to Tab A (Sustainment Overlay) to Appendix 1 (Logistics) to Annex F (Sustainment) and Appendix 3 (Health Service Support) to Annex F (Sustainment) as required.*

d. (U) <u>Foreign Nation and Host-Nation Support</u>. *Provide host-nation support information. Refer to Annex P (Host-Nation Support) as required.*

e. (U) <u>Resource Availability</u>. *Identify significant competing demands for sustainment resources where expected requirements may exceed resources.*

f. (U) <u>Miscellaneous</u>. *Provide any general miscellaneous information not covered in this annex.*

5. (U) <u>Command and Signal</u>.

a. (U) <u>Command</u>.

(1) (U) <u>Location of the Commander and Key Leaders</u>. *State the location of the commander and sustainment area leaders.*

(2) (U) <u>Succession of Command</u>. *State the succession of command if not covered in the unit's standard operating procedures (SOPs).*

(3) (U) <u>Liaison Requirements</u>. *State the sustainment liaison requirements not covered in the base order.*

b. (U) <u>Control</u>.

(1) (U) <u>Command Posts</u>. *Describe the employment of sustainment-specific command posts (CPs), including the location of each CP and its time of opening and closing.*

(2) (U) <u>Reports</u>. *List sustainment-specific reports not covered in SOPs. Refer to Annex R (Reports) as required.*

c. (U) <u>Signal</u>. *Address any sustainment-specific communications requirements. Refer to Annex H (Signal) as required.*

ACKNOWLEDGE: *Include only if attachment is distributed separately from the base order.*

[Commander's last name]
[Commander's rank]

The commander or authorized representative signs the original copy of the attachment. If the representative signs the original, add the phrase "For the Commander." The signed copy is the historical copy and remains in the headquarters' files.

OFFICIAL:

[Authenticator's name]
[Authenticator's position]

[page number]
[CLASSIFICATION]

Figure D-6. Sample Annex F (Sustainment) format (continued)

[CLASSIFICATION]

ANNEX F (SUSTAINMENT) TO OPERATION PLAN/ORDER [number] [(code name)]— [issuing headquarters] [(classification of title)]

Use only if the commander does not sign the original attachment. If the commander signs the original, no further authentication is required. If the commander does not sign, the signature of the preparing staff officer requires authentication and only the last name and rank of the commander appear in the signature block.

ATTACHMENTS: *List lower-level attachments (appendixes, tabs, and exhibits).*

Appendix 1–Logistics
Appendix 2–Personnel Services Support
Appendix 3–Health Service Support

DISTRIBUTION: *Show only if distributed separately from the base order or higher-level attachments.*

[page number]
[CLASSIFICATION]

Figure D-6. Sample Annex F (Sustainment) format (continued)

ANNEX G (ENGINEER) FORMAT AND INSTRUCTIONS

D-37. This annex provides fundamental considerations, formats, and instructions for developing Annex G (Engineer) to the base plan or order. This annex follows the five-paragraph attachment format.

D-38. Commanders and staffs use Annex G (Engineer) to describe how the engineer plan supports the concept of operations described in the base plan or order. The engineer officer develops Annex G (Engineer).

D-39. This annex follows the five-paragraph (situation, mission, execution, sustainment, and command and signal) format of the base plan or order. Engineers use this annex to define engineer support to the maneuver commander's intent, coordinating instructions to subordinate commanders, and essential tasks for mobility, countermobility, and survivability. This annex is not intended to function as the internal order for an engineer organization, where the engineer commander will articulate intent, concept of operations, and coordinating instructions to subordinate, supporting, and supported commanders. This annex seeks to clarify engineer support to the base plan or order. Guidance to maneuver units on obstacle responsibilities should be listed in the body of the base plan or order, not in this annex. (See figure D-7 on pages D-41 through D-45.)

[CLASSIFICATION]

Place the classification at the top and bottom of every page of the attachments. Place the classification marking at the front of each paragraph and subparagraph in parentheses. Refer to AR 380-5 for classification and release marking instructions.

Copy ## of ## copies
Issuing headquarters
Place of issue
Date-time group of signature
Message reference number

Include the full heading if attachment is distributed separately from the base order or higher-level attachment.

ANNEX G (ENGINEER) TO OPLAN/OPORD [number] [(code name)]—[issuing headquarters] [(classification of title)]

(U) References: *List documents essential to understanding this attachment.*

a. *List maps and charts first. Map entries include series number, country, sheet names or numbers, edition, and scale.*

b. *List other references in subparagraphs labeled as shown.*

c. *Doctrinal references for this annex are ATTP 3-34.23, ATTP 3-34.80, ATP 3-37.34, FM 3-34, FM 3-34.5, FM 3-34.170, FM 3-34.400, FM 5-102, FM 6-0, and TM 3-34.85.*

(U) Time Zone Used Throughout the OPLAN/OPORD: *Write the time zone established in the base plan or order.*

[page number]
[CLASSIFICATION]

Figure D-7. Sample Annex G (Engineer) format

[CLASSIFICATION]

ANNEX G (ENGINEER) TO OPLAN/OPORD [number] [(code name)]—[issuing headquarters] [(classification of title)]

1. (U) Situation. *Include information affecting engineer support that paragraph 1 of the OPLAN or OPORD does not cover or that needs expansion.*

a. (U) <u>Area of Interest</u>. *Describe the area of interest as it relates to engineer operations. Refer to Annex B (Intelligence) as required.*

b. (U) <u>Area of Operations</u>. *Refer to Appendix 2 (Operation Overlay) to Annex C (Operations) as required.*

(1) (U) <u>Terrain</u>. *Describe the aspects of terrain that impact engineer operations. Refer to Annex B (Intelligence) as required.*

(2) (U) <u>Weather</u>. *Describe the aspects of weather that impact engineer operations. Refer to Annex B (Intelligence) as required.*

c. (U) <u>Enemy Forces</u>. *List known and templated locations and activities of enemy engineer units for one echelon up and two echelons down. List enemy maneuver and other capabilities that will impact engineer operations. State expected enemy courses of action and employment of enemy engineer assets. Give a detailed description of enemy engineer units, assets, and any known obstacles. Refer to Annex B (Intelligence) as required.*

d. (U) <u>Friendly Forces</u>. *Outline the higher headquarters' engineer operation plan. List designation, location, and outline of plan of higher, adjacent, and other engineer assets that support or impact the issuing headquarters or require coordination and additional support.*

e. (U) <u>Interagency, Intergovernmental, and Nongovernmental Organizations</u>. *Identify and describe other organizations in the area of operations that may impact the conduct of engineer operations or implementation of engineer-specific equipment and tactics. Refer to Annex V (Interagency Coordination) as required.*

f. (U) <u>Civil Considerations</u>. *Describe the critical aspects of the civil situation that impact engineer operations. Refer to Annex B (Intelligence) and Annex K (Civil Affairs Operations) as required.*

g. (U) <u>Attachments and Detachments</u>. *List all engineer assets with a command support relationship with higher headquarters. List any units detached or under the operational control of other headquarters. Refer to Annex A (Task Organization) as required.*

h. (U) <u>Assumptions</u>. *List any engineer-specific assumptions that support the annex development.*

2. (U) <u>Mission</u>. *State the engineer mission in support of the base plan or order.*

3. (U) <u>Execution</u>.

a. (U) <u>Scheme of Engineer Support</u>. *Describe how engineer operations support the commander's intent and concept of operations. Establish the priorities of engineer support to units for each phase of the operation. Refer to the base plan or order and Annex C (Operations) as required.*

(1) (U) <u>Assured Mobility</u>. *Describe the plan to maintain freedom of movement and maneuver. Refer to Appendix 1 (Mobility/Countermobility) to Annex G (Engineer) as required.*

(a) (U) <u>Mobility Support</u>. *State the scheme of mobility operations to include task and purpose. This includes breaching (proofing, marking lanes, providing guides, and maintaining and clearing routes), relative location (route or objective), priority for reduction assets used (use plows first, then mine-clearing line charge), priority of clearance assets, and unit responsible. For gap crossing operations, refer to Appendix 4 (Gap Crossing Operations) to Annex C (Operations).*

[page number]
[CLASSIFICATION]

Figure D-7. Sample Annex G (Engineer) format (continued)

[CLASSIFICATION]

ANNEX G (ENGINEER) TO OPLAN/OPORD [number] [(code name)]—[issuing headquarters] [(classification of title)]

(b) (U) Countermobility Support. *State the scheme of countermobility operations including task and purpose, unit responsible for task, priority of effort, intent, target number assignments (by unit) and planned grid coordinates. Operations requiring obstacle emplacement will also be required to include a Tab A (Obstacle Overlay) to Appendix 1 (Mobility/Countermobility) to Annex G (Engineer).*

(2) (U) Survivability. *Describe how survivability operations support the commander's intent and concept of operations. Establish the priorities of survivability support to units for each phase of the operation. Refer to the base plan or order, Annex C (Operations) and Appendix 2 (Survivability) to Annex G (Engineer) as required.*

(3) (U) General Engineering. *Describe how general engineering assets support the commander's intent and concept of operations. Establish the priorities of support to subordinate units for each phase of the operation. Refer to the base plan or order and Annex C (Operations) and refer to Appendix 3 (General Engineering) to Annex G (Engineer) as required.*

(4) (U) Geospatial Engineering. *Describe how geospatial engineering capabilities will support the operation. Expand the scheme of engineer operations in Annex G (Engineer) with any additional information that clarifies the geospatial engineering tasks, purposes, and priorities in support of each phase of the scheme of maneuver. The four primary functions of geospatial engineering (generate, analyze, manage, and disseminate) may be used to structure this narrative. Refer to Appendix 4 (Geospatial Engineering) to Annex G (Engineer) as required.*

(5) (U) Environmental Considerations. *Summarize the commander's concept of environmental actions required to support the OPLAN, OPORD, or concept plan. Identify issues and actions that should be addressed during all phases of the operation. Refer to Appendix 5 (Environmental Considerations) to Annex G (Engineer) as required.*

(6) (U) Engineer Reconnaissance. *State the scheme of engineer reconnaissance by task and purpose for engineer tactical and technical reconnaissance including infrastructure reconnaissance requirements.*

b. (U) Tasks to Subordinate Units. *List engineering tasks to specific units that are not assigned in the base plan or order. List tasks specific to engineering and mobility, countermobility, and survivability assets only as necessary to ensure unity of effort. Specific and detailed task descriptions should be done in each respective appendix as applicable.*

c. (U) Coordinating Instructions. *List only instructions applicable to two or more subordinate units not covered in the base plan or order. Provide additional coordinating instructions for the following:*

(1) (U) *Identify and list the times or events when obstacle control measures become effective.*

(2) (U) *List supported unit information requirements focused on mobility, countermobility, and survivability that must be considered by subordinate engineer staff officers or that the supported unit requires. This includes engineer-related commander's critical information requirements and perhaps the requests for information that have already been submitted to higher.*

(3) (U) *Explain and describe the countermobility and survivability timelines.*

4. (U) Sustainment. *Identify sustainment priorities for engineer key tasks and specify additional sustainment instructions as necessary, and, at a minimum, address engineer Class IV and V locations. Refer to Annex F (Sustainment) as required.*

[page number]
[CLASSIFICATION]

Figure D-7. Sample Annex G (Engineer) format (continued)

[CLASSIFICATION]

ANNEX G (ENGINEER) TO OPLAN/OPORD [number] [(code name)]—[issuing headquarters] [(classification of title)]

a. (U) <u>Logistics</u>. *Use subparagraphs to identify priorities and specific instructions for engineer logistics support. Refer to Annex F (Sustainment) and Annex P (Host-Nation Support) as required.*

(1) (U) <u>Command-Regulated Classes of Supply</u>. *Identify command-regulated classes of supply. Highlight supported unit allocations that affect engineer support (such as Class IV barrier material allocated to other efforts).*

(2) (U) <u>Supply Distribution Plan</u>. *Establish Class IV and Class V (obstacle) supply distribution plan. State method of supply for each class and for each supported unit subordinate element. List supply linkup points. Identify and list all allocations of Class IV and Class V by support unit element by obstacle control measure or combination. Summarize in a matrix or table as necessary.*

(3) (U) <u>Transportation</u>. *List any transportation coordination to include supported troop movements, Class IV building materials, and Class V materials.*

b. (U) <u>Personnel</u>. *Use subparagraphs to identify priorities and specific instructions for human resources support, financial management, legal support, and religious support. Refer to Annex F (Sustainment) as required.*

c. (U) <u>Health Service Support</u>. *Identify availability, priorities, and instructions for medical care. Refer to Annex F (Sustainment) as required.*

5. (U) <u>Command and Signal</u>.

a. (U) <u>Command</u>.

(1) (U) <u>Location of the Commander and Key Leaders</u>. *State the location of the commander and key engineer leaders. Designate the headquarters that controls the mobility, countermobility, and survivability effort within work lines on an area basis. Clearly identify release authority for special munitions, such as the Intelligent Munitions System (Scorpion).*

(2) (U) <u>Succession of Command</u>. *State the succession of command or leadership if not covered in the unit's standard operating procedures (SOPs).*

(3) (U) <u>Liaison Requirements</u>. *State engineer liaison requirements not covered in the unit's SOPs.*

b. (U) <u>Control</u>.

(1) (U) <u>Command Posts</u>. *Describe the employment of command posts (CPs), including the location of each CP and state the primary controlling CP for specific tasks or phases of the operation.*

(2) (U) <u>Reports</u>. *Identify critical engineer reporting requirements of subordinates if not covered in SOPs. Refer to Annex R (Reports) as required.*

c. (U) <u>Signal</u>. *Describe the concept of signal support as it pertains to engineer support operations. Refer to Annex H (Signal) as required.*

[page number]
[CLASSIFICATION]

Figure D-7. Sample Annex G (Engineer) format (continued)

[CLASSIFICATION]

ANNEX G (ENGINEER) TO OPLAN/OPORD [number] [(code name)]—[issuing headquarters] [(classification of title)]

ACKNOWLEDGE: *Include only if attachment is distributed separately from the base order.*

[Commander's last name]
[Commander's rank]

The commander or authorized representative signs the original copy of the attachment. If the representative signs the original, add the phrase "For the Commander." The signed copy is the historical copy and remains in the headquarters' files.

OFFICIAL:

[Authenticator's name]
[Authenticator's position]

Use only if the commander does not sign the original attachment. If the commander signs the original, no further authentication is required. If the commander does not sign, the signature of the preparing staff officer requires authentication and only the last name and rank of the commander appear in the signature block.

ATTACHMENTS: *List lower-level attachment (appendixes, tabs, and exhibits). If a particular attachment is not used, place "not used" beside the attachment number. Unit SOPs will dictate attachment development and format. Common attachments include the following:*

Appendix 1–Mobility/Countermobility
Appendix 2–Survivability
Appendix 3–General Engineering
Appendix 4–Geospatial Engineering
Appendix 5–Environmental Considerations

DISTRIBUTION: *Show only if distributed separately from the base order or higher-level attachments.*

[page number]
[CLASSIFICATION]

Figure D-7. Sample Annex G (Engineer) format (continued)

ANNEX H (SIGNAL) FORMAT AND INSTRUCTIONS

D-40. This annex provides fundamental considerations, formats, and instructions for developing Annex H (Signal) to the base plan or order. This annex follows the five-paragraph attachment format.

D-41. Commanders and staffs use Annex H (Signal) to describe how signal supports the concept of operations described in the base plan or order. The G-6 (S-6) develops Annex H (Signal). (See figure D-8 on pages D-46 through D-50.)

[CLASSIFICATION]

Place the classification at the top and bottom of every page of the attachments. Place the classification marking at the front of each paragraph and subparagraph in parentheses. Refer to AR 380-5 for classification and release marking instructions.

Copy ## of ## copies
Issuing headquarters
Place of issue
Date-time group of signature
Message reference number

Include the full heading if attachment is distributed separately from the base order or higher-level attachment.

ANNEX H (SIGNAL) TO OPERATION PLAN/ORDER [number] [(code name)]—[issuing headquarters] [(classification of title)]

(U) References: *List documents essential to understanding the attachment.*

a. *List maps and charts first. Map entries include series number, country, sheet names or numbers, edition, and scale.*

b. *List other references in subparagraphs labeled as shown.*

c. *Doctrinal references for signal support to operations include FM 6-0, FM 6-02.40, FM 6-02.43, FM 6-02.53, FM 6-02.70, and FM 6-02.71.*

(U) Time Zone Used Throughout the Order: *Write the time zone established in the base plan or order.*

1. (U) <u>Situation</u>. *Include information affecting signal operations that paragraph 1 of the OPLAN or OPORD does not cover or that needs expansion.*

a. (U) <u>Area of Interest</u>. *Describe the area of interest as it relates to signal support to operations. Refer to Annex B (Intelligence) as required.*

b. (U) <u>Area of Operations</u>. *Describe the area of operations as it relates to signal support to operations. Refer to Appendix 2 (Operation Overlay) to Annex C (Operations).*

(1) (U) <u>Terrain</u>. *Describe the aspects of terrain that impact signal support to operations. Refer to Annex B (Intelligence) as required.*

(2) (U) <u>Weather</u>. *Describe all critical weather aspects that impact signal support to operations such as rain, flooding, windstorms, and snow, that also may impact network availability or reliability in the area of operations. Refer to Annex B (Intelligence) as required.*

[page number]
[CLASSIFICATION]

Figure D-8. Sample Annex H (Signal) format

[CLASSIFICATION]

ANNEX H (SIGNAL) TO OPERATION PLAN/ORDER [number] [(code name)]—[issuing headquarters] [(classification of title)]

c. (U) <u>Enemy Forces</u>. *List known and templated locations and activities of enemy units for one echelon above and two echelons below. List enemy maneuver and other area capabilities that will impact friendly signal support to operations. State expected enemy courses of action. Refer to Annex B (Intelligence) as required.*

d. (U) <u>Friendly Forces</u>. *Briefly identify the signal mission of friendly forces and the objectives, goals, and missions of civilian organizations that impact support to operations. Refer to Annex A (Task Organization) and Annex C (Operations) as required.*

(1) (U) <u>Higher Headquarters' Signal Operations Mission</u>. *Identify and state the signal mission of the higher headquarters.*

(2) (U) <u>Signal Support Operations Impact of Adjacent Units</u>. *Identify and state the missions of adjacent units and other units whose actions have a significant impact on the issuing headquarters' support to operations.*

e. (U) <u>Interagency, Intergovernmental, and Nongovernmental Organizations</u>. *Identify and state the objectives or goals and primary tasks of those non-Department of Defense organizations that may impact the conduct of support to operations or implementation of signal-specific equipment and tactics in the area of operations. Refer to Annex V (Interagency Coordination) as required.*

f. (U) <u>Signal Support to Cyber Electromagnetic Activities</u>. *List considerations related to the planning, integration, coordination, and synchronization of network operations and defense functions with other cyber electromagnetic activities.*

g. (U) <u>Civil Considerations</u>. *Describe the critical aspects of the civil situation that impact voice and data network operations using the memory aid ASCOPE (areas, structures, capabilities, organizations, people, and events). Refer to Annex B (Intelligence) and Annex K (Civil Affairs Operations) as required.*

h. (U) <u>Attachments and Detachments</u>. *List units attached or detached only as necessary to clarify task organization that impact signal support to operations. Refer to Annex A (Task Organization) as required.*

i. (U) <u>Assumptions</u>. *List key assumptions that pertain to support to operations which support development of the annex.*

2. (U) <u>Mission</u>. *State the mission of signal in support of the base plan or order.*

3. (U) <u>Execution</u>.

a. (U) <u>Scheme of Signal Support to Operations</u>. *Describe how signal support to operations supports the commander's intent and concept of operations described in the base plan or order. Establish the priorities of support to units for each phase of the operation. Refer to Annex C (Operations) as required.*

(1) (U) <u>Scheme of Defensive Cyberspace Operations.</u> *Describe how defensive cyberspace operations supports the commander's intent and concept of operations described in the base plan or order. Outline defensive cyberspace operations that protect against, monitor for, detect (find), analyze (fix), and respond (finish) to cyber threats on Nonsecure Internet Protocol Router Network, SECRET Internet Protocol Router Network, and Joint Worldwide Intelligence Communications System. Refer to Appendix 1 (Defensive Cyberspace Operations) to Annex H (Signal) as required.*

(2) (U) <u>Scheme of Network Operations</u>. *Describe how the commander's intent and concept of operations are supported through actions taken to gain and maintain access to the cyber domain via the execution of resource allocation, configuration, continuous monitoring of performance and effectiveness, event handling, and security functions that operate and sustain the Nonsecure Internet Protocol Router Network, SECRET Internet Protocol Router Network, and Joint Worldwide Intelligence Communications System. Refer to Appendix 2 (Network Operations) to Annex H (Signal) as required.*

[page number]
[CLASSIFICATION]

Figure D-8. Sample Annex H (Signal) format (continued)

[CLASSIFICATION]

ANNEX H (SIGNAL) TO OPERATION PLAN/ORDER [number] [(code name)]—[issuing headquarters] [(classification of title)]

(3) (U) <u>Scheme of Voice, Video, and Data Routing</u>. *Describe how the routing and movement of voice, video, and data network traffic via primary and alternate routes support the commander's intent and concept of operations described in the base plan or order. Establish the priorities of support to units for each phase of the operation. Provide a detailed network diagram including the internet protocol scheme of the network being established. Refer to Appendix 3 (Voice, Video, and Data Network Diagrams) to Annex H (Signal) as required.*

(4) (U) <u>Scheme of Satellite Communications</u>. *Describe how satellite communications support the commander's intent and concept of operations described in the base plan or order. Establish the priorities of support to units for each phase of the operation. Provide a chart for all required frequencies, access times, access dates, and in the case of IP-based satellite communications systems, provide the internet protocol scheme for the modem, as well as the Nonsecure Internet Protocol Router Network and SECRET Internet Protocol Router Network routers. (This subparagraph will serve as a reference to units along with attachments.). Refer to Appendix 4 (Satellite Communications) to Annex H (Signal) as required.*

(5) (U) <u>Scheme of Foreign Data Exchanges</u>. *Describe how foreign data exchanges support the commander's intent and concept of operations described in the base plan or order. Outline procedures to prevent unauthorized disclosure and release of classified information on the SECRET Internet Protocol Router Network and the Joint Worldwide Intelligence Communications System. Outline the information to be disclosed to, released to, or received from foreign entities and the planned approach, including safeguarding steps to be taken. Refer to Appendix 5 (Foreign Data Exchanges) to Annex H (Signal) as required.*

(6) (U) <u>Spectrum Management Operations</u>. *Describe how spectrum management operations support the commander's intent and concept of operations described in the base plan or order. Outline the effects the commander wants to achieve while prioritizing tasks for spectrum management operations. List objectives and the primary tasks to achieve those objectives. Refer to Appendix 6 (Spectrum Management Operations) to Annex H (Signal) as required.*

(7) (U) <u>Scheme of Information Services</u>. *Describe how information services on Nonsecure Internet Protocol Router Network, SECRET Internet Protocol Router Network, and Joint Worldwide Intelligence Communications System will be provided and integrated to support the commander's intent and concept of operations as described in the base plan or order. Explain how information will be staged and the dissemination of that information controlled to facilitate data collection, processing, storage, discovery, and access by the user. Refer to Appendix 7 (Information Services) to Annex H (Signal) as required.*

b. (U) <u>Tasks to Subordinate Units</u>. *List signal support to operations tasks assigned to subordinate signal units not contained in the base order. Each task must include* who *(the subordinate unit assigned the task),* what *(the task itself),* when, where, *and* why *(purpose). Include tasks for supporting interagency, intergovernmental, and nongovernmental organizations. Use a separate subparagraph for each unit. List units in task organization sequence. Place tasks that affect two or more units in paragraph 3c (Coordinating Instructions).*

c. (U) <u>Coordinating Instructions</u>. *List only instructions applicable to two or more subordinate units not covered in the base plan or order.*

4. (U) <u>Sustainment</u>. *Identify priorities of sustainment for signal support to operations key tasks and specify additional instructions as required in the paragraph below. Refer to Annex F (Sustainment) as required.*

a. (U) <u>Logistics</u>. *Use subparagraphs to identify priorities and specific instructions for signal logistics support. Refer to Annex F (Sustainment) and Annex P (Host-Nation Support) as required.*

[page number]
[CLASSIFICATION]

Figure D-8. Sample Annex H (Signal) format (continued)

[CLASSIFICATION]

ANNEX H (SIGNAL) TO OPERATION PLAN/ORDER [number] [(code name)]—[issuing headquarters] [(classification of title)]

b. (U) <u>Personnel</u>. *Use subparagraphs to identify priorities and specific instructions for human resources support, financial management, legal support, and religious support. Refer to Annex F (Sustainment) as required.*

c. (U) <u>Health Service Support</u>. *Identify availability, priorities, and instruction for medical care. Refer to Annex F (Sustainment) as required.*

5. (U) <u>Command and Signal</u>.

a. (U) <u>Command</u>.

(1) (U) <u>Location of the Commander and Key Leaders</u>. *State the location of the commander and key signal unit commanders and staff officers.*

(2) (U) <u>Succession of Command</u>. *State the succession of command if not covered in the unit's standard operating procedures (SOPs).*

(3) (U) <u>Liaison Requirements</u>. *State the signal liaison requirements not covered in unit SOPs.*

b. (U) <u>Control</u>.

(1) (U) <u>Command Posts</u>. *Describe the employment of signal command posts (CPs), including the location of each CP and its time of opening and closing.*

(2) (U) <u>Reports</u>. *List reports not covered in SOPs. Describe signal support to operations reporting requirements for subordinate units. Refer to Annex R (Reports) as required.*

c. (U) <u>Signal</u>. *List signal operating instructions for signal support to operations as needed, as well as primary and alternate means of communications with both military and nonmilitary organizations conducting signal support to operations. Consider operations security requirements.*

(1) (U) *Describe the networks to monitor for reports.*

(2) (U) *Address any support to operations communications or digitization connectivity requirements or coordination necessary to meet functional responsibilities (consider telephone listing).*

ACKNOWLEDGE: *Include only if attachment is distributed separately from the base order.*

[Commander's last name]
[Commander's rank]

The commander or authorized representative signs the original copy of the attachment. If the representative signs the original, add the phrase "For the Commander." The signed copy is the historical copy and remains in the headquarters' files.

OFFICIAL:

[Authenticator's name]
[Authenticator's position]

Use only if the commander does not sign the original attachment. If the commander signs the original, no further authentication is required. If the commander does not sign, the signature of the preparing staff officer requires authentication and only the last name and rank of the commander appear in the signature block.

[page number]
[CLASSIFICATION]

Figure D-8. Sample Annex H (Signal) format (continued)

[CLASSIFICATION]

ANNEX H (SIGNAL) TO OPERATION PLAN/ORDER [number] [(code name)]—[issuing headquarters] [(classification of title)]

ATTACHMENTS: *List lower-level attachment (appendixes, tabs, and exhibits). If a particular attachment is not used, place "not used" beside the attachment number. Unit SOPs will dictate attachment development and format. Common attachments include the following:*

Appendix 1–Defensive Cyberspace Operations
Appendix 2–Information Network Operations
Appendix 3–Voice, Video, and Data Network Diagrams
Appendix 4–Satellite Communications
Appendix 5–Foreign Data Exchanges
Appendix 6–Spectrum Management Operations
Appendix 7–Information Services

DISTRIBUTION: *Show only if distributed separately from the base order or higher-level attachments.*

[page number]

[CLASSIFICATION]

Figure D-8. Sample Annex H (Signal) format (continued)

Annex I (Not Used)

ANNEX J (PUBLIC AFFAIRS) FORMAT AND INSTRUCTIONS

D-42. This annex provides fundamental considerations, formats, and instructions for developing Annex J (Public Affairs) to the base plan or order. This annex follows the five-paragraph attachment format.

D-43. Commanders and staffs use Annex J (Public Affairs) to describe how public affairs activities support the concept of operations described in the base plan or order. The public affairs officer develops Annex J (Public Affairs). (See figure D-9 on pages D-52 through D-55.)

[CLASSIFICATION]

Place the classification at the top and bottom of every page of the attachments. Place the classification marking at the front of each paragraph and subparagraph in parentheses. Refer to AR 380-5 for classification and release marking instructions.

Copy ## of ## copies
Issuing headquarters
Place of issue
Date-time group of signature
Message reference number

Include the full heading if attachment is distributed separately from the base order or higher-level attachment.

ANNEX J (PUBLIC AFFAIRS) TO OPERATION PLAN/ORDER [number] [(code name)]— [(classification of title)]

(U) References: *List documents essential to understanding the attachment.*

 a. *List maps and charts first. Map entries include series number, country, sheet names or numbers, edition, and scale.*

 b. *List other references in subparagraphs labeled as shown.*

 c. *Doctrinal references for public affairs activities include FM 3-61 and JP 3-61.*

(U) Time Zone Used Throughout the OPLAN or OPORD: *Write the time zone established in the base plan or order.*

(U) **Task Organization**: *Describe the organization of forces available to the issuing headquarters and their command and support relationships. Refer to Annex A (Task Organization) if long or complicated.*

1. (U) **Situation.** *Include information affecting public affairs that paragraph 1 of the OPLAN or OPORD does not cover or that needs expansion.*

 a. (U) Area of Interest. *Describe the area of interest as it relates to public affairs. Refer to Annex B (Intelligence) as required.*

 b. (U) Area of Operations. *Refer to Appendix 2 (Operation Overlay) to Annex C (Operations).*

[page number]
[CLASSIFICATION]

Figure D-9. Sample Annex J (Public Affairs) format

ANNEX J (PUBLIC AFFAIRS) TO OPERATION PLAN/ORDER |number| |(code name)|— |(classification of title)|

(1) (U) <u>Terrain</u>. *Describe the aspects of terrain that impact public affairs activities. Refer to Annex B (Intelligence) as required.*

(2) (U) <u>Weather</u>. *Describe the aspects of weather that impact public affairs. Refer to Annex B (Intelligence) as required.*

c. (U) <u>Enemy Forces</u>. *Identify enemy forces' general communications and media capabilities. Describe the enemy's disposition, location, strength, and probable public affairs courses of actions, including disinformation, rumors, and propaganda. Refer to Appendix B (Intelligence) as required.*

(1) (U) <u>Enemy Communications and Media Capabilities</u>. *Identify enemy forces' general communications and media capabilities, including television, radio, and print mediums as well as online and social media capabilities.*

(2) (U) <u>Enemy Courses of Action</u>. *Describe enemy's employment of communications and media capabilities that would impact friendly operations and public affairs operations.*

d. (U) <u>Friendly Forces</u>. *Outline the higher headquarters' plan (and public affairs annex) and adjacent unit public affairs plans. Provide information on friendly coalition forces, which may impact the public affairs mission. Note public affairs resources supporting the unit (who, where, when) and higher, allied, and adjacent headquarters.*

(1) (U) <u>Higher Headquarters Public Affairs Mission</u>. *State the public affairs mission of the higher headquarters.*

(2) (U) <u>Public Affairs Mission of Adjacent Units</u>. *Identify and state the public affairs missions of adjacent units and other units whose actions have a significant impact on the issuing headquarters.*

e. (U) <u>Interagency, Intergovernmental, and Nongovernmental Organizations</u>. *Identify and describe other organizations in the area of operations that may impact the conduct of operations of public affairs operations or implementation of public affairs activities.*

f. (U) <u>Civil Considerations</u>. *Describe critical aspects of the civil situation that impact public affairs operations. Refer to Annex K (Civil Affairs Operations) as required.*

g. (U) <u>Attachments and Detachments</u>. *Identify all augmenting public affairs units supporting this command and all attached or assigned subordinate units. Include effective dates, if applicable.*

h. (U) <u>Media</u>. *Identify media in the area (who, where, and pools) including U.S., international, and host nation.*

i. (U) <u>Assumptions</u>. *List any additional assumptions or information not included in the general situation that will impact the public affairs mission.*

2. (U) Mission. *State the mission of public affairs in support of the base plan or order.*

3. (U) Execution.

a. (U) <u>Scheme of Public Affairs</u>. *Describe how public affairs supports the commander's intent and concept of operations. Summarize how the commander visualizes executing the public affairs plan. Include public affairs priorities: Intent (access, information, welfare, morale, and will to win), concept (who, where, what, why, and when), specifics (tasks to a subordinate; who is to do what, where, and when, including nonpublic affairs activities), and actions with media (credential, train, and transport).*

|page number|
|CLASSIFICATION|

Figure D-9. Sample Annex J (Public Affairs) format (continued)

[CLASSIFICATION]

ANNEX J (PUBLIC AFFAIRS) TO OPERATION PLAN/ORDER [number] [(code name)]— [(classification of title)]

(1) (U) <u>Outline of Public Affairs Objectives.</u> *Describe clearly defined public affairs objectives that the commander intends to achieve.*

(2) (U) <u>Outline of Public Affairs Tasks.</u> *Identify and assign supporting public affairs tasks to each objective. Assign specific tasks to elements of the command charged with public affairs tasks. Establish priorities of support for each phase of the operation.*

(b) (U) <u>Tasks to Subordinate Units.</u> *Identify and list public affairs tasks assigned to subordinate units not contained in the base order including maneuver and augmenting public affairs units. Also identify unit public affairs representatives' requirements.*

(c) (U) <u>Coordinating Instructions.</u> *Give details on coordination, task organization, and groupings. List instructions that apply to two or more subordinate elements or units. Refer to supporting appendixes (public affairs running estimate) not referenced elsewhere (public affairs guidance, media in country, media en route with U.S. forces, media contact report, handover checklist, task organization, and public affairs synchronization requirements).*

4. (U) <u>Sustainment</u>. *Identify priorities of sustainment for public affairs key tasks and specify additional instructions as required by the paragraph below. Refer to Annex F (Sustainment) as required.*

a. (U) <u>Logistics.</u> *Use subparagraphs to identify priorities and specific instructions for maintenance, transportation, supply, field services, distribution, contracting, and general engineering support. Outline requirements for establishing a media operations center (if required) and embedded journalists. Refer to Annex F (Sustainment) and Annex P (Host-Nation Support) as required.*

b. (U) <u>Personnel.</u> *Use subparagraphs to identify priorities and specific instructions for human resources support, financial management, legal support, and religious support. Refer to Annex F (Sustainment) as required.*

5. (U) <u>Command and Signal</u>.

a. (U) <u>Command.</u> *State the location of key public affairs leaders (to include media operations center location and public affairs contact information).*

b. (U) <u>Control.</u> *State the public affairs liaison requirements not covered in the base order.*

c. (U) <u>Signal.</u> *Address any public affairs specific communication requirements (such as commercial internet or Defense Visual Information Distribution Systems) and reports. Refer to Annex H (Signal) as required.*

ACKNOWLEDGE: *Include only if attachment is distributed separately from the base order.*

[Commander's last name]
[Commander's rank]

The commander or authorized representative signs the original copy of the attachment. If the representative signs the original, add the phrase "For the Commander." The signed copy is the historical copy and remains in the headquarters' files.

[page number]
[CLASSIFICATION]

Figure D-9. Sample Annex J (Public Affairs) format (continued)

|CLASSIFICATION|

ANNEX J (PUBLIC AFFAIRS) TO OPERATION PLAN/ORDER |number| |(code name)|— |(classification of title)|

OFFICIAL:

[Authenticator's name]
[Authenticator's position]

Use only if the commander does not sign the original attachment. If the commander signs the original, no further authentication is required. If the commander does not sign, the signature of the preparing staff officer requires authentication and only the last name and rank of the commander appear in the signature block.

ATTACHMENTS: *List lower-level attachment (appendixes, tabs, and exhibits).*

Appendix 1–Public Affairs Running Estimate
Appendix 2–Public Affairs Guidance

DISTRIBUTION: *Show only if distributed separately from the base order or higher-level attachments.*

|page number|
|CLASSIFICATION|

Figure D-9. Sample Annex J (Public Affairs) format (continued)

ANNEX K (CIVIL AFFAIRS OPERATIONS) FORMAT AND INSTRUCTIONS

D-44. This annex provides fundamental considerations, formats, and instructions for developing Annex K (Civil Affairs Operations) to the base plan or order. This annex follows the five-paragraph attachment format.

D-45. Commanders and staffs use Annex K (Civil Affairs Operations) to describe how civil affairs operations, in coordination with other military and civil organizations, support the concept of operations described in the base plan or order. The G-9 (S-9) is responsible for developing Annex K (Civil Affairs Operations). (See figure D-10 on pages D-56 through D-61.)

[CLASSIFICATION]

Place the classification at the top and bottom of every page of the attachments. Place the classification marking at the front of each paragraph and subparagraph in parentheses. Refer to AR 380-5 for classification and release marking instructions.

Copy ## of ## copies
Issuing headquarters
Place of issue
Date-time group of signature
Message reference number

Include the full heading if attachment is distributed separately from the base order or higher-level attachment.

ANNEX K (CIVIL AFFAIRS OPERATIONS) TO OPLAN/OPORD [number] [(code name)]— [issuing headquarters] [(classification of title)]

(U) References: *List documents essential to understanding this attachment.*

    a. List maps and charts first. Map entries include series number, country, sheet names or numbers, edition, and scale.

    b. List other references in subparagraphs such as the civil affairs operations annex of higher headquarters, relevant civilian agency operations guides and standard documents, relevant plans of participating civilian organizations, coordinated transition plans, international treaties and agreements, and civil information management plans.

    c. Doctrinal references for civil affairs operations include FM 3-57, FM 6-0, and JP 3-57.

(U) Time Zone Used Throughout the OPLAN/OPORD: *Write the time zone established in the base plan or order.*

1. (U) <u>Situation</u>. *Include information affecting civil affairs operations that paragraph 1 of the OPLAN or OPORD does not cover or that needs expansion.*

    a. (U) <u>Area of Interest</u>. Describe the area of interest as it relates to civil affairs operations. Refer to Annex B (Intelligence) as required.

    b. (U) <u>Area of Operations</u>. Refer to Appendix 2 (Operation Overlay) to Annex C (Operations).

[page number]
[CLASSIFICATION]

Figure D-10. Sample Annex K (Civil Affairs Operations) format

---

[CLASSIFICATION]

ANNEX K (CIVIL AFFAIRS OPERATIONS) TO OPLAN/OPORD [number] [(code name)]— [issuing headquarters] [(classification of title)]

(1) (U) <u>Terrain</u>. *Describe the aspects of terrain that impact civil affairs operations such as population centers, likely movement corridors of dislocated civilians, and terrain that channels dislocated civilians. Refer to Annex B (Intelligence) as required.*

(2) (U) <u>Weather</u>. *Describe the aspects of weather that impact civil affairs operations such as seasonal events (rain, flooding, wind storms, and snow) that may impact commercial mobility, agricultural production, farmer to market access, and populace and resources control in the area of operations. Refer to Annex B (Intelligence) as required.*

c. (U) <u>Enemy Forces</u>. *List known and templated locations and activities of enemy civil affairs operations units for one echelon up and two echelons down. Identify enemy forces and appraise their general capabilities and impacts on the indigenous population and civil affairs operations. State expected enemy courses of action and employment of enemy civil affairs operations assets. Refer to Annex B (Intelligence) as required.*

d. (U) <u>Friendly Forces</u>. *Outline the higher headquarters' civil affairs operation plan. Briefly identify the mission of friendly forces and the objectives, goals, and mission of civilian organization that impact civil affairs operations. List designation, location, and outline of plan of higher, adjacent, and other civil affairs organizations and assets that support or impact the issuing headquarters or require coordination and additional support.*

(1) (U) <u>Higher Headquarters' Civil Affairs Operations Mission</u>. *Identify and state the civil affairs operations mission of the higher headquarters.*

(2) (U) <u>Civil Affairs Operations Missions of Adjacent Units</u>. *Identify and state the civil affairs operations missions of adjacent units and other units whose actions have a significant impact on the issuing headquarters.*

e. (U) <u>Interagency, Intergovernmental, and Nongovernmental Organizations</u>. *Identify and state the objectives or goals and primary tasks of those non-Department of Defense organizations that have a significant role within the civil situation in the area of operations or implementation of civil affairs operations-specific equipment and tactics. Refer to Annex V (Interagency Coordination) as required.*

(1) (U) <u>Interagency Organizations</u>. *Identify and state the objectives and primary tasks of those interagency organizations that impact the unit's civil affairs operations mission. Briefly describe the capabilities and capacity of each organization if not listed in Annex V (Interagency Coordination).*

(2) (U) <u>Intergovernmental Organizations</u>. *Identify and state the objectives and primary tasks of those intergovernmental organizations that impact the unit's civil affairs operations mission. Briefly describe the capabilities and capacities of each organization.*

(3) (U) <u>Nongovernmental Organizations</u>. *Identify and state the objectives and primary tasks of those nongovernmental organizations that impact the unit's civil affairs operations mission. Briefly describe the capabilities and capacities of each organization.*

f. (U) <u>Civil Considerations</u>. *Describe the critical aspects of the civil situation that impact civil affairs operations using the memory aid ASCOPE (areas, structures, capabilities, organizations, people, and events). Refer to Annex B (Intelligence) as required.*

(1) (U) <u>Areas</u>. *List the key civilian areas such as political boundaries; locations of government centers; social, political, religious, or criminal enclaves; agricultural and mining regions; trade routes; possible sites for the temporary settlement of dislocated civilians in the area of interest. Describe how these civilian areas affect the mission and how military operations may affect these areas.*

[page number]
[CLASSIFICATION]

Figure D-10. Sample Annex K (Civil Affairs Operations) format (continued)

[CLASSIFICATION]

ANNEX K (CIVIL AFFAIRS OPERATIONS) TO OPLAN/OPORD [number] [(code name)]— [issuing headquarters] [(classification of title)]

(2) (U) <u>Structures</u>. *List the locations of existing civil structures (critical infrastructure) such as ports, air terminals, transportation networks, bridges, communications towers, power plants, and dams. Identify churches, mosques, national libraries, hospitals, and other cultural sites generally protected by international law or other agreements. Other infrastructure includes governance and public safety structures (national, regional, and urban government facilities, record archives, judiciary, police, fire, and emergency medical services) and economic and environmental structures (banking, stock and commodity exchanges, toxic industrial facilities, and pipelines). Identify facilities with practical applications—such as jails, warehouses, schools, television stations, radio stations, and print plants— which may be useful for military purposes.*

(3) (U) <u>Capabilities</u>. *Describe civil capabilities by assessing the population's capabilities of sustaining itself through public safety, emergency services, and food and agriculture. Include whether the population needs assistance with public works and utilities, public health, public transportation, economics, and commerce. Refer to the civil affairs preliminary area assessment.*

(4) (U) <u>Organizations</u>. *Identify and list civil organizations that may or may not be affiliated with government agencies, such as religious groups, ethnic groups, multinational corporations, fraternal organizations, patriotic or service organizations, intergovernmental organizations, or nongovernmental organizations. Do not repeat those listed in Annex V (Interagency Coordination) or paragraph 1e (Interagency, Intergovernmental, and Nongovernmental Organizations) of this annex. Include host-nation organizations capable of forming the nucleus for humanitarian assistance programs, interim-governing bodies, civil defense efforts, and other activities.*

(5) (U) <u>People</u>. *List key personnel and their linkage to the population, leaders, figureheads, clerics, and subject matter experts such as plant operators and public utility managers. Categorize groups of civilians using local nationals (town and city dwellers, farmers and other rural dwellers, and nomads), local civil authorities (elected and traditional leaders at all levels of government), expatriates, tribal or clan figureheads and religious leaders, third-nation government agency representatives, foreign employees of intergovernmental organizations or nongovernmental organizations, contractors (American citizens, local nationals, and third-nation citizens providing contract services), the media (journalists from print, radio, and visual media), and dislocated civilians (refugees, displaced persons, evacuees, migrants, and stateless persons).*

*(**Note**: This list may extend to personnel outside of the area of operations whose actions, opinions, and influence can affect the commander's area of operations.)*

(6) (U) <u>Events</u>. *Determine what events, military and civilian, are occurring and analyze the events for their political, economic, psychological, environmental, moral, and legal implications. Categorize civilian events that may affect military missions. Events may include harvest seasons, elections, riots, voluntary and involuntary evacuations, holidays, school years, and religious periods.*

g. (U) <u>Attachments and Detachments</u>. *List units attached to or detached from the issuing headquarters only as necessary to clarify task organization that impact civil affairs operations. Refer to Annex A (Task Organization) as required.*

h. (U) <u>Assumptions</u>. *List key assumptions that pertain to civil affairs operations that were used to form the civil affairs operations running estimate and develop the OPLAN or OPORD and this annex.*

2. (U) <u>Mission</u>. *State the mission of civil affairs operations in support of the base plan or order.*

3. (U) <u>Execution</u>.

a. (U) <u>Scheme of Civil Affairs Operations</u>. *Describe how civil affairs operations support the commander's intent and concept of operations described in the base plan or order. Outline the effects the commander wants civil affairs operations to achieve while prioritizing civil affairs tasks. Identify and list civil-military objectives and the primary tasks to achieve those objectives.*

[page number]
[CLASSIFICATION]

Figure D-10. Sample Annex K (Civil Affairs Operations) format (continued)

[CLASSIFICATION]

ANNEX K (CIVIL AFFAIRS OPERATIONS) TO OPLAN/OPORD [number] [(code name)]— [issuing headquarters] [(classification of title)]

(1) (U) <u>Execution Matrix</u>. *Provide the execution matrix. Refer to Appendix 1 (Execution Matrix) to Annex K (Civil Affairs Operations).*

(2) (U) <u>Populace and Resources Control Plan</u>. *Provide the populace and resources control plan. Refer to Appendix 2 (Populace and Resources Control Plan) to Annex K (Civil Affairs Operations).*

(3) (U) <u>Civil Information Management Plan</u>. *Provide the civil information management plan. Refer to Appendix 3 (Civil Information Management Plan) to Annex K (Civil Affairs Operations).*

b. (U) <u>Tasks to Subordinate Units</u>. *State the civil affairs operations tasks assigned to each unit that report directly to the headquarters issuing the order. Each task must include who (the subordinate unit assigned the task), what (the task itself), when, where, and why (purpose). Include interagency, intergovernmental organization, or nongovernmental organization supporting tasks. Use a separate subparagraph for each unit. List units in task organization sequence. Place tasks that affect two or more units in paragraph 3c (Coordinating Instructions) of this annex.*

c. (U) <u>Coordinating Instructions</u>. *List only instructions applicable to two or more subordinate units not covered in the base plan or order.*

(1) (U) <u>Environmental Considerations</u>. *Review environmental planning guidance and, if available, the Environmental Management Support Plan for implied civil affairs operations tasks that support environmental activities. For example, establishing and supporting camps for dislocated civilians may require air and water purification, hazardous waste and material disposal, sanitation facilities and personal hygiene facilities, and identification of hazards such as pesticides, toxic chemicals, and historic or cultural resources for preservation. Refer to Annex G (Engineer) as required.*

(2) (U) <u>Stability tasks</u>. *Describe how civil affairs operations support the command's identified minimum-essential stability tasks—civil control, civil security, and restoration of essential services. Units responsible for an area of operations must execute the minimum-essential tasks with available resources if no civilian agency or organization is capable. Address course of action support to governance and economic stability if required by mission taskings of the higher headquarters.*

4. (U) <u>Sustainment</u>. *Identify priorities of sustainment for civil affairs operations key tasks and specify additional instructions as required. Refer to Annex F (Sustainment) as required.*

a. (U) <u>Logistics</u>. *Identify unique sustainment requirements, procedures, and guidance to support civil affairs teams and operations. Specify procedures for specialized technical logistics support from external organizations as necessary. Use subparagraphs to identify priorities and specific instructions for civil affairs operations logistics support. Refer to Annex F (Sustainment) and Annex P (Host-Nation Support) as required.*

b. (U) <u>Personnel</u>. *Identify unique personnel requirements and concerns associated with civil affairs operations, including global sourcing support and contracted linguist requirements. Use subparagraphs to identify priorities and specific instructions for human resources support, financial management, legal support, and religious support. Refer to Annex F (Sustainment) as required.*

c. (U) <u>Health Service Support</u>. *Identify availability, priorities, and instructions for medical care. Refer to Annex F (Sustainment) as required. Provide additional information on the following:*

(1) (U) *Identify and list locations, capabilities, and capacity of nonmilitary medical treatment facilities that can or will support civil affairs operations.*

(2) (U) *Identify and list unique problems, challenges, and legal considerations of providing health service support to the indigenous population.*

[page number]
[CLASSIFICATION]

Figure D-10. Sample Annex K (Civil Affairs Operations) format (continued)

[CLASSIFICATION]

ANNEX K (CIVIL AFFAIRS OPERATIONS) TO OPLAN/OPORD [number] [(code name)]—[issuing headquarters] [(classification of title)]

(3) (U) *Identify and list host-nation medical support capabilities if not addressed in Annex P (Host-Nation Support).*

(4) (U) *Identify and list areas requiring capacity building activities, such as in veterinary services or agriculture realms.*

5. (U) Command and Signal.

a. (U) Command.

(1) (U) Location of the Commander and Key Leaders. *List the location of the commander and key civil affairs leaders.*

(2) (U) Succession of Command. State the succession of command if not covered in the unit's standard operating procedures (SOPs).

(3) (U) Liaison Requirements. *State civil affairs liaison requirements not covered in the unit's SOPs.*

b. (U) Control.

(1) (U) Command Posts. *Describe the employment of command posts (CPs), including the location of each CP and its time of opening and closing, as appropriate. State the primary controlling CP for specific tasks or phases of the operation (for example, "Civil-military operations center (CMOC) will be co-located with division main CP").*

(a) (U) *Location and alternate locations of civil affairs command post or CMOC.*

(b) (U) *Location and alternate locations of higher, adjacent, and subordinate CMOCs.*

(c) (U) *Location of key civil affairs operations leaders.*

(2) (U) Reports. *List reports not covered in SOPs. Describe civil affairs operations reporting requirements for subordinate units. Refer to Annex R (Reports) as required.*

c. (U) Signal. *List signal operating instructions for civil affairs operations as needed, as well as primary and alternate means of communications with both military and nonmilitary organizations conducting civil affairs operations. Consider operations security requirements. Refer to Annex H (Signal) as required.*

(1) (U) *Describe the networks to monitor for reports.*

(2) (U) *Address any civil affairs operations specific communications or digitization connectivity requirements or coordination necessary to meet functional responsibilities (consider telephone listing). Provide instructions regarding maintenance and update of the civil information management database.*

ACKNOWLEDGE: *Include only if attachment is distributed separately from the base order.*

[Commander's last name]
[Commander's rank]

The commander or authorized representative signs the original copy of attachment. If the representative signs the original, add the phrase "For the Commander." The signed copy is the historical copy and remains in the headquarters' files.

OFFICIAL:

[Authenticator's name]
[Authenticator's position]

[page number]
[CLASSIFICATION]

Figure D-10. Sample Annex K (Civil Affairs Operations) format (continued)

[CLASSIFICATION]

ANNEX K (CIVIL AFFAIRS OPERATIONS) TO OPLAN/OPORD [number] [(code name)]— [issuing headquarters] [(classification of title)]

Use only if the commander does not sign the original attachment. If the commander signs the original, no further authentication is required. If the commander does not sign, the signature of the preparing staff officer requires authentication and only the last name and rank of the commander appear in the signature block.

ATTACHMENTS: *List lower-level attachment (appendixes, tabs, and exhibits).*

Appendix 1–Execution Matrix
Appendix 2–Populace and Resources Control Plan
Appendix 3–Civil Information Management Plan

DISTRIBUTION: *Show only if distributed separately from the base order or higher-level attachment.*

[page number]
[CLASSIFICATION]

Figure D-10. Sample Annex K (Civil Affairs Operations) format (continued)

ANNEX L (INFORMATION COLLECTION) FORMAT AND INSTRUCTIONS

D-46. This annex provides fundamental considerations, formats, and instructions for developing Annex L (Information Collection) in Army plans and orders. It provides a format for the annex that can be modified to meet the requirements of the base order and operations and an example information collection plan. This annex follows the five-paragraph attachment format.

D-47. The information collection annex clearly describes how information collection activities support the offensive, defensive, and stability or defense support of civil authorities operations throughout the conduct of the operations described in the base order. It synchronizes activities in time, space, and purpose to achieve objectives and accomplish the commander's intent for reconnaissance, surveillance, and intelligence operations (including military intelligence disciplines). The G-3 (S-3), in conjunction with the G-2 (S-2), is responsible for this annex. (See figure D-11 on pages D-62 through D-66.)

[CLASSIFICATION]

Place the classification at the top and bottom of every page of the attachments. Place the classification marking at the front of each paragraph and subparagraph in parentheses. Refer to AR 380-5 for classification and release marking instructions.

Copy ## of ## copies
Issuing headquarters
Place of issue
Date-time group of signature
Message reference number

Include the full heading if attachment is distributed separately from the base order or higher-level attachment.

ANNEX L (INFORMATION COLLECTION) TO OPERATION PLAN/ORDER [number] [(code name)]—[issuing headquarters] [(classification of title)]

(U) References: *List documents essential to understanding Annex L.*

a. *List maps and charts first. Map entries include series number, country, sheet names or numbers, edition, and scale.*

b. *List other references in subparagraphs labeled as shown.*

c. *Doctrinal references for this annex include FM 2-0 and FM 6-0.*

(U) Time Zone Used Throughout the Plan/Order: *Write the time zone established in the base plan or order.*

1. (U) <u>Situation</u>.

a. (U) <u>Area of Interest</u>. *Refer to Annex B (Intelligence) or Appendix 2 (Operation Overlay) to Annex C (Operations).*

b. (U) <u>Area of Operations</u>. *Refer to Appendix 2 (Operation Overlay) to Annex C (Operations).*

(1) (U) <u>Terrain</u>. *Describe the aspects of terrain that impact information collection. Refer to Annex B (Intelligence) as required.*

(2) (U) <u>Weather</u>. *Describe the aspects of weather that impact information collection. Refer to Annex B (Intelligence) as required.*

c. (U) <u>Enemy Forces</u>. *Refer to Annex B (Intelligence) as required.*

d. (U) <u>Friendly Forces</u>. *Refer to base order, Annex A (Task Organization) and Annex C (Operations) as required.*

[page number]
[CLASSIFICATION]

Figure D-11. Sample Annex L (Information Collection) format

[CLASSIFICATION]

ANNEX L (INFORMATION COLLECTION) TO OPERATION PLAN/ORDER [number] [(code name)]—[issuing headquarters] [(classification of title)]

e. (U) <u>Interagency, Intergovernmental, and Nongovernmental Organizations</u>. *Identify and describe other organizations in the area of operations that may impact the conduct of operations or implementation of information collection-specific equipment and tactics. Refer to Annex V (Interagency Coordination) as required.*

f. (U) <u>Civil Considerations</u>. *Describe the critical aspects of the civil situation that impact information collection activities. Refer to Appendix 1 (Intelligence Estimate) to Annex B (Intelligence) and Annex K (Civil Affairs Operations) as required.*

g. (U) <u>Attachments and Detachments</u>. *If pertinent, list units or assets attached to or detached from the issuing headquarters. State when each attachment or detachment is effective (for example, on order, on commitment of the reserve) if different from the effective time of the base plan or order. Do not repeat information already listed in Annex A (Task Organization).*

h. (U) <u>Assumptions</u>. *List any information collection-specific assumptions that support the annex development.*

2. (U) <u>Mission</u>. *State the mission of information collection in support of the operation—a short description of the who, what (task), when, where, and why (purpose) that clearly indicates the action to be taken and the reason for doing so.*

3. (U) <u>Execution</u>.

a. (U) <u>Concept of Operations</u>. *This is a statement of the overall information collection objective. Describe how the tasks or missions of reconnaissance, surveillance, security, intelligence operations, and so forth support the commander's intent and the maneuver plan. Direct the manner in which each element of the force cooperates to accomplish the key information collection tasks and ties that to support of the operation with task and purpose statement. Must describe, at minimum, the overall scheme of maneuver and concept of fires. Refer to Appendix 1 (Information Collection Plan) to Annex L (Information Collection). The following subparagraphs are examples. Omit what is unnecessary for brevity.*

(1) (U) <u>Movement and Maneuver</u>. *Provide the scheme of movement and maneuver for collection assets and any other unit given a key information collection task, in accordance with the concept of operations in the base order (paragraph 3b) and Annex C (Operations). Describe the employment of information collection assets in relation to the rest of the force and state the method forces will enter the area of operations.*

(2) (U) <u>Intelligence</u>. *Describe the intelligence concept for supporting information collection. Refer to Annex B (Intelligence) as required.*

(3) (U) <u>Fires</u>. *Describe the concept of fires in support of information collection. Identify which information collection assets have priority of fires and the coordinating purpose of, priorities for, allocation of, and restrictions on fire support and fire support coordinating measures. Refer to Annex D (Fires) as required.*

(4) (U) <u>Protection</u>. *Describe protection support to information collection. Refer to Annex E (Protection) as required.*

(5) (U) <u>Engineer</u>. *Describe engineer support, if applicable, to information collection. Identify priority of mobility and survivability assets. Refer to Annex G (Engineer) as required.*

(6) (U) <u>Sustainment</u>. *Describe sustainment support to information collection as required. Refer to Annex F (Sustainment).*

[page number]
[CLASSIFICATION]

Figure D-11. Sample Annex L (Information Collection) format (continued)

[CLASSIFICATION]

ANNEX L (INFORMATION COLLECTION) TO OPERATION ORDER # [number] [(code name)]—[issuing headquarters] [(classification of title)]

(7) (U) <u>Signal</u>. *Describe signal support to information collection as required. Refer to Annex H (Signal).*

(8) (U) <u>Soldier and Leader Engagement</u>. *State overall concept for synchronizing information collection with Soldier and leader engagement. Refer to coordinating instructions in Annex C (Operations).*

(9) (U) <u>Assessment</u>. *If required, describe the priorities for assessment for the information collection plan and identify the measures of effectiveness used to assess end state conditions and objectives. Refer to Annex M (Assessment) as required.*

b. (U) <u>Tasks to Subordinate Units</u>. *State the information collection task assigned to each unit not identified in the base order. Refer to Appendix 1 (Information Collection Plan) to Annex L (Information Collection) as required.*

(1) (U) <u>Information Collection Support Tasks for Maneuver Units</u>.

(a) (U) <u>Tasks to Maneuver Unit 1</u>.

(b) (U) <u>Tasks to Maneuver Unit 2</u>.

(c) (U) <u>Tasks to Maneuver Unit 3</u>.

(2) (U) <u>Information Collection Support Tasks for Support Units</u>. *Direct units to observe and report in accordance with Appendix 1 (Information Collection Plan) to Annex L (Information Collection).*

(a) (U) <u>Military Intelligence</u>. *Refer to Annex B (Intelligence) as required.*

(b) (U) <u>Engineer</u>. *Refer to Annex G (Engineer) as required.*

(c) (U) <u>Fires</u>. *Refer to Annex D (Fires) as required.*

(d) (U) <u>Signal</u>. *Refer to Annex H (Signal) as required.*

(e) (U) <u>Sustainment</u>. *Refer to Annex F (Sustainment) as required.*

(f) (U) <u>Protection</u>. *Refer to Annex E (Protection) as required.*

(g) (U) <u>Civil Affairs</u>. *Refer to Annex K (Civil Affairs Operations) as required.*

c. (U) <u>Coordinating Instructions</u>. *List only instructions applicable or not covered in unit standard operating procedures (SOPs).*

(1) (U) <u>Time or Condition When the Plan Becomes Effective</u>.

(2) (U) <u>Priority Intelligence Requirements</u>. *List priority intelligence requirements (PIRs) here, the information collection tasks associated with them, and the latest time information is of value for each PIR.*

(3) (U) <u>Essential Elements of Friendly Information</u>. *List essential elements of friendly information (EEFIs) here.*

(4) (U) <u>Fire Support Coordinating Measures</u>. *List fire support coordinating or control measures. Establish no fire areas.*

[page number]

[CLASSIFICATION]

Figure D-11. Sample Annex L (Information Collection) format (continued)

[CLASSIFICATION]

ANNEX L (INFORMATION COLLECTION) TO OPERATION PLAN/ORDER [number] [(code name)]—[issuing headquarters] [(classification of title)]

(5) (U) Intelligence Handover Lines with Adjacent Units. *Identify handover guidance and parameters; refer to necessary graphics or attachments as required.*

(6) (U) Limits of Advance, Limits of Reconnaissance, and Quick Reaction Force Response Instructions. *Identify as required, referencing graphical depictions in attachments or instructions as needed.*

(7) (U) Airspace Coordinating Measures. *List airspace control measures.*

(8) (U) Intelligence Coordination Measures. *List information such as restrictions on international borders or other limitations and the coordination or special instructions that apply. Identify what unit is responsible for coordinating information collection activities in relation to the area of operations.*

(9) (U) Rules of Engagement. *Refer to Appendix 11 (Rules of Engagement) to Annex C (Operations) as required.*

(10) (U) Risk Reduction Control Measures. *State any reconnaissance, surveillance, and security-specific guidance such as fratricide prevention measures not included in SOPs, referring to Annex E (Protection) as required.*

(11) (U) Environmental Considerations. *Refer to Appendix 5 (Environmental Considerations) to Annex G (Engineer) as required.*

(12) (U) Other Coordinating Instructions. *List only instructions applicable to two or more subordinate units not covered in the base plan or order.*

4. (U) **Sustainment**. *Describe any sustainment requirements, subparagraphs may include:*

a. (U) Logistics. *Identify unique sustainment requirements, procedures, and guidance to support information collection. Specify procedures for specialized technical logistics support from external organizations as necessary. Use subparagraphs to identify priorities and specific instructions for information collection logistics support. Refer to Annex F (Sustainment) and Annex P (Host-Nation Support) as required.*

b. (U) Personnel. *Identify unique personnel requirements and concerns, associated with information collection, including global sourcing support and contracted linguist requirements. Use subparagraphs to identify priorities and specific instructions for human resources support, financial management, legal support, and religious support. Refer to Annex F (Sustainment) as required.*

c. (U) Health Service Support. *Provide information including the health threat (endemic and epidemic diseases, state of health of the enemy forces, medical capabilities of the enemy force and the civilian population), and medical evacuation routes, barriers, and significant terrain features. Refer to Appendix 3 (Health System Support) to Annex F (Sustainment) as required.*

5. (U) **Command and Signal**.

a. (U) Command.

(1) (U) Location of the Commander and Key Leaders. *List the location of the commander and key intelligence collection leaders and staff officers.*

(2) (U) Succession of Command. *State the succession of command if not covered in the unit's SOPs.*

(3) (U) Liaison Requirements. *State intelligence collection liaison requirements not covered in the unit's SOPs.*

[page number]
[CLASSIFICATION]

Figure D-11. Sample Annex L (Information Collection) format (continued)

[CLASSIFICATION]

ANNEX L (INFORMATION COLLECTION) TO OPERATION PLAN/ORDER [number] [(code name)]—[issuing headquarters] [(classification of title)]

b. (U) <u>Control</u>.

(1) (U) <u>Command Posts</u>. *Describe the employment of command posts (CPs), including the location of each CP and its time of opening and closing, as appropriate. State the primary controlling CP for specific tasks or phases of the operation.*

(2) (U) <u>Reports</u>. *List reports not covered in SOPs. Describe information collection reporting requirements for subordinate units. Refer to Annex R (Reports) as required.*

c. (U) <u>Signal</u>. *List signal operating instructions for intelligence collection as needed. Consider operations security requirements. Address any intelligence collection specific communications and digitization connectivity requirements. Refer to Annex H (Signal) as required.*

ACKNOWLEDGE: *Include only if attachment is distributed separately from the base plan or order.*

[Commander's last name]
[Commander's rank]

The commander or authorized representative signs the original copy. If the representative signs the original, add the phrase "For the Commander." The signed copy is the historical copy and remains in the headquarters' files.

OFFICIAL:

[Authenticator's name]
[Authenticator's position]

Use only if the commander does not sign the original attachment. If the commander signs the original, no further authentication is required. If the commander does not sign, the signature of the preparing staff officer requires authentication and only the last name and rank of the commander appear in the signature block.

ATTACHMENTS: *List lower-level attachment (appendixes, tabs, and exhibits).*

Appendix 1–Information Collection Plan
Appendix 2–Information Collection Overlay

DISTRIBUTION: (if distributed separately from the base order).

[page number]
[CLASSIFICATION]

Figure D-11. Sample Annex L (Information Collection) format (continued)

ANNEX M (ASSESSMENT) FORMAT AND INSTRUCTIONS

D-48. This annex provides fundamental considerations, formats, and instructions for developing Annex M (Assessment) to the base plan or order. This annex uses the five-paragraph attachment format.

D-49. Commanders and staffs use Annex M (Assessment) to as a means to quantify and qualify mission success or task accomplishment. The G-3 (S-3) or G-5 (S-5) is responsible for the development of Annex M (Assessment).

D-50. This annex describes the assessment concept of support objectives. This annex includes a discussion of the overall assessment concept of support, with the specific details in element subparagraphs and attachments. (See figure D-12 on pages D-67 through D-70.)

[CLASSIFICATION]

Place the classification at the top and bottom of every page of the attachments. Place the classification marking at the front of each paragraph and subparagraph in parentheses. Refer to AR 380-5 for classification and release marking instructions.

Copy ## of ## copies
Issuing headquarters
Place of issue
Date-time group of signature
Message reference number

Include the full heading if attachment is distributed separately from the base order or higher-level attachment.

ANNEX M (ASSESSMENT) TO OPERATION PLAN/ORDER [number] [(code name)]—[issuing headquarters] [(classification of title)]

(U) References: *List documents essential to understanding the attachment.*

    a. *List maps and charts first. Map entries include series number, country, sheet names or numbers, edition, and scale.*

    b. *List other references in subparagraphs labeled as shown. List available assessment products that are produced external to this unit. This includes classified and open-source assessment products of the higher headquarters, adjacent units, key government organizations (such as the Department of State), and any other relevant military or civilian organizations.*

    c. *Doctrinal references for assessment include ADRP 5-0 and FM 6-0.*

(U) Time Zone Used Throughout the Plan/Order: *Write the time zone established in the base plan or order.*

1. (U) <u>Situation</u>. *See the base order or use the following subparagraphs. Include information affecting assessment that paragraph 1 of the OPLAN or OPORD does not cover or that needs expansion.*

    a. (U) <u>Area of Interest</u>. *Describe the area of interest as it relates to assessment. Refer to Annex B (Intelligence) as required.*

[page number]
[CLASSIFICATION]

Figure D-12. Sample Annex M (Assessment) format

[CLASSIFICATION]

ANNEX M (ASSESSMENT) TO OPERATION PLAN/ORDER [number] [(code name)]—[issuing headquarters] [(classification of title)]

b. (U) <u>Area of Operations</u>. *Refer to Appendix 2 (Operation Overlay) to Annex C (Operations).*

(1) (U) <u>Terrain</u>. *Describe the aspects of terrain that impact assessment. Refer to Annex B (Intelligence) as required.*

(2) (U) <u>Weather</u>. *Describe the aspects of weather that impact assessment. Refer to Annex B (Intelligence) as required.*

c. (U) <u>Enemy Forces</u>. *List known and templated locations and activities of enemy assessment units for one echelon up and two echelons down. List enemy maneuver and other area capabilities that will impact friendly operations. State expected enemy courses of action and employment of enemy assessment assets. Refer to Annex B (Intelligence) as required.*

d. (U) <u>Friendly Forces</u>. *Outline the higher headquarters' assessment plan. List designation, location, and outline of plans of higher, adjacent, and other assessment organizations and assets that support or impact the issuing headquarters or require coordination and additional support.*

e. (U) <u>Interagency, Intergovernmental, and Nongovernmental Organizations</u>. *Identify and describe other organizations in the area of operations that may impact assessment. Refer to Annex V (Interagency Coordination) as required.*

f. (U) <u>Civil Considerations</u>. *Describe the aspects of the civil situation that impact assessment. Refer to Annex B (Intelligence) and Annex K (Civil Affairs Operations) as required.*

g. (U) <u>Attachments and Detachments</u>. *List units attached or detached only as necessary to clarify task organization. Refer to Annex A (Task Organization) as required.*

h. (U) <u>Assumptions</u>. *List any assessment-specific assumptions that support the annex development.*

2. (U) <u>Mission</u>. *State the mission of assessment in support of the base plan or order.*

3. (U) <u>Execution</u>.

a. (U) <u>Scheme of Operational Assessment</u>. *State the overall concept for assessing the operation. Include priorities of assessment, quantitative and qualitative indicators, and the general concept for how the recommendations produced by the assessment process will reach decisionmakers at the relevant time and place.*

(1) (U) <u>Nesting with Higher Headquarters</u>. *Provide the concept of nesting of unit assessment practices with lateral and higher headquarters (include military and interagency organizations, where applicable). Use Appendix 1 (Nesting of Assessment Efforts) to Annex M (Assessment) to provide a diagram or matrix that depicts the nesting of headquarters assessment procedures.*

(2) (U) <u>Information Requirements (Data Collection Plan)</u>. *Information requirements for assessment are synchronized through the information collection process and may be commander's critical information requirements. Provide a narrative that describes the plan to collect the data needed to inform the status on metrics and indicators developed. The data collection plan should include a consideration to minimize impact on subordinate unit operations. Provide diagrams or matrixes that depict the hierarchy of assessment objectives with the underlying measures of effectiveness, measures of performance, indicators, and metrics. Provide measures of effectiveness with the underlying data collection requirements and responsible agency for collecting the data.*

(3) (U) <u>Battle Rhythm</u>. *Establish the sequence of regularly occurring assessment activities. Explicitly state frequency of data collection for each data element. Include requirements to higher units, synchronization with lateral units, and products provided to subordinate units.*

[page number]
[CLASSIFICATION]

Figure D-12. Sample Annex M (Assessment) format (continued)

|CLASSIFICATION|

ANNEX M (ASSESSMENT) TO OPERATION PLAN/ORDER |number| |(code name)|—|issuing headquarters| |(classification of title)|

(4) (U) <u>Reframing Criteria</u>. *Identify key assumptions, events, or conditions that staffs will periodically assess to refine understanding of the existing problem and, if appropriate, trigger a reframe.*

b. (U) <u>Tasks to Subordinate Units</u>. *Identify the unit, agency, or staff section assigned responsibility for collecting data, conducting analysis, and generating recommendations for each condition or measure of effectiveness. Refer to paragraph 3a(2) (Information Requirements) of this annex as necessary.*

c. (U) <u>Coordinating Instructions</u>. *List only instructions applicable to two or more subordinate units not covered in the base plan or order. Use Appendix 3 (Assessment Working Group) to Annex M (Assessment) to include quad charts that provide details about meeting location, proponency, members, agenda, and inputs or outputs.*

4. (U) <u>Sustainment</u>. Identify priorities of sustainment assessment key tasks and specify additional instructions as required. Refer to Annex F (Sustainment) as required.

a. (U) <u>Logistics</u>. *Identify unique sustainment requirements, procedures, and guidance to support assessment teams. Use subparagraphs to identify priorities and specific instructions for assessment logistics support. Refer to Annex F (Sustainment) and Annex P (Host-Nation Support) as required.*

b. (U) <u>Personnel</u>. *Use subparagraphs to identify priorities and specific instructions for human resources support, financial management, legal support, and religious support. Refer to Annex F (Sustainment) as required.*

c. (U) <u>Health Service Support</u>. *Identify availability, priorities, and instructions for medical care. Refer to Annex F (Sustainment) as required.*

5. (U) <u>Command and Signal</u>.

a. (U) <u>Command</u>. *State the location of key assessment cells. State assessment liaison requirements not covered in the unit's standard operating procedures (SOPs).*

(1) (U) <u>Location of the Commander and Key Leaders</u>. *State the location of the commander and key assessment leaders.*

(2) (U) <u>Succession of Command</u>. *State the succession of command if not covered in the unit's SOPs.*

(3) (U) <u>Liaison Requirements</u>. *State the assessment liaison requirements not covered in the unit's SOPs.*

b. (U) <u>Control</u>.

(1) (U) <u>Command Posts</u>. *Describe the employment of assessment-specific command posts (CPs), including the location of each CP and its time of opening and closing.*

(2) (U) <u>Reports</u>. *List assessment-specific reports not covered in SOPs. Refer to Annex R (Reports) as required.*

c. (U) <u>Signal</u>. *Address any assessment-specific communications requirements. Refer to Annex H (Signal) as required.*

|page number|
|CLASSIFICATION|

Figure D-12. Sample Annex M (Assessment) format (continued)

[CLASSIFICATION]

ANNEX M (ASSESSMENT) TO OPERATION PLAN/ORDER [number] [(code name)]—[issuing headquarters] [(classification of title)]OFFICIAL:

ACKNOWLEDGE: *Include only if attachment is distributed separately from the base order.*

[Commander's last name]
[Commander's rank]

The commander or authorized representative signs the original copy of the attachment. If the representative signs the original, add the phrase "For the Commander." The signed copy is the historical copy and remains in the headquarters' files.

[Authenticator's name]
[Authenticator's position]

Use only if the commander does not sign the original attachment. If the commander signs the original, no further authentication is required. If the commander does not sign, the signature of the preparing staff officer requires authentication and only the last name and rank of the commander appear in the signature block.

ATTACHMENTS: *List lower-level attachment (appendixes, tabs, and exhibits).*

Appendix 1–Nesting of Assessment Efforts
Appendix 2–Assessment Framework
Appendix 3–Assessment Working Group

DISTRIBUTION: *Show only if distributed separately from the base order or higher-level attachments.*

[page number]
[CLASSIFICATION]

Figure D-12. Sample Annex M (Assessment) format (continued)

ANNEX N (SPACE OPERATIONS) FORMAT AND INSTRUCTIONS

D-51. This annex provides fundamental considerations, formats, and instructions for developing Annex N (Space Operations) to the base plan or order. This annex uses the five-paragraph attachment format.

D-52. Commanders and staffs use Annex N (Space Operations) to describe how space operations support the concept of operations described in the base plan or order. The space operations officer develops the Annex N (Space Operations).

D-53. This annex is used to coordinate early with the staff, to include the G-2 (S-2), G-6 (S-6), air defense artillery officer, and the special technical operations element to synchronize efforts and avoid duplication of information. While the G-2 (S-2) may want to produce and include the enemy space assessment portion in Annex B (Intelligence), there are products space professionals may uniquely contribute. This annex requests space orders of battle through the Joint Space Operations Center prior to deployment. (See figure D-13 on pages D-71 through D-75.)

[CLASSIFICATION]

Place the classification at the top and bottom of every page of the attachments. Place the classification marking at the front of each paragraph and subparagraph in parentheses. Refer to AR 380-5 for classification and release marking instructions.

Copy ## of ## copies
Issuing headquarters
Place of issue
Date-time group of signature
Message reference number

Include heading if attachment is distributed separately from the base order or higher-level attachment.

ANNEX N (SPACE OPERATIONS) TO OPLAN/OPORD [number] [(code name)]—[issuing headquarters] [(classification of title)]

(U) References: *List documents essential to understanding the attachment.*

a. List maps and charts first. Map entries include series number, country, sheet names or numbers, edition, and scale.

b. List other references in subparagraphs labeled as shown.

c. Doctrinal references for space operations include FM 3-14, FM 6-0, JP 3-14, and U.S. National Space Policy.

(U) Time Zone Used Throughout the Order: *Write the time zone established in the base plan or order.*

1. (U) <u>Situation</u>. *Include information affecting space operations that paragraph 1 of the OPLAN or OPORD does not cover or that needs expansion.*

[page number]
[CLASSIFICATION]

Figure D-13. Sample Annex N (Space Operations) format

[CLASSIFICATION]

ANNEX N (SPACE OPERATIONS) TO OPLAN/OPORD [number] [(code name)]—[issuing headquarters] [(classification of title)]

a. (U) <u>Area of Interest</u>. *Describe the area of interest as it relates to space operations. Refer to Annex B (Intelligence) as required.*

b. (U) <u>Area of Operations</u>. *Refer to Appendix 2 (Operation Overlay) to Annex C (Operations).*

(1) (U) <u>Terrain</u>. *Describe the aspects of terrain that impact space operations such as terrain masking. Refer to Annex B (Intelligence) as required.*

(2) (U) <u>Weather</u>. *Describe the aspects of space and terrestrial weather that impact space operations. Refer to Annex B (Intelligence) as required.*

c. (U) <u>Enemy Forces</u>. *List known locations and activities of enemy space capable assets and units. List enemy space capabilities that can impact friendly operations. State expected enemy courses of action and employment of enemy and commercial space assets. Refer to Annex B (Intelligence) as required.*

d. (U) <u>Friendly Forces</u>. *Outline the higher headquarters' plan for space operations and space support teams including but not limited to space support elements, Army space support teams, and an organic space weapons officer. List designation, location, and outline of plans of higher, adjacent, and other space operations-related assets that support or impact the issuing headquarters or require coordination and additional support. For example, the space coordinating authority and specified processes established for the area of responsibility.*

e. (U) <u>Interagency, Intergovernmental, and Nongovernmental Organizations</u>. *Identify and describe other organizations in the area of operations that may impact the conduct of space operations or implementation of space-specific equipment, tactics, and capabilities. Consider all multinational, civil, and nongovernmental organizations such as civilian relief agencies and other customers and providers of space-based capabilities. Refer to Annex V (Interagency Coordination) as required.*

f. (U) <u>Civil Considerations</u>. *Describe the aspects of the civil situation that impact space operations. Refer to Annex B (Intelligence) and Annex K (Civil Affairs Operations) as required.*

g. (U) <u>Attachments and Detachments</u>. *List units attached or detached only as necessary to clarify task organization. Refer to Annex A (Task Organization) as required.*

h. (U) <u>Assumptions</u>. *List space operations-specific assumptions that support the annex development.*

2. (U) <u>Mission</u>. *State the mission of space operations in support of the base plan or order.*

3. (U) <u>Execution</u>.

a. (U) <u>Scheme of Space Operations</u>. *Describe how space capabilities support the commander's intent and concept of operations. Establish the priorities of space support to units for each phase of the operation. For example, electromagnetic interference resolution and defended asset list. Also address unique space reliances or vulnerabilities related to unit systems and capabilities. Refer to Annex C (Operations) as required.*

(1) (U) <u>Space Force Enhancement</u>. *Identify space activities required to support the operation plan, including the following specific areas as applicable:*

[page number]
[CLASSIFICATION]

Figure D-13. Sample Annex N (Space Operations) format (continued)

[CLASSIFICATION]

ANNEX N (SPACE OPERATIONS) TO OPLAN/OPORD [number] [(code name)]—[issuing headquarters] [(classification of title)]

(a) (U) <u>Satellite Communication</u>. *Describe the space operations communications plan. Ensure defensive space priorities for satellite communication links are established and coordinated based on operational priorities. Refer to Annex H (Signal) as required.*

(b) (U) <u>Remote Sensing/Environmental Monitoring</u>. *Identify and list meteorological, oceanographic, geodetic, and other environmental support information provided by space assets that affect space, air, surface, or subsurface activities and assets. Refer to Annex G (Engineer) as required.*

(c) (U) <u>Position, Navigation, and Timing</u>. *Provide navigational capabilities that would aid the transit of ships, aircraft, personnel, or ground vehicles and determine the course and distance traveled or position location. Provide global positioning system (GPS) accuracy to support GPS-aided munitions.*

(d) (U) <u>Information Collection</u>. *Provide information pertaining to friendly and enemy forces in or external to the operational areas that would aid in operations and force positioning. Refer to Annex L (Information Collection) as required.*

(e) (U) <u>Theater Missile Warning</u>. *Provide information on the notification of enemy ballistic missile or space-weapon attacks evaluated from available sensor and intelligence sources and the possible affect on the operational area. Provide notification of friendly ballistic missile launches and the impacts on the operational areas that would require early warning of affected friendly forces and an estimated point of impact for each launch. Establish provisions, in coordination with the air defense artillery officer, to disseminate information quickly throughout the operational areas. Refer to Annex B (Intelligence), Annex D (Fires), and Annex E (Protection) as required.*

(2) (U) <u>Space Control</u>. *Provide information on space-related activities, whether performed by space, air, or surface assets that ensure friendly forces and deny enemy forces the unrestricted use of space and space assets. Identify targetable enemy assets and limitations of targeting. Address all capabilities and effects related to offensive or defensive space control and space situational understanding requirements.*

(3) (U) <u>Nuclear Detonation</u>. *Provide information on the notification of detected nuclear detonations that might affect the operation and require evaluation as to yield and location. Refer to Annex B (Intelligence) as required.*

(4) (U) <u>Cyber Electromagnetic Activities</u>. *Integrate cyber electromagnetic activities to optimally synchronize their effects. Refer to Annex C (Operations) as required.*

(5) (U) <u>Special Technical Programs</u>. *Provide information on the organization and synchronization of the integrated Army and integrated joint special technical operations and alternate compensatory control measures plans in support of the commander's objectives. Refer to Annex S (Special Technical Operations) as required.*

(6) (U) <u>Mission Command</u>. *Provide information and an assessment on friendly space reliances upon satellite communications, missile warning, and network architectures. Determine how organic unit systems and equipment rely upon these communications paths (architectures).*

[page number]
[CLASSIFICATION]

Figure D-13. Sample Annex N (Space Operations) format (continued)

[CLASSIFICATION]

ANNEX N (SPACE OPERATIONS) TO OPLAN/OPORD [number] [(code name)]—[issuing headquarters] [(classification of title)]

b. (U) <u>Tasks to Subordinate Units</u>. *List space tasks assigned to specific subordinate units not contained in the base plan or order. Refer to any tasks in base order.*

c. (U) <u>Coordinating Instructions</u>. *List only instructions applicable to two or more subordinate units not covered in the base plan or order. Document coordination and reachback support requests in accordance with space coordinating authority guidance such as "Space Coordinating Plans" and other directives for the area of responsibility; include unique equipment sustainment and technical points of contact.*

4. (U) <u>Sustainment</u>. *Identify priorities of sustainment for space operations key tasks and specify additional instructions as required. Refer to Annex F (Sustainment) as required.*

a. (U) <u>Logistics</u>. *Identify unique sustainment requirements, procedures, and guidance to support space operations teams and operations. Specify procedures for specialized technical logistics support from external organizations as necessary. Use subparagraphs to identify priorities and specific instructions for space operations logistics support. Refer to Annex F (Sustainment) and Annex P (Host-Nation Support) as required.*

b. (U) <u>Personnel</u>. *Use subparagraphs to identify priorities and specific instructions for human resources support, financial management, legal support, and religious support. Refer to Annex F (Sustainment) as required.*

c. (U) <u>Health Service Support</u>. *Identify availability, priorities, and instructions for medical care. Refer to Annex F (Sustainment) as required.*

5. (U) <u>Command and Signal</u>.

a. (U) **<u>Command</u>.**

(1) (U) <u>Location of the Commander and Key Leaders</u>. *State the location of the commander and key space leaders such as the space coordinating authority, director of space forces, Joint Space Operations Center, electronic warfare officers, and other key reachback leaders.*

(2) (U) <u>Succession of Command</u>. *State the succession of command if not covered in the unit's SOPs.*

(3) (U) <u>Liaison Requirements</u>. *State the space liaison requirements not covered in the unit's standard operating procedures (SOPs), such as air component coordination element or multinational space officers.*

b. (U) <u>Control</u>.

(1) (U) <u>Command Posts</u>. *Describe the employment of space command, control, and functional chains including their location and contact information.*

(2) (U) <u>Reports</u>. *List space related reports not covered in SOPs. Refer to any space coordinating authority concept of operations or guidance and Annex R (Reports) as required.*

c. (U) <u>Signal</u>. *Address any space-specific communications requirements such as secure chat communications applications. These often require a lengthy approval process to tunnel through existing networks and should be specified well in advance. Refer to Annex H (Signal) as required.*

ACKNOWLEDGE: *Include only if attachment is distributed separately from the base order.*

[Commander's last name]
[Commander's rank]

[page number]
[CLASSIFICATION]

Figure D-13. Sample Annex N (Space Operations) format (continued)

[CLASSIFICATION]

ANNEX N (SPACE OPERATIONS) TO OPLAN/OPORD [number] [(code name)]—[issuing headquarters] [(classification of title)]

The commander or authorized representative signs the original copy of attachment. If the representative signs the original, add the phrase "For the Commander." The signed copy is the historical copy and remains in the headquarters' files.

OFFICIAL:

[Authenticator's name]
[Authenticator's position]

Use only if the commander does not sign the original attachment. If the commander signs the original, no further authentication is required. If the commander does not sign, the signature of the preparing staff officer requires authentication and only the last name and rank of the commander appear in the signature block.

ATTACHMENT: *List lower-level attachments (appendixes, tabs, and exhibits).*

DISTRIBUTION: *Show only if distributed separately from the base order or higher-level attachments.*

[page number]
[CLASSIFICATION]

Figure D-13. Sample Annex N (Space Operations) format (continued)

Annex O (Not Used)

ANNEX P (HOST-NATION SUPPORT) FORMAT AND INSTRUCTIONS

D-54. This annex provides fundamental considerations, formats, and instructions for developing Annex P (Host-Nation Support) to the base plan or order. This annex uses the five-paragraph attachment format.

D-55. Commanders and staffs use Annex P (Host-Nation Support) to describe how sustainment operations support the concept of operations described in the base plan or order. The G-4 (S-4) is the staff officer responsible for Annex P (Host-Nation Support).

D-56. Host-nation support is the civil and military assistance provided by the host nation to the forces located in or transiting through that host nation's territory. Efficient use of available host-nation support can greatly aid forces and augment the deployed sustainment structure. (See figure D-14 on pages D-77 through D-81.)

[CLASSIFICATION]

Place the classification at the top and bottom of every page of the attachments. Place the classification marking at the front of each paragraph and subparagraph in parentheses. Refer to AR 380-5 for classification and release marking instructions.

Copy ## of ## copies
Issuing headquarters
Place of issue
Date-time group of signature
Message reference number

Include heading if attachment is distributed separately from the base order or higher-level attachment.

ANNEX P (HOST-NATION SUPPORT) TO OPLAN/OPORD [number] [(code name)]—[issuing headquarters] [(classification of title)]

(U) References: *List documents essential to understanding the attachment.*

   a. *List maps and charts first. Map entries include series number, country, sheet names or numbers, edition, and scale.*

   b. *List other references in subparagraphs labeled as shown.*

   c. *Doctrinal references for host-nation support include FM 3-16 and FM 6-0.*

(U) Time Zone Used Throughout the Order: *Write the time zone established in the base plan or order.*

1. (U) Situation. *Include information affecting host-nation support that paragraph 1 of the OPLAN or OPORD does not cover or that needs expansion.*

   a. (U) Area of Interest. *Describe the area of interest as it relates to host-nation support. Refer to Annex B (Intelligence) as required.*

[page number]
[CLASSIFICATION]

Figure D-14. Sample Annex P (Host-Nation Support) format

[CLASSIFICATION]

ANNEX P (HOST-NATION SUPPORT) TO OPLAN/OPORD [number] [(code name)]—[issuing headquarters] [(classification of title)]

b. (U) <u>Area of Operations</u>. *Refer to Appendix 2 (Operation Overlay) to Annex C (Operations).*

(1) (U) <u>Terrain</u>. *Describe the aspects of terrain that impact host-nation support operations. Refer to Annex B (Intelligence) as required.*

(2) (U) <u>Weather</u>. *Describe the aspects of weather that impact host-nation support operations. Refer to Annex B (Intelligence) as required.*

c. (U) <u>Enemy Forces</u>. *List known and templated locations and activities of enemy host-nation support for one echelon up and two echelons down. List enemy maneuver and other area capabilities that will impact friendly host-nation support operations. State expected enemy courses of action and employment of enemy host-nation support assets. Refer to Annex B (Intelligence) as required.*

d. (U) <u>Friendly Forces</u>. *Outline the higher headquarters' host-nation support plan. List designation, location, and outline of plans of higher, adjacent, and other host-nation support assets that support or impact the issuing headquarters or require coordination and additional support.*

e. (U) <u>Interagency, Intergovernmental, and Nongovernmental Organizations</u>. *Identify and describe other organizations in the area of operations that may impact the conduct of host-nation support operations or implementation of host-nation support-specific equipment and tactics. Refer to Annex V (Interagency Coordination) as required.*

f. (U) <u>Civil Considerations</u>. *Describe the aspects of the civil situation that impact host-nation support operations. Refer to Annex B (Intelligence) and Annex K (Civil Affairs Operations) as required.*

g. (U) <u>Attachments and Detachments</u>. *List units attached or detached only as necessary to clarify task organization. Refer to Annex A (Task Organization) as required.*

h. (U) <u>Assumptions</u>. *List any host-nation support-specific assumptions that support the annex development. State assumptions concerning host-nation support and the operational impact if the assumptions are inaccurate.*

i. (U) <u>Host-Nation Support Agreements</u>. *List host-nation support agreements, unreliable or doubtful agreements, and presumed host-nation support agreements.*

2. (U) <u>Mission</u>. *State the mission of host-nation support in support of the base plan or order.*

3. (U) <u>Execution</u>.

a. (U) <u>Scheme of Host-Nation Support</u>. *Describe how the commander's intent and concept of operations is supported by host-nation support. Cover the overall status of negotiations and agreements, including customs requirements, by country or treaty organization, presumed host-nation support, and the reliability of host-nation support. Identify peacetime and pre-conflict military information support operations that would develop support in foreign countries for the provision of host-nation support. Establish the priorities of support to units for each phase of the operation. Refer to Annex C (Operations) as required.*

b. (U) <u>Host Nation Support Considerations</u>. *The subparagraphs below are not an all inclusive list. Each host-nation agreement is unique. Refer to Annex F (Sustainment) as required.*

(1) (U) <u>Accommodations</u>. *Describe host-nation accommodation considerations for the following: billeting; offices; stores and warehouses; workshops, vehicle parks, gun parks; medical; hardstands; fuel; weapons and ammunition; transportation including aircraft; firing ranges; training areas and facilities; recreational areas and facilities; and laundry and dry-cleaning facilities.*

[page number]
[CLASSIFICATION]

Figure D-14. Sample Annex P (Host-Nation Support) format (continued)

|CLASSIFICATION|

ANNEX P (HOST-NATION SUPPORT) TO OPLAN/OPORD |number| |(code name)|—|issuing headquarters| |(classification of title)|

(2) (U) <u>Ammunition and Weapons</u>. *Describe host-nation considerations for ammunition and weapons security, storage, and collection or delivery.*

(3) (U) <u>Communications</u>. *Describe host-nation considerations for local and international communications and security.*

(4) (U) <u>Finance</u>. *Describe host-nation considerations and payment for accommodations, supplies, communications, equipment, local labor, maintenance, medical treatment facilities, movement facilities, emergency facilities, and personnel facilities.*

(5) (U) <u>Fuel</u>. *Describe host-nation fuel considerations for aircraft, vehicles, ships, methods of delivery, storage, interoperability of refueling equipment, and common use of refueling installations.*

(6) (U) <u>Local Labor</u>. *Describe host-nation local labor considerations for method of hiring, method of payment, and administration.*

(7) (U) <u>Maintenance</u>. *Describe host-nation maintenance considerations for accommodations, vehicles, ships, equipment, roads, fixed and rotary wing aircraft, provision of assembly areas, damage control, emergency facilities for visitors' vehicles and equipment, and evacuation of disabled vehicles and equipment.*

(8) (U) <u>Medical</u>. *Describe host-nation medical considerations for medical treatment facilities, emergency facilities, reciprocal national health agreements, and availability of medical equipment and supplies, standards of care, public health facilities, accessibility to care, and medical and casualty evacuation.*

(9) (U) <u>Movement</u>. *Describe host-nation movement considerations for airheads (facilities, alternates, equipment, and refueling), ports (facilities, alternates, ships, draft, bunkering/fueling, and repair), road and rail movement (personnel, equipment, security, and traffic control), and pipeline movement.*

(10) (U) <u>Rations</u>. *Describe host-nation rations considerations for fresh food, packaged foods, and potable water.*

(11) (U) <u>Supplies and Equipment</u>. *Describe host-nation supplies and equipment considerations for common use items other than ammunition, fuel, or rations.*

(12) (U) <u>Translation</u>. *Describe host-nation translation considerations for interpreters, linguists, language specialists, and document translation.*

(13) (U) <u>Transportation Equipment</u>. *Describe host-nation transportation equipment considerations for use of host-nation military vehicles, equipment, ships, and aircraft; locally hired vehicles and equipment, ships, and aircraft; and the policy on drivers and handlers of the military and locally hired vehicles.*

(14) (U) <u>Water</u>. *Describe host-nation water considerations for production and purification capability (municipal and other water treatment systems), distribution capability (trucks, pipeline, and hose line), storage capability, receipt and issue capability, available water sources (wells, surface, and subsurface), and host-nation water quality standards.*

c. (U) <u>Tasks to Subordinate Units</u>. *List host-nation support tasks assigned to specific subordinate units not contained in the base order. Identify the office of primary responsibility for each type of host-nation support managed separately within the command.*

d. (U) <u>Coordinating Instructions</u>. *List only instructions applicable to two or more subordinate units not covered in the base plan or order.*

|page number|
|CLASSIFICATION|

Figure D-14. Sample Annex P (Host-Nation Support) format (continued)

[CLASSIFICATION]

ANNEX P (HOST-NATION SUPPORT) TO OPLAN/OPORD [number] [(code name)]—[issuing headquarters] [(classification of title)]

4. (U) Sustainment. *Identify priorities of sustainment for host-nation support key tasks and specify additional instructions as required. Outline support limitations that are due to lack of host-nation water agreements, operational impact, status of any current negotiations, and prospects for availability of the required support on a emergency basis. Refer to Annex F (Sustainment) as required.*

a. (U) Logistics. *Identify unique sustainment requirements, procedures, and guidance to support host-nation support teams and operations. Specify procedures for specialized technical logistics support from external organizations as necessary. Use subparagraphs to identify priorities and specific instructions for host-nation logistics support. Refer to Annex F (Sustainment) as required.*

b. (U) Personnel. *Identify host-nation support unique personnel requirements and concerns, including global sourcing support and contracted linguist requirements. Use subparagraphs to identify priorities and specific instructions for human resources support, financial management, legal support, and religious support. Refer to Annex F (Sustainment) as required.*

c. (U) Health Service Support. *Identify availability, priorities, and instructions for medical care. Determine if locally available medical supplies and equipment meet U.S. and Food and Drug Administration standards for use with U.S. Forces. Determine if the host-nation blood supply is tested and considered safe (if not, where will blood products be obtained), and determine the availability of medical equipment repairers. Refer to Annex F (Sustainment) as required.*

5. (U) Command and Signal.

a. (U) Command.

(1) (U) Location of the Commander and Key Leaders. *State the location of the commander and key host-nation support leaders.*

(2) (U) Succession of Command. *State the succession of command if not covered in the unit's standard operating procedures (SOPs).*

(3) (U) Liaison Requirements. *State the host-nation support liaison requirements not covered in the base order.*

b. (U) Control.

(1) (U) Command Posts. *Describe the employment of host-nation support-specific command posts (CPs), including the location of each CP and its time of opening and closing.*

(2) (U) Reports. *List host-nation support-specific reports not covered in SOPs. Refer to Annex R (Reports) as required.*

c. (U) Signal. *Address any host-nation support-specific communications requirements or reports. Refer to Annex H (Signal) as required.*

ACKNOWLEDGE: *Include only if attachment is distributed separately from the base order.*

[Commander's last name]
[Commander's rank]

The commander or authorized representative signs the original copy of attachment. If the representative signs the original, add the phrase "For the Commander." The signed copy is the historical copy and remains in the headquarters' files.

[page number]
[CLASSIFICATION]

Figure D-14. Sample Annex P (Host-Nation Support) format (continued)

[CLASSIFICATION]

ANNEX P (HOST-NATION SUPPORT) TO OPLAN/OPORD [number] [(code name)]—[issuing headquarters] [(classification of title)]

OFFICIAL:

[Authenticator's name]
[Authenticator's position]

Use only if the commander does not sign the original attachment. If the commander signs the original, no further authentication is required. If the commander does not sign, the signature of the preparing staff officer requires authentication and only the last name and rank of the commander appear in the signature block.

ATTACHMENT: *List lower-level attachments (appendixes, tabs, and exhibits).*

DISTRIBUTION: *Show only if distributed separately from the base order or higher-level attachments.*
[page number]
[CLASSIFICATION]

Figure D-14. Sample Annex P (Host-Nation Support) format (continued)

ANNEX Q (KNOWLEDGE MANAGEMENT) FORMAT AND INSTRUCTIONS

D-57. This annex provides a format for the knowledge management annex. This annex describes how knowledge management supports the commander's intent and concept of operations. It also describes how knowledge management creates shared understanding through the alignment of people, processes, and tools within the organizational structure. The knowledge management officer is responsible for this annex. This annex uses the five-paragraph attachment format. (See figure D-15 on pages D-82 through D-84.)

[CLASSIFICATION]

Place the classification at the top and bottom of every page of the attachments. Place the classification marking at the front of each paragraph and subparagraph in parentheses. Refer to AR 380-5 for classification and release marking instructions.

Copy ## of ## copies
Issuing headquarters
Place of issue
Date-time group of signature
Message reference number

Include the full heading if attachment is distributed separately from the base order or higher-level attachment.

ANNEX Q (KNOWLEDGE MANAGEMENT) TO OPERATION PLAN/ORDER [number] [(code name)]—[issuing headquarters] [(classification of title)]

(U) References: *List documents essential to understanding the attachment.*

    a. *List maps and charts first. Map entries include series number, country, sheet names or numbers, edition, and scale.*

    b. *List other references in subparagraphs labeled as shown.*

    c. *Doctrinal References for this annex include the following: ADRP 3-0, ADRP 5-0, ADRP 6-0, FM 6-0, and FM 6-01.1.*

(U) Time Zone Used Throughout the Plan/Order: *Write the time zone established in the base plan or order.*

1. (U) Situation. *Include information affecting the functional area that paragraph 1 of the OPLAN or OPORD does not cover or needs to be expanded.*

    a. (U) Area of Interest. *Describe the area of interest as it relates to knowledge management. Refer to Annex B (Intelligence) as required.*

    b. (U) Area of Operations. *Refer to Appendix 2 (Operation Overlay) to Annex C (Operations) as required.*

[page number]
[CLASSIFICATION]

Figure D-15. Sample Annex Q (Knowledge Management) format

[CLASSIFICATION]
ANNEX Q (KNOWLEDGE MANAGEMENT) TO OPERATION PLAN/ORDER [number] [(code name)]—[issuing headquarters] [(classification of title)]

c. (U) <u>Enemy Forces</u>. *Refer to Annex B (Intelligence) as required.*

d. (U) <u>Friendly Forces</u>. *Outline the knowledge management and information management structure, including higher headquarters. This will include the joint force commander involved with the operation.*

e. (U) <u>Interagency, Intergovernmental, and Nongovernmental Organizations</u>. *Identify and describe other organizations in the area of operations that may impact knowledge management (data sharing and collaboration capabilities). Refer to Annex V (Interagency Coordination) as required.*

f. (U) <u>Civil Considerations</u>. *Refer to Annex K (Civil Affairs Operations) as required.*

g. (U) <u>Attachments and Detachments</u>. *List units and capabilities attached or detached only as necessary to clarify task organization and knowledge management and information management. Refer to Annex A (Task Organization) as required.*

h. (U) <u>Assumptions</u>. *List any knowledge management integration assumptions that support the annex development.*

2. (U) <u>Mission</u>. *State the mission of knowledge management in support of the base plan or order.*

3. (U) <u>Execution</u>.

a. (U) <u>Scheme of Knowledge Management Support</u>. *Describe how knowledge management supports the commander's intent and concept of operations. Describe how knowledge management will create shared understanding through the alignment of people, processes, and tools within the organizational structure and culture in order to increase collaboration and interaction between leaders and subordinates, enabling decisions through improved flexibility, adaptability, integration, and synchronization to achieve a position of relative advantage. Describe how knowledge management enhances shared understanding, learning, and decisionmaking during the phases of the operation. Specify the authority exercised at each echelon for each phase of the operation. Describe the roles and relationships between knowledge management elements in the organization and how they will coordinate with joint, combined, and intergovernmental knowledge management elements. Describe how units' knowledge management elements and assets are integrated into the unit battle rhythm, operations process, and during execution.*

b. (U) <u>Tasks to Subordinate Units</u>. *List knowledge management critical tasks assigned to subordinate units not contained in the base plan or order. This may include tasks to combat units and other functional organizations.*

c. (U) <u>Coordinating Instructions</u>. *List only instructions applicable to two or more subordinate units not covered in the base order that affect knowledge management procedures (for example, commander's critical information requirements).*

4. (U) <u>Sustainment</u>. *Identify and list sustainment priorities for knowledge management key tasks and specify additional sustainment instructions as necessary, to include contractor support. Refer to Annex F (Sustainment) as required.*

a. (U) <u>Logistics</u>. *Identify unique sustainment requirements, procedures, and guidance to support knowledge management. Specify procedures for specialized technical logistics support from external organizations as necessary. Use subparagraphs to identify priorities and specific instructions for knowledge management logistics support. Refer to Annex F (Sustainment) and Annex P (Host-Nation Support) as required.*

[page number]
[CLASSIFICATION]

Figure D-15. Sample Annex Q (Knowledge Management) format (continued)

|CLASSIFICATION|

ANNEX Q (KNOWLEDGE MANAGEMENT) TO OPERATION PLAN/ORDER |number| |(code name)|—|issuing headquarters| |(classification of title)|

b. (U) <u>Personnel</u>. *Identify knowledge management unique personnel requirements and concerns, including global sourcing support and contracted linguist requirements. Use subparagraphs to identify priorities and specific instructions for human resources support, financial management, legal support, and religious support. Refer to Annex F (Sustainment) as required.*

c. (U) <u>Health Service Support</u>. *Identify availability, priorities, and instructions for medical care. Identify medical-unique automation requirements for medical records and other medical documentation and support requirements for medical units. Refer to Annex F (Sustainment) as required.*

5. (U) <u>Command and Signal</u>.

a. (U) <u>Command</u>.

(1) (U) <u>Location of the Commander and Key Leaders</u>. *State the location of the commander and key knowledge management leaders. Indentify who is authorized to make knowledge management decisions for the commander.*

(2) (U) <u>Succession of Command</u>. *State the succession of command if not covered in the unit's standard operating procedures (SOPs).*

(3) (U) <u>Liaison Requirements</u>. *State the knowledge management liaison requirements not covered in the base order.*

b. (U) <u>Control</u>.

(1) (U) <u>Command Posts</u>. *Describe the employment of knowledge management-specific command posts (CPs), including the location of each CP and its time of opening and closing.*

(2) (U) <u>Reports</u>. *List knowledge management support-specific reports not covered in SOPs. Refer to Annex R (Reports) as required.*

c. (U) <u>Signal</u>. *Address any knowledge management support-specific communications requirements or reports. Refer to Annex H (Signal) as required.*

ACKNOWLEDGE: *Include only if attachment is distributed separately from the base order.*

[Commander's last name]

[Commander's rank]

The commander or authorized representative signs the original copy of the attachment. If the representative signs the original, add the phrase "For the Commander." The signed copy is the historical copy and remains in the headquarters' files.

OFFICIAL:

[Authenticator's name]
[Authenticator's position]

Use only if the commander does not sign the original attachment. If the commander signs the original, no further authentication is required. If the commander does not sign, the signature of the preparing staff officer requires authentication and only the last name and rank of the commander appear in the signature block.

ATTACHMENTS: *List lower-level attachment (appendixes, tabs, and exhibits).*

Appendix 1–Knowledge Management Decision Support Matrix
Appendix 2–Common Operational Picture Configuration Matrix
Appendix 3–Mission Command Information Systems Integration Matrix
Appendix 4–Content Management
Appendix 5–Battle Rhythm

DISTRIBUTION: *Show only if distributed separately from the base order or higher-level attachments.*

|page number|
|CLASSIFICATION|

Figure D-15. Sample Annex Q (Knowledge Management) format (continued)

ANNEX R (REPORTS) ANNEX FORMAT AND INSTRUCTIONS

D-58. This annex provides fundamental considerations, formats, and instructions for developing Annex R (Reports) to the base plan or order. This annex does not follow the five-paragraph attachment format. Unit SOPs will dictate the development and format for this annex.

D-59. Commanders and staffs use Annex R (Reports) to list and catalog all unit reports and their respective formats. The G-3 (S-3) or G-5 (S-5), in coordination with the knowledge management officer, develops Annex R (Reports). (See figure D-16 on pages D-85 through D-86.)

[CLASSIFICATION]

Place the classification at the top and bottom of every page of the attachments. Place the classification marking at the front of each paragraph and subparagraph in parentheses. Refer to AR 380-5 for classification and release marking instructions.

Copy ## of ## copies
Issuing headquarters
Place of issue
Date-time group of signature
Message reference number

Include heading if attachment is distributed separately from the base order or higher-level attachment.

ANNEX R (REPORTS) TO OPERATION PLAN/ORDER [number] [(code name)]—[issuing headquarters] [(classification of title)]

(U) References: *List documents essential to understanding the attachment.*

a. *List maps and charts first. Map entries include series number, country, sheet names or numbers, edition, and scale.*

b. *List other references in subparagraphs labeled as shown.*

c. *Doctrinal references for this annex include FM 6-0 and FM 6-99.*

(U) Time Zone Used Throughout the Order: *Write the time zone established in the base plan or order.*

(U) Reports. *List all reports (formats, submission standards and times) not covered in unit standard operating procedures. Specify reporting requirements for all assigned, attached, operational control, and tactical control command relationships.*

ACKNOWLEDGE: *Include only if attachment is distributed separately from the base order.*

[Commander's last name]
[Commander's rank]

[page number]
[CLASSIFICATION]

Figure D-16. Sample Annex R (Reports) format

[CLASSIFICATION]

ANNEX R (REPORTS) TO OPERATION PLAN/ORDER [number] [(code name)]—[issuing headquarters] [(classification of title)]

The commander or authorized representative signs the original copy of attachment. If the representative signs the original, add the phrase "For the Commander." The signed copy is the historical copy and remains in the headquarters' files.

OFFICIAL:

[Authenticator's name]
[Authenticator's position]

Use only if the commander does not sign the original attachment. If the commander signs the original, no further authentication is required. If the commander does not sign, the signature of the preparing staff officer requires authentication and only the last name and rank of the commander appear in the signature block.

ATTACHMENTS: *List lower-level attachment (appendixes, tabs, and exhibits).*

DISTRIBUTION: *Show only if distributed separately from the base order or higher-level attachments.*

[page number]
[CLASSIFICATION]

Figure D-16. Sample Annex R (Reports) format (continued)

ANNEX S (SPECIAL TECHNICAL OPERATIONS) FORMAT AND INSTRUCTIONS

D-60. This annex provides fundamental considerations, formats, and instructions for developing Annex S (Special Technical Operations) to the base plan or order. This annex follows the five-paragraph attachment format.

D-61. Commanders and staffs use Annex S (Special Technical Operations) to expand the plan or order and provide the mission, scheme, and tasks to units for special technical operations. The special technical operations officer is the staff officer responsible for developing Annex S (Special Technical Operations). Due to classification, this annex may be produced separately from the base order and other annexes, with access restricted to personnel authorized to view its content. (See figure D-17 on pages D-87 through D-90.)

[CLASSIFICATION]

Place the classification at the top and bottom of every page of the attachments. Place the classification marking at the front of each paragraph and subparagraph in parentheses. Refer to AR 380-5 for classification and release marking instructions.

Copy ## of ## copies
Issuing headquarters
Place of issue
Date-time group of signature
Message reference number

Include heading if attachment is distributed separately from the base order or higher-level attachment.

ANNEX S (SPECIAL TECHNICAL OPERATIONS) TO OPLAN/OPORD [number] [(code name)]—[issuing headquarters] [(classification of title)]

(U) References: *List documents essential to understanding the attachment.*

   a. List maps and charts first. Map entries include series number, country, sheet names or numbers, edition, and scale.

   b. List other references in subparagraphs labeled as shown.

   c. Doctrinal references for this annex include the CJCSM 3122 series and FM 6-0.

(U) Time Zone Used Throughout the Order: *Write the time zone established in the base plan or order.*

1. (U) <u>Situation</u>. *Include information affecting special technical operations that paragraph 1 of the OPLAN or OPORD does not cover or that needs expansion.*

   a. (U) <u>Area of Interest</u>. *Describe the area of interest as it relates to special technical operations. Refer to Annex B (Intelligence) as required.*

   b. (U) <u>Area of Operations</u>. *Refer to Appendix 2 (Operation Overlay) to Annex C (Operations).*

[page number]
[CLASSIFICATION]

Figure D-17. Sample Annex S (Special Technical Operations) format

|CLASSIFICATION|

ANNEX S (SPECIAL TECHNICAL OPERATIONS) TO OPLAN/OPORD |number| |(code name)|—|issuing headquarters| |(classification of title)|

(1) (U) <u>Terrain</u>. *Describe the aspects of terrain that impact special technical operations. Refer to Annex B (Intelligence) as required.*

(2) (U) <u>Weather</u>. *Describe the aspects of weather that impact special technical operations. Refer to Annex B (Intelligence) as required.*

c. (U) <u>Enemy Forces</u>. *List known and templated locations and activities of enemy special technical operations units for one echelon up and two echelons down. List enemy maneuver and other area capabilities that will impact friendly operations. State expected enemy courses of action and employment of enemy special technical operations assets. Refer to Annex B (Intelligence) as required.*

d. (U) <u>Friendly Forces</u>. *Outline the higher headquarters' special technical operation plan. List designation, location, and outline of plans of higher, adjacent, and other special technical operations assets that support or impact the issuing headquarters or require coordination and additional support.*

e. (U) <u>Interagency, Intergovernmental, and Nongovernmental Organizations</u>. *Identify and describe other organizations in the area of operations that may impact the conduct of special technical operations. Refer to Annex V (Interagency Coordination) as required.*

f. (U) <u>Civil Considerations</u>. *Describe the aspects of the civil situation that impact special technical operations. Refer to Annex B (Intelligence) and Annex K (Civil Affairs Operations) as required.*

g. (U) <u>Attachments and Detachments</u>. *List units attached or detached only as necessary to clarify task organization. Refer to Annex A (Task Organization) as required.*

h. (U) <u>Assumptions</u>. *List any special technical operations-specific assumptions that support the annex development.*

2. (U) <u>Mission</u>. *State the mission of special technical operations in support of the base plan or order.*

3. (U) <u>Execution</u>.

a. (U) <u>Scheme of Special Technical Operations</u>. *Describe how the special technical operations support the commander's intent and concept of operations. List and describe the commander's objective for each special technical operations target set or functional area in separately numbered subparagraphs. Establish the priorities of support to units for each phase of the operation. Refer to Annex C (Operations) as required.*

(1) (U) <u>Capabilities Integration Matrix</u>. *Refer to Appendix 1 (Special Technical Operations Capabilities Integration Matrix) to Annex S (Special Technical Operations) as required.*

(2) (U) <u>Objective for Functional Area I</u>. *Describe commander's objective for this functional area. Refer to Appendix 2 (Functional Area I Program and Objectives) to Annex S (Special Technical Operations) as required.*

(3) (U) <u>Objective for Functional Area II</u>. *Describe commander's objective for this functional area. Refer to Appendix 3 (Functional Area II Program and Objectives) to Annex S (Special Technical Operations) as required.*

b. (U) <u>Tasks to Subordinate Units</u>. *List special technical operations tasks assigned to specific subordinate units not contained in the base order.*

c. (U) <u>Coordinating Instructions</u>. *List only instructions applicable to two or more subordinate units not covered in the base order.*

page number|
|CLASSIFICATION|

Figure D-17. Sample Annex S (Special Technical Operations) format (continued)

[CLASSIFICATION]

ANNEX S (SPECIAL TECHNICAL OPERATIONS) TO OPLAN/OPORD [number] [(code name)]—[issuing headquarters] [(classification of title)]

4. (U) <u>Sustainment</u>. *Identify priorities of sustainment for special technical operations key tasks and specify additional instructions as required. Provide general instructions concerning the movement, support, and maintenance of special technical operations capabilities. Provide additional information on equipment to support special technical operation planning and operations. Provide any additional guidance on special technical operations-specific administrative matters. Refer to Annex F (Sustainment) as required.*

a. (U) <u>Logistics</u>. *Identify unique sustainment requirements, procedures, and guidance to support special technical operations teams and operations. Specify procedures for specialized technical logistics support from external organizations as necessary. Use subparagraphs to identify priorities and specific instructions for special technical operations logistics support. Refer to Annex F (Sustainment) and Annex P (Host-Nation Support) as required.*

b. (U) <u>Personnel</u>. *Use subparagraphs to identify priorities and specific instructions for human resources support, financial management, legal support, and religious support. Refer to Annex F (Sustainment) as required.*

c. (U) <u>Health Service Support</u>. *Identify availability, priorities, and instructions for medical care. Refer to Annex F (Sustainment) as required.*

5. (U) <u>Command and Signal</u>.

a. (U) <u>Command</u>.

(1) (U) <u>Location of the Commander and Key Leaders</u>. *State the location of the commander and key special technical operations leaders. Provide guidance on specific approval authorities for deployment and employment of special technical operations capabilities.*

(2) (U) <u>Succession of Command</u>. *State the succession of command if not covered in the unit's standard operating procedures.*

(3) (U) <u>Liaison Requirements</u>. *State the special technical operations liaison requirements not covered in the base order.*

b. (U) <u>Control</u>.

(1) (U) <u>Command Posts</u>. *Describe the employment of special technical operations-specific command posts (CPs), including the location of each CP and its time of opening and closing.*

(2) (U) <u>Reports</u>. *List special technical operations-specific reports not covered in standard operating procedures. Refer to Annex R (Reports) as required.*

c. (U) <u>Signal</u>. *Address any special technical operations-specific communications requirements or reports. Provide guidance on the communication methods authorized to transmit planning, coordination, deconfliction, deployment, and employment information for special technical operations capabilities included in this annex. Refer to Annex H (Signal) as required.*

ACKNOWLEDGE: *Include only if attachment is distributed separately from the base order.*

[Commander's last name]
[Commander's rank]

[page number]
[CLASSIFICATION]

Figure D-17. Sample Annex S (Special Technical Operations) format (continued)

[CLASSIFICATION]

ANNEX S (SPECIAL TECHNICAL OPERATIONS) TO OPLAN/OPORD [number] [(code name)]—[issuing headquarters] [(classification of title)]

The commander or authorized representative signs the original copy of attachment. If the representative signs the original, add the phrase "For the Commander." The signed copy is the historical copy and remains in the headquarters' files.

OFFICIAL:

[Authenticator's name]
[Authenticator's position]

Use only if the commander does not sign the original attachment. If the commander signs the original, no further authentication is required. If the commander does not sign, the signature of the preparing staff officer requires authentication and only the last name and rank of the commander appear in the signature block.

ATTACHMENT: *List lower-level attachments (appendixes, tabs, and exhibits).*

Appendix 1–Special Technical Operations Capabilities Integration Matrix
Appendix 2–Functional Area I Program and Objectives
Appendix 3–Functional Area II Program and Objectives

DISTRIBUTION: *Show only if distributed separately from the base order or higher-level attachments.*

[page number]
[CLASSIFICATION]

Figure D-17. Sample Annex S (Special Technical Operations) format (continued)

Annex T (Spare)

ANNEX U (INSPECTOR GENERAL) FORMAT AND INSTRUCTIONS

D-62. This annex provides fundamental considerations, formats, and instructions for developing Annex U (Inspector General) to the base plan or order. This annex follows the five-paragraph attachment format.

D-63. The inspector general uses Annex U (Inspector General) to describe and outline the inspector general support to the concept of operations described in the base plan or order. Staffs include this annex when they need to expand the inspector general functions beyond the base plan or order. The inspector general is responsible for developing Annex U (Inspector General). (See figure D-18 on pages D-92 through D-95.)

[CLASSIFICATION]

Place the classification at the top and bottom of every page of the attachments. Place the classification marking at the front of each paragraph and subparagraph in parentheses. Refer to AR 380-5 for classification and release marking instructions.

Copy ## of ## copies
Issuing headquarters
Place of issue
Date-time group of signature
Message reference number

Include heading if attachment is distributed separately from the base order or higher-level attachment.

ANNEX U (INSPECTOR GENERAL) TO OPLAN/OPORD [number] [(code name)]—[issuing headquarters] [(classification of title)]

(U) References: *List documents essential to understanding the attachment.*

    a. *List maps and charts first. Map entries include series number, country, sheet names or numbers, edition, and scale.*

    b. *List other references in subparagraphs labeled as shown.*

    c. *Policy references for this annex include AR 1-201 and AR 20-1. A doctrinal reference for this annex is FM 6-0.*

(U) Time Zone Used Throughout the Order: *Write the time zone established in the base plan or order.*

1. (U) Situation. *Include information affecting inspector general operations that paragraph 1 of the OPLAN or OPORD does not cover or that needs expansion.*

    a. (U) Area of Interest. *Describe the area of interest as it relates to inspector general operations. Refer to Annex B (Intelligence) as required.*

    b. (U) Area of Operations. *Refer to Appendix 2 (Operation Overlay) to Annex C (Operations).*

        (1) (U) Terrain. *Describe the aspects of terrain that impact inspector general operations. Refer to Annex B (Intelligence) as required.*

[page number]
[CLASSIFICATION]

Figure D-18. Sample Annex U (Inspector General) format

[CLASSIFICATION]

ANNEX U (INSPECTOR GENERAL) TO OPLAN/OPORD [number] [(code name)]—[issuing headquarters] [(classification of title)]

(2) (U) <u>Weather</u>. *Describe the aspects of weather that impact inspector general operations. Refer to Annex B (Intelligence) as required.*

c. (U) <u>Enemy Forces</u>. *Describe the possible or anticipated impact of enemy activities and courses of action on inspector general operations. Refer to Annex B (Intelligence) as required.*

d. (U) <u>Friendly Forces</u>. *Outline the higher headquarters' inspector general plan. List designation, location, and outline of plan of higher, adjacent, and other inspector general assets that support or impact the issuing headquarters or require coordination and additional support.*

e. (U) <u>Interagency, Intergovernmental, and Nongovernmental Organizations</u>. *Identify and describe other organizations in the area of operations that may impact the conduct of inspector general operations. Refer to Annex V (Interagency Coordination) as required.*

f. (U) <u>Civil Considerations</u>. *Describe the aspects of the civil situation that impact inspector general operations. Refer to Annex B (Intelligence) and Annex K (Civil Affairs Operations) as required.*

g. (U) <u>Attachments and Detachments</u>. *List units attached or detached only as necessary to clarify task organization. Refer to Annex A (Task Organization) as required.*

h. (U) <u>Assumptions</u>. *List inspector general-specific assumptions that support the annex development.*

2. (U) Mission. *State the mission of the inspector general in support of the base plan or order. For example, "On order, the inspector general provides the full range of inspector general functions (inspections, assistance, investigations, teaching, and training) in support of assigned and attached units of (unit name) for the duration of this operation."*

3. (U) Execution.

a. (U) <u>Scheme of Inspector General Support</u>. *Describe how the inspector general supports the commander's intent and concept of operations. Establish the priorities of support to units, or the concept for inspector general employment, for each phase of the operation. Focus on the commander's guidance, mission, and intent, and emphasize how inspector general operations reduce friction that affects readiness and warfighting capability. List any general areas the commander has asked the inspector general to assess in any travels. Refer to Annex C (Operations) as required.*

(1) (U) <u>Inspections</u>. *Outline inspection plan by phase based on the commanding general's guidance and the compressed inspection plan for unanticipated inspection topics when directed. Inspection plans should focus on high-payoff issues for the commander related to each phase of the operation (such as mobilization, deployment, employment, and sustainment). Include command guidance on requirements for the Organizational Inspection Program in theater, to include command inspections, staff inspections, inspector general inspections, intelligence oversight inspections, and audits. Include request and tasking procedures for subject-matter experts to serve as temporary assistant inspectors general. List upcoming outside agency assessments—Government Accountability Office and Department of Defense—that may impact the command's resources.*

[page number]
[CLASSIFICATION]

Figure D-18. Sample Annex U (Inspector General) format (continued)

[CLASSIFICATION]

ANNEX U (INSPECTOR GENERAL) TO OPLAN/OPORD [number] [(code name)]—[issuing headquarters] [(classification of title)]

(2) (U) <u>Assistance and Investigations</u>. *Develop assistance coverage plan for subordinate units with considerations for geographically dispersed units and split-based operations. Description of coverage should include unit visitation plans and plans for use of acting inspectors general for assistance. Emphasize the inspector general's role of underwriting the chain of command in addressing issues and allegations, including handling of law of war violations. The inspector general assistance plan should also address support for units under the operational control or direct-supporting role of the inspector general's organization (such as assistance support on an area-support basis).*

(3) (U) <u>Teaching and Training</u>. *Detail plans for deliberate teaching and training tools, such as deployment and reception briefs, inspector general bulletins and newsletters, and new commander orientations.*

b. (U) <u>Tasks to Subordinate Units</u>. *List inspector general tasks assigned to specific subordinate units not contained in the base order, and areas of responsibility for inspectors general and acting inspector general elements geographically separated from the command inspector general.*

c. (U) <u>Coordinating Instructions</u>. *List only instructions applicable to two or more subordinate units not covered in the base order. Include instructions for coordination between inspector general elements conducting split-based operations and coordination for reachback assistance from nondeployed supporting inspectors general at home station. List coordination and reporting requirements to the higher command inspector general and other inspector general technical channels. List the unit's reporting process for intelligence oversight procedure 15 reports, law of war violations, whistle-blower reprisals, and other Department of Defense-level critical information requirements. List the standard "before you see the inspector general" checklist.*

4. (U) <u>Sustainment</u>. *Identify priorities of sustainment for inspector general key tasks and specify additional instructions as required. Refer to Annex F (Sustainment) as required.*

a. (U) <u>Logistics</u>. *Identify unique sustainment requirements, procedures, and guidance to support inspector general teams and operations. Specify procedures for specialized technical logistics support from external organizations as necessary. Use subparagraphs to identify priorities and specific instructions for inspector general logistics support. Refer to Annex F (Sustainment) and Annex P (Host-Nation Support) as required.*

b. (U) <u>Personnel</u>. *Identify inspector general-unique personnel requirements and concerns, including global sourcing support and contracted linguist requirements. Use subparagraphs to identify priorities and specific instructions for human resources support, financial management, legal support, and religious support. Refer to Annex F (Sustainment) as required.*

c. (U) <u>Health Service Support</u>. *Identify availability, priorities, and instructions for medical care. Identify inspector general-unique inspection requirements for medical specialty personnel, medical logistics personnel, and medical equipment maintenance personnel. Identify availability, priorities, and instructions for medical care. Refer to Annex F (Sustainment) as required.*

5. (U) <u>Command and Signal</u>.

a. (U) <u>Command</u>.

(1) (U) <u>Location of the Commander and Key Leaders</u>. *Identify current or future command post locations or map coordinate locations of inspectors general. Identify the inspector general chain of command if not addressed in the unit standard operating procedures (SOPs).*

[page number]
[CLASSIFICATION]

Figure D-18. Sample Annex U (Inspector General) format (continued)

[CLASSIFICATION]

ANNEX U (INSPECTOR GENERAL) TO OPLAN/OPORD [number] [(code name)]—[issuing headquarters] [(classification of title)]

(2) (U) <u>Succession of Leadership</u>. *State the succession of leadership if not covered in the unit's SOPs.*

(3) (U) <u>Liaison Requirements</u>. *State the inspector general liaison requirements not covered in the base order.*

b. (U) <u>Control</u>.

(1) (U) <u>Command Posts</u>. *Describe the employment of inspector general-specific command posts (CPs), including the location of each CP and its time of opening and closing.*

(2) (U) <u>Reports</u>. *List inspector general-specific reports not covered in SOPs. Refer to Annex R (Reports) as required.*

c. (U) <u>Signal</u>. *Address any inspector general-specific communications requirements or reports. List signal instructions and network-centric instructions, to include call signs, phone numbers, and addresses to reach the inspector general. Address unique digitization connectivity requirements or coordination to meet functional responsibilities. Refer to Annex H (Signal) as required.*

ACKNOWLEDGE: *Include only if attachment is distributed separately from the base order.*

[Commander's last name]
[Commander's rank]

The commander or authorized representative signs the original copy of attachment. If the representative signs the original, add the phrase "For the Commander." The signed copy is the historical copy and remains in the headquarters' files.

OFFICIAL:

[Authenticator's name]
[Authenticator's position]

Use only if the commander does not sign the original attachment. If the commander signs the original, no further authentication is required. If the commander does not sign, the signature of the preparing staff officer (normally the command inspector general) requires authentication and only the last name and rank of the commander appear in the signature block.

ATTACHMENT: *List lower-level attachments (appendixes, tabs, and exhibits).*

DISTRIBUTION: *Show only if distributed separately from the base order or higher-level attachments.*

[page number]
[CLASSIFICATION]

Figure D-18. Sample Annex U (Inspector General) format (continued)

ANNEX V (INTERAGENCY COORDINATION) FORMAT AND INSTRUCTIONS

D-64. This annex provides fundamental considerations, formats, and instructions for developing Annex V (Interagency Coordination) to the base plan or order. This annex follows the five-paragraph attachment format.

D-65. Annex V (Interagency Coordination) provides military and interagency personnel with detailed information (mission, scheme, and tasks) to direct the necessary coordination and interaction between Army forces and interagency organizations. It describes how the commander intends to cooperate, provide support, and receive support from interagency organizations throughout the operation. This annex follows the five-paragraph order format; however, some subparagraphs are modified to accommodate communication with the interagency. The G-3 (S-3), in conjunction with the G-9 (S-9), develops Annex V (Interagency Coordination). (See figure D-19 on pages D-96 through D-99.)

D-66. Interagency organizations of the United States government include the following:

- Central Intelligence Agency.
- Department of Commerce.
- Department of Defense.
- Department of Energy.
- Department of Homeland Security.
- Department of Justice.
- Department of State.
- Department of the Treasury.
- Department of Transportation.
- Environmental Protection Agency.
- National Security Council.
- Peace Corps.
- United States Agency for International Development/Office of Foreign Disaster Assistance.
- United States Department of Agriculture.

[CLASSIFICATION]

Place the classification at the top and bottom of every page of the attachments. Place the classification marking at the front of each paragraph and subparagraph in parentheses. Refer to AR 380-5 for classification and release marking instructions.

Copy ## of ## copies
Issuing headquarters
Place of issue
Date-time group of signature
Message reference number

Include heading if attachment is distributed separately from the base order or higher-level attachment.

ANNEX V (INTERAGENCY COORDINATION) TO OPLAN [number] [(code name)]—[issuing headquarters] [(classification of title)]

(U) References: *List documents essential to understanding the attachment.*

 a. List maps and charts first. Map entries include series number, country, sheet names or numbers, edition, and scale.

 b. List other references in subparagraphs labeled as shown.

[page number]
[CLASSIFICATION]

Figure D-19. Sample Annex V (Interagency Coordination) format

[CLASSIFICATION]

ANNEX V (INTERAGENCY COORDINATION) TO OPLAN [number] [(code name)]—[issuing headquarters] [(classification of title)]

c. *Doctrinal references for interagency coordination include ADRP 3-07, FM 6-0, and JP 3-08.*

(U) Time Zone Used Throughout the Order: *Write the time zone established in the base plan or order.*

1. (U) <u>Situation.</u> *Include information affecting interagency coordination that paragraph 1 of the OPLAN or OPORD does not cover or that needs expansion.*

a. (U) <u>Area of Interest.</u> *Describe the area of interest as it relates to interagency coordination. Refer to Annex B (Intelligence) as required.*

b. (U) <u>Area of Operations.</u> *Refer to Appendix 2 (Operation Overlay) to Annex C (Operations).*

(1) (U) <u>Terrain.</u> *Describe the aspects of terrain that impact interagency coordination. Refer to Annex B (Intelligence) as required.*

(2) (U) <u>Weather.</u> *Describe the aspects of weather that impact interagency coordination. Refer to Annex B (Intelligence) as required.*

c. (U) <u>Political-Military Situation.</u> *Describe the political-military situation in the area of interest and area of operations. Identify U.S. national security objectives and interests applicable to the plan or order.*

d. (U) <u>Enemy Forces.</u> *Summarize the threat to interagency personnel. Identify enemy forces and appraise their general capabilities and impacts on interagency coordination operations. Refer to Annex B (Intelligence) as required.*

e. (U) <u>Friendly Forces.</u> *Outline the higher headquarters' interagency coordination plan. Identify and state the objectives or goals and primary tasks of those interagency organizations involved in the operations in subparagraphs below.*

f. (U) <u>Civil Considerations.</u> *Describe the aspects of the civil situation that impact interagency coordination. Refer to Annex B (Intelligence) and Annex K (Civil Affairs Operations) as required.*

g. (U) <u>Attachments and Detachments.</u> *List units attached or detached only as necessary to clarify task organization. Refer to Annex A (Task Organization) as required.*

h. (U) <u>Assumptions.</u> *List any interagency coordination-specific assumptions that support the annex development.*

i. (U) <u>Legal Considerations.</u> *List any legal considerations that may affect interagency participation, such as applicable international law or the authorities established under U.S. Code titles 10 and 50.*

2. (U) <u>Mission.</u> *State the mission of interagency coordination in support of the concept of operations in the base plan or order.*

3. (U) <u>Execution.</u>

a. (U) <u>Scheme of Interagency Coordination.</u> *Summarize the concept of operations in the base plan or order including an outline of the primary objectives and desired effects of each phase. Describe the concept of interagency coordination and how it supports the concept of operations. Describe the areas of responsibility from U.S. government agencies by major areas of response: humanitarian, economic, political or diplomatic, and others as required. The operational variables are another method to organize major areas of response; they are political, military, economic, social, information, infrastructure, physical environment, and time (PMESII-PT).*

[page number]
[CLASSIFICATION]

Figure D-19. Sample Annex V (Interagency Coordination) format (continued)

[CLASSIFICATION]

ANNEX V (INTERAGENCY COORDINATION) TO OPLAN [number] [(code name)]—[issuing headquarters] [(classification of title)]

       (1) (U) <u>Humanitarian</u>. *Define, in broad terms, the desired actions and responsibilities for United States government agencies in rebuilding and shaping the humanitarian structure and health of the affected nation. Coordinate these requested actions with the commander's phase development.*

       (2) (U) <u>Economic</u>. *Define, in broad terms, the desired actions and responsibilities for United States government agencies in rebuilding and shaping the economic structure and health of the affected nation. Coordinate these requested actions with the supported commander's phase.*

       (3) (U) <u>Political/Diplomatic</u>. *Define, in broad terms, the desired actions and responsibilities for United States government agencies in rebuilding and shaping the political and diplomatic structure of the affected nation. Coordinate these requested actions with the supported commander's phase development.*

       (4) (U) <u>Others</u>. *As required.*

    b. (U) <u>Tasks to Subordinate Units and Milestones</u>. *Identify tasks and required milestones of the issuing headquarters and interagency organizations during the conduct of operations.*

    c. (U) <u>Coordinating Instructions</u>. *List only instructions applicable to two or more subordinate units not covered in the base plan or order. Identify and list general instructions applicable to other United States government agencies, such as agreements with the host country and multinational forces.*

4. (U) <u>Sustainment</u>. *Identify priorities of sustainment for interagency coordination key tasks and specify additional instructions as required. Refer to Annex F (Sustainment) as required.*

    a. (U) <u>Logistics</u>. *Use subparagraphs to identify availability, priorities, and specific instructions for interagency coordination logistics support. Refer to Annex F (Sustainment) and Annex P (Host-Nation Support) as required.*

    b. (U) <u>Personnel</u>. *Use subparagraphs to identify availability, priorities, and specific instructions for human resources support, financial management, legal support, and religious support. Refer to Annex F (Sustainment) as required.*

    c. (U) <u>Health Service Support</u>. *Identify availability, priorities, and instructions for medical care. Identify specialized medical and veterinary requirements for interagency operations. Identify availability, priorities, and instructions for medical care. Refer to Annex F (Sustainment) as required.*

5. (U) <u>Command and Signal</u>.

    a. (U) <u>Command</u>. *Identify any unique command relationships established for the purpose of interagency coordination. Identify any interagency coordination forms or bodies such as an interagency coordination working group.*

       (1) (U) <u>Location of Interagency Coordination Leadership</u>. *Identify current or future locations of key interagency coordination leadership.*

       (2) (U) <u>Succession of Command</u>. *State the succession of leadership if not covered in the unit's standard operating procedures (SOPs).*

       (3) (U) <u>Liaison Requirements</u>. *State the interagency coordination liaison requirements not covered in the base order.*

[page number]
[CLASSIFICATION]

Figure D-19. Sample Annex V (Interagency Coordination) format (continued)

[CLASSIFICATION]

ANNEX V (INTERAGENCY COORDINATION) TO OPLAN [number] [(code name)]—[issuing headquarters] [(classification of title)]

b. (U) <u>Control</u>. *List the locations of key interagency leaders and contact information.*

(1) (U) <u>Command Posts</u>. *Describe the employment of interagency coordination command posts (CPs), including the location of each CP and its time of opening and closing.*

(2) (U) <u>Reports</u>. *List interagency coordination specific reports not covered in SOPs. Refer to Annex R (Reports) as required.*

c. (U) <u>Signal</u>. *Describe the communication plan used among the issuing force and interagency organizations to include the primary and alternate means of communications. Consider operations security requirements. Refer to Annex H (Signal) as required.*

ACKNOWLEDGE: *Include only if attachment is distributed separately from the base order.*

[Commander's last name]
[Commander's rank]

The commander or authorized representative signs the original copy of attachment. If the representative signs the original, add the phrase "For the Commander." The signed copy is the historical copy and remains in the headquarters' files.

OFFICIAL:

[Authenticator's name]
[Authenticator's position]

Either the commander or coordinating staff officer responsible for the functional area may sign attachments.

ATTACHMENT: *List lower-level attachments (appendixes, tabs, and exhibits).*

DISTRIBUTION: *Show only if distributed separately from the base order or higher-level attachments.*

[page number]
[CLASSIFICATION]

Figure D-19. Sample Annex V (Interagency Coordination) format (continued)

ANNEX W (OPERATIONAL CONTRACT SUPPORT) FORMAT AND INSTRUCTIONS

D-67. This annex provides fundamental considerations, formats, and instructions for developing Annex W (Operational Contract Support) to the base plan or order. This annex follows the five-paragraph attachment format.

D-68. Commanders and staffs use Annex W (Operational Contract Support) to describe how operational contract support is integrated into the overall concept of operations as well as other support functions described in the base plan or order and applicable annex. The G-4 (S-4) is the staff officer responsible for this annex.

D-69. Order writers describe the operational contract support concept of support objectives. A complex operational contract support concept of support may require a schematic to show the operational contract support objectives and task relationships. Order writers then include a discussion of the overall operational contract support organizational structure, command guidance, and subordinate unit responsibilities with the specific details in element subparagraphs, tabs, appendixes, or exhibits. Order writers include operational contract support-related assumptions and constraints, as appropriate.

D-70. A detailed description of specific contract support requirements and guidance will be addressed in the appropriate appendix, tab, or exhibit. (See figure D-20 on pages D-100 through D-103.)

[CLASSIFICATION]

Place the classification at the top and bottom of every page of the attachments. Place the classification marking at the front of each paragraph and subparagraph in parentheses. Refer to AR 380-5 for classification and release marking instructions.

Copy ## of ## copies
Issuing headquarters
Place of issue
Date-time group of signature
Message reference number

Include the full heading if attachment is distributed separately from the base order or higher-level attachment.

ANNEX W (OPERATIONAL CONTRACT SUPPORT) TO OPERATION PLAN/ORDER [number] [(code name)]-[issuing headquarters] [(classification of title)]

(U) References: *List documents essential to understanding the attachment.*

    a. List maps and charts first. Map entries include series number, country, sheet names or numbers, edition, and scale.

    b. List other references in subparagraphs labeled as shown. At a minimum, include higher level headquarters Annex W (Operational Contract Support) and any operational contract support related standard operating procedures (SOPs).

    c. Doctrinal and policy references for this attachment include AR 715-9, ATTP 4-10, and FM 6-0.

(U) Time Zone Used Throughout the Order: *Write the time zone established in the base plan or order.*

1. (U) Situation. *Include information affecting operational contract support that Paragraph 1 of the OPLAN or OPORD does not cover or needs expansion.*

    a. (U) Area of Interest. *Describe the area of interest as it relates to operational contract support, including general business climate and information on existing U.S. government contracts. Refer to Annex B (Intelligence) as required.*

[page number]
[CLASSIFICATION]

Figure D-20. Sample Annex W (Operational Contract Support) format

[CLASSIFICATION]

ANNEX W (OPERATIONAL CONTRACT SUPPORT) TO OPERATION PLAN/ORDER [number] [(code name)]-[issuing headquarters] [(classification of title)]

b. (U) <u>Area of Operations</u>. *Refer to Appendix 2 (Operation Overlay) to Annex C (Operations) as required.*

(1) (U) <u>Terrain</u>. *Describe the aspects of terrain that impact operational contract support operations. Refer to Annex B (Intelligence) as required.*

(2) (U) <u>Weather</u>. *Describe the aspects of weather that impact operational contract support operations. Refer to Annex B (Intelligence) as required.*

c. (U) <u>Enemy Forces</u>. *List enemy maneuver and other capabilities that could impact friendly operational contract support operations. Refer to Annex B (Intelligence) as required.*

d. (U) <u>Friendly Forces</u>. *List supporting contracting (contracting support brigade) and contract support units (team logistics civil augmentation program forward, Defense Contract Management Agency) as necessary to clarify operational contract support related task organization.*

e. (U) <u>Multinational Military, Interagency, Intergovernmental, and Nongovernmental Organizations</u>. *Identify and describe other organizations in the area of operations that may impact the conduct of operational contract support with special emphasis on avoiding undue competition for locally available commercial supplies and services. Refer to Annex V (Interagency Coordination) as required.*

f. (U) <u>Civil Considerations</u>. *Describe the aspects of the civil situation that impact operational contract support operations. Refer to Annex B (Intelligence) and Annex K (Civil Affairs Operations) as required.*

g. (U) <u>Attachments and Detachments</u>. *List attached or direct support contract support units as necessary to clarify operational contract support related task organization. Refer to Annex A (Task Organization) as required.*

h. (U) <u>Assumptions and Constraints</u>. *List any operational contract support-specific assumptions and constraints that support the annex development.*

2. (U) <u>Mission</u>. *State the mission of operational contract support in support of the base plan or order.*

3. (U) <u>Execution.</u>

a. (U) <u>Scheme of Operational Contract Support</u>. *Describe how operational contract support will be used to support the commander's intent and concept of operations. Provide initial guidance on suitability, availability, acceptability and any restrictions on contracted support by major support or sustainment function or commodity. Include any contract priorities of support that are different than priorities of support described in the base plan. Refer to Annex C (Operations), Annex F (Sustainment), or Annex P (Host-Nation Support) as required.*

b. (U) <u>Tasks to Subordinate Units</u>. *List operational contract support tasks assigned to specific subordinate units not contained in the base order to include activity responsibilities by major support or sustainment functions. This includes base life support, transportation, and linguists. Include any mission specific contract management assist responsibilities (contracting officer representative requirements support to award fee boards, and any other mission specific operational contract support related tasks and reporting requirements.*

[page number]
[CLASSIFICATION]

Figure D-20. Sample Annex W (Operational Contract Support) format (continued)

[CLASSIFICATION]

ANNEX W (OPERATIONAL CONTRACT SUPPORT) TO OPERATION PLAN/ORDER [number] [(code name)]-[issuing headquarters] [(classification of title)]

c. (U) <u>Coordinating Instructions</u>. *List only instructions applicable to two or more subordinate units not covered in the base plan or order. Include any operational contract support-specific reports not covered in SOPs.*

4. (U) <u>Sustainment</u>. *Identify and list sustainment priorities for operational contract support key tasks and specify additional sustainment instructions as necessary, including contractor support. Refer to Annex F (Sustainment) as required.*

a. (U) <u>Logistics</u>. *Identify unique sustainment requirements, procedures, and guidance to support operational contract support teams and operations. Specify procedures for specialized technical logistics support from external organizations as necessary. Use subparagraphs to identify priorities and specific instructions for operational contract support logistics. Refer to Annex F (Sustainment) and Annex P (Host-Nation Support) as required.*

b. (U) <u>Personnel</u>. *Identify operational contract support unique personnel requirements and concerns, including global sourcing support and contracted linguist requirements. Use subparagraphs to identify priorities and specific instructions for human resources support, financial management, legal support, and religious support. Refer to Annex F (Sustainment) as required.*

c. (U) <u>Health Service Support</u>. *Identify availability, priorities, and instructions for medical care. Identify medical contract support requirements which will be coordinated through the medical logistics management center. Identify availability, priorities, and instructions for medical care. Refer to Annex F (Sustainment) as required.*

5. (U) <u>Command and Signal</u>.

a. (U) <u>Command</u>. *Identify any unique command relationships established for the purpose of interagency coordination. Identify any interagency coordination forms or bodies such as an interagency coordination working group.*

(1) (U) <u>Location of the Commander and Key Leaders</u>. *Identify current or future locations of the commander and key operational contract support leadership.*

(2) (U) <u>Succession of Command</u>. *State the succession of leadership if not covered in the unit's standard operating procedures.*

(3) (U) <u>Liaison Requirements</u>. *State the liaison requirements not covered in the base order.*

b. (U) <u>Command and Support Relationships</u>. *State the specific command or support relationship between the major contracting and contract support units.*

c. (U) <u>Liaison Requirements</u>. *State any operational contract support liaison requirements not covered in the unit's SOPs.*

d. (U) <u>Control</u>. *List the locations of key interagency leaders and contact information.*

(1) (U) <u>Command Posts</u>. *Describe the employment of operational contract support command posts, including the location of each command post and its time of opening and closing.*

(2) (U) <u>Reports</u>. *List operational contract support-specific reports not covered in SOP. Refer to Annex R (Reports) as required.*

e. (U) <u>Signal</u>. *Describe the communication plan used among the issuing force and operational contract support organizations to include the primary and alternate means of communications. Consider operations security requirements. Refer to Annex H (Signal) as required.*

[page number]
[CLASSIFICATION]

Figure D-20. Sample Annex W (Operational Contract Support) format (continued)

[CLASSIFICATION]
ANNEX W (OPERATIONAL CONTRACT SUPPORT) TO OPERATION PLAN/ORDER [number] [(code name)]-[issuing headquarters] [(classification of title)]

ACKNOWLEDGE: *Include only if attachment is distributed separately from the base order.*

[Commander's last name]
[Commander's rank]

The commander or authorized representative signs the original copy of the attachment. If the representative signs the original, add the phrase "For the Commander." The signed copy is the historical copy and remains in the headquarters' files.

OFFICIAL:

[Authenticator's name]
[Authenticator's position]

Use only if the commander does not sign the original attachment. If the commander signs the original, no further authentication is required. If the commander does not sign, the signature of the preparing staff officer requires authentication and only the last name and rank of the commander appear in the signature block.

ATTACHMENTS: *List lower-level attachment (appendixes, tabs, and exhibits).*

DISTRIBUTION: *Show only if distributed separately from the base order or higher-level attachments.*

[page number]
[CLASSIFICATION]

Figure D-20. Sample Annex W (Operational Contract Support) format (continued)

Annex X (Spare)

Annex Y (Spare)

ANNEX Z (DISTRIBUTION) FORMAT AND INSTRUCTIONS

D-71. This annex provides fundamental considerations, formats, and instructions for developing Annex Z (Distribution) to the base plan or order. This annex does not follow the five-paragraph attachment format. Unit SOPs dictate the development and format.

D-72. Commanders and staffs use Annex Z (Distribution) to track the distribution of the operation plan and order and attachments. The G-3 (S-3), in coordination with the knowledge management officer, is responsible for developing Annex Z (Distribution).

D-73. An important information management task is determining what organizations receive copies of the unit's operation plan and order. Normally, the distribution list is located at the end of the base plan or order. If the distribution plan is lengthy or complicated, use Annex Z (Distribution). (See figure D-21 on page D-106 through D-107.)

[CLASSIFICATION]

Place the classification at the top and bottom of every page of the attachments. Place the classification marking at the front of each paragraph and subparagraph in parentheses. Refer to AR 380-5 for classification and release marking instructions.

Copy ## of ## copies
Issuing headquarters
Place of issue
Date-time group of signature
Message reference number

Include heading if attachment is distributed separately from the base order or higher-level attachment.

ANNEX Z (DISTRIBUTION) TO OPLAN/OPORD [number] [(code name)]—[issuing headquarters] [(classification of title)]References:

(U) References: *List documents essential to understanding Annex Z.*

    a. *List maps and charts first. Map entries include series number, country, sheet names or numbers, edition, and scale.*

    b. *List other references in subparagraphs labeled as shown.*

    c. *Doctrinal references include AR 25-50, AR 380-10, CJCSM 3122.05, and FM 6-0.*

Time Zone Used Throughout the Order: *Write the time zone established in the base plan or order.*

(U) Distribution: *Furnish distribution copies either for action or for information. List in detail those who are to receive the plan or order. When referring to a standard distribution list, also show distribution to reinforcing, supporting, and adjacent units, since that list does not normally include these units. Refer to Annex A (Task Organization) as a guide to major subordinate commands involved in the operation and the base operation order for description of adjacent units. When units from multinational forces or host-nation forces are involved, ensure distribution is in accordance with theater foreign disclosure policies and AR 380-10.*

[page number]
[CLASSIFICATION]

Figure D-21. Sample Annex Z (Distribution) format

|CLASSIFICATION|

ANNEX Z (DISTRIBUTION) TO OPLAN/OPORD |number| |(code name)|—|issuing headquarters| |(classification of title)| References:

Distribution lists for paper copies should include the following information:

Duty Position, Unit, Location, Copy Number(s)

Example: CDR, C/1-503/173 ABN, Patrol Base Rock, #10-11

Electronic distribution and posting on a secure unit Web-portal (such as SECRET Internet Protocol Router Network) may also be used. Ensure all recipients have required privileges to access Web-portal and acknowledge in accordance with instructions provided in the base order.

ACKNOWLEDGE: *Include only if attachment is distributed separately from the base order.*

[Commander's last name]
[Commander's rank]

The commander or authorized representative signs the original copy of attachment. If the representative signs the original, add the phrase "For the Commander." The signed copy is the historical copy and remains in the headquarters' files.

OFFICIAL:

[Authenticator's name]
[Authenticator's position]

Either the commander or coordinating staff officer responsible for distribution may sign attachments.

DISTRIBUTION: *Show only if distributed separately from the base order or higher-level attachments.*

|page number|
|CLASSIFICATION|

Figure D-21. Sample Annex Z (Distribution) format (continued)

This page intentionally left blank.

Glossary

The glossary lists acronyms and terms with Army or joint definitions. Where Army and joint definitions differ, (Army) precedes the definition. Terms for which FM 6-0 is the proponent are marked with an asterisk (*). The proponent publication for other terms is listed in parentheses after the definition.

SECTION I – ACRONYMS AND ABBREVIATIONS

ABCT	armored brigade combat team
ACOS	assistant chief of staff
ADCON	administrative control
ADP	Army doctrine publication
ADRP	Army doctrine reference publication
AO	area of operations
AR	Army regulation
ASCC	Army Service component commander
ASCOPE	areas, structures, capabilities, organizations, people, and events
ATTP	Army tactics, techniques, and procedures
BCT	brigade combat team
CBRN	chemical, biological, radiological, and nuclear
CCIR	commander's critical information requirement
CJCSM	Chairman of the Joint Chiefs of Staff manual
CMOC	civil-military operations center
COA	course of action
+COCOM	combatant command
COP	common operational picture
COS	chief of staff
CP	command post
DA	Department of the Army
EEFI	essential element of friendly information
FFIR	friendly force information requirement
FM	field manual
FOUO	for official use only
FRAGORD	fragmentary order
G-1	assistant chief of staff, personnel
G-2	assistant chief of staff, intelligence
G-3	assistant chief of staff, operations
G-4	assistant chief of staff, logistics
G-5	assistant chief of staff, plans
G-6	assistant chief of staff, signal
G-8	assistant chief of staff, fininacial management
G-9	assistant chief of staff, civil affairs operations

GS	general support
GSR	general support-reinforcing
IPB	intelligence preparation of the battlefield
JP	joint publication
KMO	knowledge management officer
LNO	liaison officer
MDMP	military decisionmaking process
METT-TC	mission, enemy, terrain and weather, troops and support available, time available, and civil considerations
MISO	military information support operations
MOE	measure of effectiveness
MOP	measure of performance
NATO	North Atlantic Treaty Organization
NCO	noncomissioned officer
OAKOC	observation and fields of fire, avenues of approach, key terrain, obstacles, and cover and concealment
OPCON	operational control
OPLAN	operation plan
OPORD	operation order
OPSEC	operations security
PIR	priority intelligence requirement
PMESII-PT	political, military, economic, social, information, infrastructure, physical environment, and time
RSOI	reception, staging, onward movement, and integration
S-1	personnel staff officer
S-2	intelligence staff officer
S-3	operations staff officer
S-4	logistics staff officer
S-5	plans staff officer
S-6	signal staff officer
S-9	civil affairs operations staff officer
SBU	Sensitive But Unclassified
SOP	standard operating procedure
TACON	tactical control
TF	task force
TLP	troop leading procedures
TOE	table of organization and equipment
TTP	tactics, techniques, and procedures
U.S.	United States
WARNORD	warning order
XO	executive officer

SECTION II – TERMS

administrative control

Direction or exercise of authority over subordinate or other organizations in respect to administration and support. (JP 1)

after action review

A guided analysis of an organization's performance, conducted at appropriate times during and at the conclusion of a training event or operation with the objective of improving future performance. It includes a facilitator, event participants, and other observers. (ADRP 7-0)

assessment

Determination of the progress toward accomplishing a task, creating a condition, or achieving an objective. (JP 3-0)

assign

To place units or personnel in an organization where such placement is relatively permanent, and/or where such organization controls and administers the units or personnel for the primary function, or greater portion of the functions, of the unit or personnel. (JP 3-0)

assumption

A supposition on the current situation or a presupposition on the future course of events, either or both assumed to be true in the absence of positive proof, necessary to enable the commander in the process of planning to complete an estimate of the situation and make a decision on the course of action. (JP 5-0)

attach

The placement of units or personnel in an organization where such placement is relatively temporary. (JP 3-0)

avenue of approach

An air or ground route of an attacking force of a given size leading to its objective or to key terrain in its path. (JP 2-01.3)

*backbrief

A briefing by subordinates to the commander to review how subordinates intend to accomplish their mission.

*battle rhythm

A deliberate daily cycle of command, staff, and unit activities intended to synchronize current and future operations.

*be-prepared mission

A mission assigned to a unit that might be executed.

*board

A grouping of predetermined staff representatives with delegated decision authority for a particular purpose or function.

branch

The contingency options built into the base plan used for changing the mission, orientation, or direction of movement of a force to aid success of the operation based on anticipated events, opportunities, or disruption caused by enemy actions and reactions. (JP 5-0)

campaign plan

A joint operation plan for a series of related major operations aimed at achieving strategic or operational objectives within a given time and space. (JP 5-0)

C-day

The unnamed day on which a deployment operation commences or is to commence. (JP 5-0)

civil considerations

The influence of manmade infrastructure, civilian institutions, and activities of the civilian leaders, populations, and organizations within an area of operations on the conduct of military operations. (ADRP 5-0)

+close support

That action of the supporting force against targets or objectives that are sufficiently near the supported force as to require detailed integration or coordination of the supporting action (JP 3-31).

combat power

(Army) The total means of destructive, constructive, and information capabilities that a military unit or formation can apply at a given time. (ADRP 3-0)

commander's critical information requirement

An information requirement identified by the commander as being critical to facilitating timely decision making. (JP 3-0)

commander's intent

A clear and concise expression of the purpose of the operation and the desired military end state that supports mission command, provides focus to the staff, and helps subordinate and supporting commanders act to achieve the commander's desired results without further orders, even when the operation does not unfold as planned. (JP 3-0)

*command group

The commander and selected staff members who assist the commander in controlling operations away from a command post.

*command post

A unit headquarters where the commander and staff perform their activities.

*command post cell

A grouping of personnel and equipment organized by warfighting function or by planning horizon to facilitate the exercise of mission command.

common operational picture

(Army) A single display of relevant information within a commander's area of interest tailored to the user's requirements and based on common data and information shared by more than one command. (ADRP 6-0)

concealment

Protection from observation or surveillance. (ADRP 1-02)

concept plan

In the context of joint operation planning level 3 planning detail, an operation plan in an abbreviated format that may require considerable expansion or alteration to convert it into a complete operation plan or operation order. (JP 5-0)

*constraint

(Army) A restriction placed on the command by a higher command. A constraint dictates an action or inaction, thus restricting the freedom of action of a subordinate commander.

cover

(Army) Protection from the effects of fires. (ADRP 1-02)

*critical event

An event that directly influences mission accomplishment.

data

(Army) Unprocessed signals communicated between any nodes in an information system, or sensing from the environment detected by a collector of any kind (human, mechanical, or electronic). (ADRP 6-0)

D-day

The unnamed day on which a particular operation commences or is to commence. (JP 3-02)

decision point

A point in space and time when the commander or staff anticipates making a key decision concerning a specific course of action. (JP 5-0)

decisive terrain

Key terrain whose seizure and retention is mandatory for successful mission accomplishment. (FM 3-90-1)

demonstration

In military deception, a show of force in an area where a decision is not sought that is made to deceive an adversary. It is similar to a feint but no actual contact with the adversary is intended. (JP 3.13.4)

+direct liaison authorized

That authority granted by a commander (any level) to a subordinate to directly consult or coordinate an action with a command or agency within or outside of the granting command (JP 1).

+direct support

(Army) A support relationship requiring a force to support another specific force and authorizing it to answer directly to the supported force's request for assistance. (ADRP 5-0) (joint) A mission requiring a force to support another specific force and authorizing it to answer directly to the supported force's request for assistance (JP 3-09.3).

display

In military deception, a static portrayal of an activity, force, or equipment intended to deceive the adversary's visual observation. (JP 3.13.4)

*early-entry command post

A lead element of a headquarters designed to control operations until the remaining portions of the headquarters are deployed and operational.

essential element of friendly information

(Army) A critical aspect of a friendly operation that, if known by the enemy, would subsequently compromise, lead to failure, or limit success of the operation and therefore should be protected from enemy detection. (ADRP 5-0)

*essential task

(Army) A specified or implied task that must be executed to accomplish the mission.

evaluating

Using criteria to judge progress toward desired conditions and determining why the current degree of progress exists. (ADRP 5-0)

execution

Putting a plan into action by applying combat power to accomplish the mission. (ADP 5-0)

feint

In military deception, an offensive action involving contact with the adversary conducted for the purpose of deceiving the adversary as to the location and/or time of the actual main offensive action. (JP 3.13.4)

F-hour

The effective time of announcement by the Secretary of Defense to the Military Departments of a decision to mobilize Reserve units. (JP 3-02)

field of fire

The area that a weapon or group of weapons may cover effectively from a given position. (FM 3-90-1)

fragmentary order

An abbreviated form of an operation order issued as needed after an operation order to change or modify that order or to execute a branch or sequel to that order. (JP 5-0)

friendly force information requirement

Information the commander and staff need to understand the status of friendly force and supporting capabilities. (JP 3-0)

general support

That support which is given to the supported force as a whole and not to any particular subdivision thereof. (JP 3-09.3)

general support-reinforcing

(Army) A support relationship assigned to a unit to support the force as a whole and to reinforce another similar-type unit. (ADRP 5-0)

H-hour

The specific hour on D-day at which a particular operation commences. (JP 3-02)

*implied task

(Army) A task that must be performed to accomplish a specified task or mission but is not stated in the higher headquarters' order.

indicator

(Army) In the context of assessment, an item of information that provides insight into a measure of effectiveness or measure of performance. (ADRP 5-0)

information management

(Army) The science of using procedures and information systems to collect, process, store, display, disseminate, and protect data, information, and knowledge products. (ADRP 6-0)

information requirement

(Army) Any information element the commander and staff require to successfully conduct operations. (ADRP 6-0)

key terrain

Any locality, or area, the seizure or retention of which affords a marked advantage to either combatant. (JP 2-01.3)

knowledge

Information that has been analyzed to provide meaning or value or evaluated as to implications for the operation. (FM 6-01.1)

knowledge creation

The process of developing new knowledge or combining, restructuring, or repurposing existing knowledge in response to identified knowledge gaps. (FM 6-01.1)

knowledge management

The process of enabling knowledge flow to enhance shared understanding, learning, and decisionmaking. (ADRP 6-0)

knowledge transfer

Movement of knowledge—including knowledge based on expertise or skilled judgment—from one person to another. (FM 6-01.1)

L-hour

The specific hour on C-day at which a deployment operation commences or is to commence. (JP 5-0)

liaison

That contact or intercommunication maintained between elements of military forces or other agencies to ensure mutual understanding and unity of purpose and action. (JP 3-08)

*main command post

> A facility containing the majority of the staff designed to control current operations, conduct detailed analysis, and plan future operations.

M-day

> The term used to designate the unnamed day on which full mobilization commences or is due to commence. (JP 3-02)

measure of effectiveness

> A criterion used to assess changes in system behavior, capability, or operational environment that is tied to measuring the attainment of an end state, achievement of an objective, or creation of an effect. (JP 3-0)

measure of performance

> A criterion used to assess friendly actions that is tied to measuring task accomplishment. (JP 3-0)

military decisionmaking process

> An iterative planning methodology to understand the situation and mission, develop a course of action, and produce an operation plan or order. (ADP 5-0)

mission command

> (Army) The exercise of authority and direction by the commander using mission orders to enable disciplined initiative within the commander's intent to empower agile and adaptive leaders in the conduct of unified land operations. (ADP 6-0)

mission command warfighting function

> The related tasks and systems that develop and integrate those activities enabling a commander to balance the art of command and the science of control in order to integrate the other warfighting functions. (ADRP 3-0)

mission orders

> Directives that emphasize to subordinates the results to be attained, not how they are to achieve them. (ADP 6-0)

mission statement

> A short sentence or paragraph that describes the organization's essential task(s), purpose, and action containing the elements of who, what, when, where, and why. (JP 5-0)

monitoring

> Continuous observation of those conditions relevant to the current operation. (ADRP 5-0)

+mutual support

> That support which units render each other against an enemy, because of their assigned tasks, their position relative to each other and to the enemy, and their inherent capabilities (JP 3-31).

N-day

> The unnamed day an active duty unit is notified for deployment or redeployment. (JP 3-02)

observation

> The condition of weather and terrain that permits a force to see the friendly, enemy, and neutral personnel and systems, and the key aspects of the environment. (ADRP 1-02)

obstacle

> Any natural or man-made obstruction designed or employed to disrupt, fix, turn, or block the movement of an opposing force, and to impose additional losses in personnel, time, and equipment on the opposing force. (JP 3-15)

*on-order mission

> A mission to be executed at an unspecified time.

operational approach

A description of the broad actions the force must take to transform current conditions into those desired at end state. (JP 5-0)

operational control

The authority to perform those functions of command over subordinate forces involving organizing and employing commands and forces, assigning tasks, designating objectives, and giving authoritative direction necessary to accomplish the mission. (JP 1)

operation order

A directive issued by a commander to subordinate commanders for the purpose of effecting the coordinated execution of an operation. (JP 5-0)

operation plan

1. Any plan for the conduct of military operations prepared in response to actual and potential contingencies. 2. A complete and detailed joint plan containing a full description of the concept of operations, all annexes applicable to the plan, and a time-phased force and deployment data. (JP 5-0)

organic

Assigned to and forming an essential part of a military organization as listed in its table of organization for the Army, Air Force, and Marine Corps, and are assigned to the operating forces for the Navy. (JP 1)

*P-hour

The specific hour on D-day at which a parachute assault commences with the exit of the first Soldier from an aircraft over a designated drop zone. P hour may or may not coincide with H-hour.

planning

The art and science of understanding a situation, envisioning a desired future, and laying out effective ways of bringing that future about. (ADP 5-0)

planning horizon

A point in time commanders use to focus the organization's planning efforts to shape future events. (ADRP 5-0)

priority intelligence requirement

An intelligence requirement, stated as a priority for intelligence support, that the commander and staff need to understand the adversary or other aspects of the operational environment. (JP 2-01)

R-day

Redeployment day. The day on which redeployment of major combat, combat support, and combat service support forces begins in an operation. (JP 3-02)

reinforcing

A support relationship requiring a force to support another supporting unit. (ADRP 5-0)

risk management

The process of identifying, assessing, and controlling risks arising from operational factors and making decisions that balance risk cost with mission benefits. (JP 3-0)

running estimate

The continuous assessment of the current situation used to determine if the current operation is proceeding according to the commander's intent and if planned future operations are supportable. (ADP 5-0)

ruse

In military deception, a trick of war designed to deceive the adversary, usually involving the deliberate exposure of false information to the adversary's intelligence collection system. (JP 3-13.4)

S-day

The day the President authorizes Selective Reserve callup (not more than 200,000). (JP 3-02)

sequel

The subsequent major operation or phase based on the possible outcomes (success, stalemate, or defeat) of the current major operation or phase. (JP 5-0)

situational understanding

The product of applying analysis and judgment to relevant information to determine the relationships among the operations and mission variables to facilitate decisionmaking. (ADP 5-0)

*specified task

(Army) A task specifically assigned to a unit by its higher headquarters.

*staff section

A grouping of staff members by area of expertise under a coordinating, special, or personal staff officer.

supporting plan

An operation plan prepared by a supporting commander, a subordinate commander, or an agency to satisfy the requests or requirements of the supported commander's plan. (JP 5-0)

*tactical command post

A facility containing a tailored portion of a unit headquarters designed to control portions of an operation for a limited time.

tactical control

The authority over forces that is limited to the detailed direction and control of movements or maneuvers within the operational area necessary to accomplish missions or tasks assigned. (JP 1)

tactical mission task

A specific activity performed by a unit while executing a form of tactical operation or form of maneuver. It may be expressed as either an action by a friendly force or effects on an enemy force. (FM 7-15)

task-organizing

The act of designing an operating force, support staff, or sustainment package of specific size and composition to meet a unique task or mission. (ADRP 3-0)

T-Day

The effective day coincident with Presidential declaration of national emergency and authorization of partial mobilization (not more than 1,000,000 personnel exclusive of the 200,000 callup). (JP 3-02)

troop leading procedures

A dynamic process used by small-unit leaders to analyze a mission, develop a plan, and prepare for an operation. (ADP 5-0)

warning order

A preliminary notice of an order or action that is to follow. (JP 5-0)

W-Day

Declared by the President, W-day is associated with an adversary decision to prepare for war (unambiguous strategic warning). (JP 3-02)

*working group

(Army) A grouping of predetermined staff representatives who meet to provide analysis, coordinate, and provide recommendations for a particular purpose or function.

This page intentionally left blank.

References

All URLs accessed on 10 April 2014.

REQUIRED PUBLICATIONS

These documents must be available to intended users of this publication.

ADRP 1-02. *Terms and Military Symbols*. 24 September 2013.

JP 1-02. *Department of Defense Dictionary of Military and Associated Terms*. 8 November 2010

RELATED PUBLICATIONS

These documents contain relevant supplemental information.

JOINT PUBLICATIONS

Most joint publications are available online: <http://www.dtic.mil/doctrine/new_pubs/jointpub.htm.>

CJCSM 3122.05. *Operating Procedures for Joint Operation Planning and Execution (JOPES)– Information Systems (IS) Governance*. 15 December 2011.

JP 1. *Doctrine for the Armed Forces of the United States*. 25 March 2013.

JP 2-01. *Joint and National Intelligence Support to Military Operations*. 5 January 2012.

JP 2-01.3. *Joint Intelligence Preparation of the Operational Environment*. 16 June 2009.

JP 3-0. *Joint Operations*. 11 August 2011.

JP 3-02. *Amphibious Operations*. 10 August 2009.

JP 3-08. *Interorganizational Coordination During Joint Operations*. 24 June 2011.

JP 3-09.3. *Close Air Support*. 8 July 2009.

JP 3-13. *Information Operations*. 27 November 2012.

JP 3-13.4. *Military Deception*. 26 January 2012.

JP 3-14. *Space Operations*. 29 May 2013.

JP 3-15. *Barriers, Obstacles, and Mine Warfare for Joint Operations*. 17 June 2011.

JP 3-16. *Multinational Operations*. 16 July 2013.

JP 3-33. *Joint Task Force Headquarters*. 30 July 2012.

JP 3-57. *Civil-Military Operations*. 11 September 2013.

JP 3-61. *Public Affairs*. 25 August 2010.

JP 5-0. *Joint Operation Planning*. 11 August 2011.

JP 6-0. *Joint Communication System*. 10 June 2010.

ARMY PUBLICATIONS

Most Army doctrinal publications are available online at <http://www.apd.army.mil/>.

ADP 2-0. *Intelligence*. 31 August 2012.

ADP 3-0. *Unified Land Operations*. 10 October 2011.

ADP 5-0. *The Operations Process*. 17 May 2012.

ADP 6-0. *Mission Command*. 17 May 2012.

ADRP 2-0. *Intelligence*. 31 August 2012.

ADRP 3-0. *Unified Land Operations*. 16 May 2012.

ADRP 3-07. *Stability*. 31 August 2012.

ADRP 3-09. *Fires*. 31 August 2012.

ADRP 3-37. *Protection*. 31 August 2012.

ADRP 3-90. *Offense and Defense.* 31 August 2012.

ADRP 4-0. *Sustainment.* 31 July 2012.

ADRP 5-0. *The Operations Process.* 17 May 2012.

ADRP 6-0. *Mission Command.* 17 May 2012.

ADRP 6-22. *Army Leadership.* 1 August 2012.

ADRP 7-0. *Training Units and Developing Leaders.* 23 August 2012.

AR 1-201. *Army Inspection Policy.* 4 April 2008.

AR 11-7. *Army Internal Review Program.* 22 June 2011.

AR 20-1. *Inspector General Activities and Procedures.* 29 November 2010.

AR 25-50. *Preparing and Managing Correspondence.* 17 May 2013.

AR 25-55. *The Department of the Army Freedom of Information Act Program.* 1 November 1997.

AR 27-1. *Judge Advocate Legal Services.* 30 September 1996.

AR 27-10. *Military Justice.* 3 October 2011.

AR 75-15. *Policy for Explosive Ordnance Disposal.* 17 December 2013.

AR 115-10. *Weather Support for the U.S. Army.* 6 January 2010.

AR 165-1. *Army Chaplain Corps Activities.* 3 December 2009.

AR 380-5. *Department of the Army Information Security Program.* 29 September 2000

AR 380-10. *Foreign Disclosure and Contacts with Foreign Representatives.* 4 December 2013.

AR 385-10. *The Army Safety Program.* 27 November 2013.

AR 525-13. *Antiterrorism.* 11 September 2008.

AR 525-28. *Personnel Recovery.* 5 March 2010.

AR 530-1. *Operations Security (OPSEC).* 19 April 2007.

AR 600-20. *Army Command Policy.* 18 March 2008.

AR 690-11. *Use and Management of Civilian Personnel in Support of Military Contingency Operations.* 26 May 2004.

AR 715-9. *Operational Contract Support Planning and Management.* 20 June 2011.

ATP 3-37.34. *Survivability Operations.* 28 June 2013.

ATP 4-16. *Movement Control.* 5 April 2013.

ATP 4-32. *Explosive Ordnance Disposal (EOD) Operations.* 30 September 2013.

ATTP 3-34.23. *Engineer Operations–Echelons Above Brigade Combat Team.* 8 July 2010.

ATTP 3-34.80. *Geospatial Engineering.* 29 July 2010.

ATTP 4-10. *Operational Contract Support Tactics, Techniques, and Procedures.* 20 June 2011.

DA Pamphlet 385-10. *Army Safety Program.* 23 May 2008.

FM 1-0. *Human Resources Support.* 6 April 2010.

FM 1-04. *Legal Support to the Operational Army.* 18 March 2013.

FM 1-05. *Religious Support.* 5 October 2012.

FM 1-06. *Financial Management Operations.* 4 April 2011.

FM 2-0. *Intelligence.* 23 March 2010.

FM 2-01.3. *Intelligence Preparation of the Battlefield/Battlespace.* 15 October 2009.

FM 3-01. *U.S. Army Air and Missile Defense Operations.* 25 November 2009.

FM 3-07. *Stability Operations.* 6 October 2008.

FM 3-09. *Field Artillery Operations and Fire Support.* 4 April 2014.

FM 3-11. *Multi-Service Doctrine for Chemical, Biological, Radiological, and Nuclear Operations.* 1 July 2011.

FM 3-11.21. *Multiservice Tactics, Techniques, and Procedures for Chemical, Biological, Radiological, and Nuclear Consequence Management Operations.* 1 April 2008.

FM 3-13. *Inform and Influence Activities.* 25 January 2013.

FM 3-14. *Space in Support of Army Operations.* 6 January 2010.

FM 3-16. *The Army in Multinational Operations.* 20 May 2010.

FM 3-24. *Counterinsurgency.* 15 December 2006.

FM 3-27. *Army Global Ballistic Missile Defense (GBMD) Operations.* 31 March 2014.

FM 3-28. *Civil Support Operations.* 20 August 2010.

FM 3-34. *Engineer Operations.* 2 April 2014.

FM 3-34.5. *Environmental Considerations.* 16 February 2010.

FM 3-34.170. *Engineer Reconnaissance.* 25 March 2008.

FM 3-34.400. *General Engineering.* 9 December 2008.

FM 3-36. *Electronic Warfare.* 9 November 2012.

FM 3-39. *Military Police Operations.* 26 August 2013.

FM 3-50.1. *Army Personnel Recovery.* 21 November 2011.

FM 3-52. *Airspace Control.* 8 February 2013.

FM 3-55. *Information Collection.* 3 May 2013.

FM 3-57. *Civil Affairs Operations.* 31 October 2011.

FM 3-60. *The Targeting Process.* 26 November 2010.

FM 3-61. *Public Affairs Operations.* 1 April 2014.

FM 3-90-1. *Offense and Defense, Volume 1.* 14 June 2013.

FM 4-02. *Army Health System.* 26 August 2013.

FM 4-02.7. *Multiservice Tactics, Techniques, and Procedures for Health Service Support in a Chemical, Biological, Radiological, and Nuclear Environment.* 15 July 2009.

FM 4-02.17. *Preventive Medicine Services.* 28 August 2000.

FM 4-02.18. *Veterinary Service Tactics, Techniques, and Procedures.* 30 December 2004.

FM 4-02.19. *Dental Service Support Operations.* 31 July 2009.

FM 4-02.21. *Division and Brigade Surgeon's Handbook (Digitized) Tactics, Techniques, and Procedures.* 15 November 2000.

FM 4-02.51. *Combat and Operational Stress Control.* 6 July 2006.

FM 5-19. *Composite Risk Management.* 21 August 2006.

FM 5-102. *Countermobility.* 14 March 1985

FM 6-01.1. *Knowledge Management Operations.* 16 July 2012.

FM 6-02.40. *Visual Information Operations.* 10 March 2009.

FM 6-02.43. *Signal Soldier's Guide.* 17 March 2009.

FM 6-02.53. *Tactical Radio Operations.* 5 August 2009.

FM 6-02.70. *Army Electromagnetic Spectrum Operations.* 20 May 2010.

FM 6-02.71. *Network Operations.* 14 July 2009.

FM 6-20-40. *Tactics, Techniques, and Procedures for Fire Support for Brigade Operations (Heavy).* 5 January 1990.

FM 6-20-50. *Tactics, Techniques, and Procedures for Fire Support for Brigade Operations (Light).* 5 January 1990.

FM 6-99. *U.S. Army Report and Message Formats.* 19 August 2013.

FM 7-15. *The Army Universal Task List.* 27 February 2009.

FM 27-10. *The Law of Land Warfare.* 18 July 1956.

TM 3-34.85. *Engineer Field Data*. 17 October 2013.

OTHER PUBLICATIONS

Title 10, United States Code. Armed Forces. Available at < http://thomas.loc.gov/home/thomas.php>.

U.S. National Space Policy. 31 August 2006. Available at <http://www.nss.org/resources/library/spacepolicy/2006NationalSpacePolicy.htm>.

PRESCRIBED FORMS

None

REFERENCED FORMS

Most Army forms are available online at <http://www.apd.army.mil/>.

DA Form 2028. *Recommended Changes to Publications and Blank Forms*.

Index

Entries are by paragraph number.

Entries are by paragraph number.

Entries are by paragraph number.

Entries are by paragraph number.

Entries are by paragraph number.

gather information and knowledge, 4-7–4-14

gather the tools, course of action analysis, 9-126
receipt of mission, 9-17–9-18

gather tools and assessment data, assessment step, 15-21

general support, defined, B-27

general support-reinforcing, defined, B-29

H

headquarters management, 2-43

H-hour, defined, C-53

historian, 2-87

Host-Nation Support (Annex P), D-54–D-56

human resources, 2-41

human terrain teams, 2-88–2-89

I

identify critical facts and develop assumptions, 9-41–9-43

identifying pages, C-58

implement, 14-28–14-30

implied task, defined, 9-34

implied tasks, 9-32–9-36

indentify the problem, 4-15–4-4-19

indicator, defined, 15-14

indicators, selecting and writing, 15-39–15-43

information briefing, 7-2–7-3

Information Collection (Annex L), D-46–D-47

information collection plan, 9-53–9-55

information management, 3-31–3-52
components, 3-33–3-42
defined, 3-31

information management tasks, 3-43–3-52
collect, 3-44
display, 3-47
disseminate, 3-48–3-50
process, 3-45
protect, 3-51–3-52
store, 3-46

information operations officer, 2-90

information requirement, defined, 3-34

information systems, 3-41–3-42

information variable, A-2

infrastructure variable, A-2

initial assessment, 9-20–9-22

inspector general, 2-109
Annex U, D-62–D-63

integrating, cells, 1-37–1-46

intelligence, Annex B, D-23–D-25
cell, 1-32

intelligence preparation of the battlefield, 9-30–9-31

intelligence responsibilities, war-gaming, 9-167

Interagency Coordination (Annex V), D-64–D-66

interagency operations, liaison considerations, 13-21–13-22

internal review officer, 2-110

issue a warning order, 9-81, 10-17–10-19

issue the order, 10-37–10-38

J

+joint command relationships, B-3–B-4

joint operations, liaison considerations, 13-20

+joint support relationships, B-9–B-11

K

key events, after action reviews, 16-26

key leader rehearsal, 12-21–12-26

key terrain, defined, A-15

knowledge, defined, 3-6

knowledge and information management, 3-53–3-62

knowledge and understanding, 3-1–3-9

knowledge creation, defined, 3-27

knowledge management, 3-10–3-30
Annex Q, D-57
components, 3-13–3-25
defined, 3-10

knowledge management components, organization, 3-25
people, 3-14–3-16
processes, 3-17–3-23
tools, 3-24

knowledge management officer, 2-91

knowledge management tasks, 3-26–3-30
applying knowledge, 3-29
create knowledge, 3-27
organizing knowledge, 3-28
transferring knowledge, 3-30

knowledge transfer, defined, 3-30

L

layout, command post, 1-16–1-17

L-hour, defined, C-53

liaison, 13-1–13-25
considerations, 13-19–13-25
defined, 13-1
elements, 13-7
officer, 2-92, 13-3–13-6
practices, 13-9
responsibilities, 13-10–13-18
role of, 13-1–13-9

liaison considerations, interagency operations, 13-21–13-22
joint operations, 13-20
multinational operations, 13-23–13-25

liaison officer, 13-3–13-6

liaison officers, time-saving techniques, 9-210

liaison responsibilities, after the tour, 13-17–13-18
during the tour, 13-15–13-16
receiving unit, 13-14
sending unit, 13-11–13-13

M

main command post, defined, 1-4

maintenance, 2-58

make a tentative plan, course of action comparison and selection, 10-32
course of action development, 10-22–10-30

Entries are by paragraph number.

Entries are by paragraph number.

Entries are by paragraph number.

84259924R00217

Made in the USA
San Bernardino, CA
06 August 2018